Nanotechnology Science and Technology

nova
science publishers
www.novapublishers.com

Nanotechnology Science and Technology

Nanosensors for Healthcare and Environmental Monitoring: Fabrication, Characterization, Simulation, and Applications
L. Ashok Kumar, PhD (Editor)
Ms. K. M. Janani (Editor)
2023. ISBN: 979-8-89113-316-7 (Hardcover)
2023. ISBN: 979-8-89113-373-0 (eBook)

X-Ray Photoelectron Spectroscopy: Principles, Techniques and Applications
Arpana Agrawal, PhD (Editor)
2023. ISBN: 979-8-89113-180-4 (Hardcover)
2023. ISBN: 979-8-89113-224-5 (eBook)

Magnetic Nanoparticles: Synthesis and Comprehensive Characterization
V. N. Lad, PhD (Editor)
Paritosh Agnihotri, PhD (Editor)
2023. ISBN: 979-8-88697-961-9 (Softcover)
2023. ISBN: 979-8-89113-061-6 (eBook)

Nanocomposite Hydrogels and Their Emerging Applications
Mohammad Sirousazar, PhD (Editor)
2023. ISBN: 979-8-88697-675-5 (Hardcover)
2023. ISBN: 979-8-88697-895-7 (eBook)

Applications of Gold Nanoparticles
George L. Morrow (Editor)
2022. ISBN: 979-8-88697-272-6 (Hardcover)
2022. ISBN: 979-8-88697-301-3 (eBook)

More information about this series can be found at
https://novapublishers.com/product-category/series/nanotechnology-science-and-technology/

Nanda Gopal Sahoo
and Anirban Dandapat
Editors

Properties and Applications of Nanosheets

nova
science publishers
www.novapublishers.com

Library of Congress Cataloging-in-Publication Data

ISBN: 979-8-89113-381-5

Published by Nova Science Publishers, Inc. † New York

Dedicated to

Late Professor Chapal Kumar Das

&

Late Professor Ganga Bisht

Contents

Preface

Two dimensional nanosheets have great potentials because of their diverse and unusual array of different properties including physical, chemical, optical, mechanical, electronic and magnetic properties that can be successfully employed in real life applications. The properties of nanosheets are strongly dependent on their structure, morphologies and compositions of the nanosheets. Thus, a comprehensive understanding of different properties of nanosheets is a prerequisite for utilization of different nanosheets in various fields. The unique properties of 2D nanosheets stem from their ultrathin structure, high surface area to volume ratio, extraordinary mechanical strength, and superior electronic conductivity, possess extraordinary physical and chemical properties to be applied in different fields. However, there is still lack of executable characterization techniques to obtain accurate structural information of ultrathin two dimensional nanosheets and thus it is still challenging to realize the exact structure and properties correlations in different two dimensional nanosheets. In this book, we focus on the recent progresses in studying the structure-property relationships of different two dimensional nanosheets and their applications in various fields.

The book comprises eleven chapters. Chapter 1 is a general introduction to the synthesis and properties of two dimensional nanosheets and their structure-property correlations for applying in various fields. Chapter 2 describes the fundamental properties of boron nitride nanosheets and their potentials in thermal conductivity, water purification and energy storage devices. Challenges and its solution for the synthesis of cerium oxide nanosheets and their successful applications are discussed in Chapter 3. The origin and stability of the perovskite structure, the properties and applications of perovskite nanosheets have been addressed in Chapter 4. Self-assembling systems play a pivotal role in the functioning of living organisms and in emerging areas of science and technology. Properties and applications of self-assembled two-dimensional peptide, peptide, peptoid and protein-based supramolecular nanosheets are described in Chapters 5 and 6. Graphene is certainly the most widely studied two dimensional materials due to its excellent performance in electrical, mechanical, and thermal properties.

Chapter 7 explains the effect of graphene and its derivatives as flame retardant probes for epoxy composites, along with a detailed explanation of the charring, anti-dripping, and thermal stability of the composites after incorporation of graphene-based materials (GBMs). This chapter also concludes the synergistic effects of GBMs in epoxy matrices with their flame retardant applications, as well as the challenges associated with the industrial use of graphene and its derivatives. Detailed description about the utilizations of different two dimensional nanosheets in supercapacitor, biomedical, sensing and water treatment applications are described in chapter 8-11. This book is organized in such a way as to give the reader a clear and concise thought regarding the properties and applications of two-dimensional nanosheets. This book will emerge as the reference guide that overviews up to date literature in the field of two dimensional nanomaterials.

We are grateful to NOVA Publications, USA for this great opportunity and the support they provided. We wish to thank all the contributors including authors and reviewers for giving their valuable time in this project without which this book would not have become a reality.

We are also thankful to Padma Shri Professor Mahesh Verma, Vice-Chancellor, Guru Gobind Singh Indraprastha University (GGSIPU), New Delhi, Professor D. S. Rawat, Vice-Chancellor, Kumaun University, Nainital; Mr. Dinesh Chandra, Registrar, Kumaun University; Professor Neeta Bora, Director, DSB Campus, Kumaun University; Professor P. C Sharma, Director, GGSIPU EDC, Professor Arvinder Kaur, Dean, USAR& USDI, GGSIPU, Professor Chitra Pandey, H.O.D, Dept. of Chemistry, Kumaun University; all the faculty members of Dept. of Chemistry, Kumaun University and USAR, GGSIPU for their support.

NGS is most grateful to his beloved parents and elder brother Mr. Asit Sahoo and sister in-law Mrs. Sushmita Sahoo for their guidance. NGS is also thankful to his wife Mrs. Koli Sahoo, son Nirvik and daughter Nivrita for their encouragement and silent support.

AD is also grateful to his parents, elder brother Mr. Ashim Kumar Dandapat and wife Mrs. Siuli Dandapat for their guidance, encouragement and supports. A special thanks to his little daughter Aayushi for her gracious presence.

Finally, this book serves as a comprehensive information resource for students, professors, scientists, industries, and researchers working on various nanosheets and their conversion technologies.

Professor Nanda Gopal Sahoo **Dr. Anirban Dandapat**

Chapter 1

General Properties of Two-Dimensional Nanosheets

Anita Rana[1]
Monika Matiyani[1]
Bhaskar Singh Bohra[1]
Pushpa Bhakuni Negi[2]
and Nanda Gopal Sahoo[1,*]

[1]Professor Rajendra Singh Nanoscience and Nanotechnology Centre, Department of Chemistry, D. S. B. Campus, Kumaun University, Nainital, Uttarakhand, India
[2]Department of Chemistry, Graphic Era Hill University, Bhimtal Campus, Nainital, Uttarakhand, India

Abstract

Two-dimensional nanosheets (2D NSs) with atomic thickness have drawn a lot of attention from researchers owing to their exceptional optical, mechanical, and electrical characteristics for future applications in sensing, energy storage, biomedicine, electronics, and catalysis. Because the shapes, dimensions, and sizes of nanomaterials have a significant impact on their properties and uses, 2D nanomaterial design and controlled synthesis have gained a great deal of attention. Due to the difficulties in producing atomically thick 2D NSs, especially for materials with non-layered structures, current research interests are mainly concentrated on the few layered samples. In this book chapter, we emphasized 2D NSs's distinct structure-property correlations and provide an overview of their uses in many fields. Finally, the main

* Corresponding Author's Email: ngsahoo@yahoo.co.in.

In: Properties and Applications of Nanosheets
Editors: Nanda Gopal Sahoo and Anirban Dandapat
ISBN: 979-8-89113-381-5

difficulties and prospects related to ultrathin 2D NSs are also discussed in light of the current advancements, leads avenues for chemistry study in the future.

Keywords: 2D nanosheets, sensor, biomedicine, energy storage, catalysis

1. Introduction

Two-dimensional (2D) materials with unique optical, electrical, and chemical properties, have been the subject of research. Electrical properties, two-dimensional nanosheets (2D NSs) with atomic thickness shows significant attention [1]. 2D NSs display unique features that are almost entirely missing in a bulk equivalent. Because NSs are extremely thin and have high surface, their active sites is nearly equal to their total atom count, giving them effective catalytic characteristics [2, 3]. Ultrathin 2D NSs display outstanding optical transparency and mechanical flexibility because of their atomic thickness and large lateral area, making them desirable candidates for highly transparent electronic devices. Due to their distinctive surface patterns and high specific surface area, ultrathin 2D NSs can be used in a range of research areas, including sensing, energy storage, catalysis etc [4–10]. Ultrathin 2D NSs would also have more core atoms exposed to the surface due to their atomic thickness, resulting in the development of many flaws and apparent changes in the electrical properties compared to the similar bulk samples [11]. Additionally, 2D NSs could be used to establish clear structure-property and get a detailed knowledge of the connections between the intrinsic properties of materials and their microstructures (like surfaces, interfaces, and defects) in solid state nano-chemistry [12-15]. Due to their acceptable bandgaps (between 1.0 and 2.0 eV), 2D transition metal dichalcogenides have intriguing optical properties [16, 17]. Additionally, it demonstrated that the lengths and thicknesses clearly affect their electrochemical performances [18]. The NSs have well potential for widespread application in many different domains [19-21]. This is due to their ultrathin structure.

Numerous techniques for producing 2D NSs with atomic thickness have been investigated up to this point to satisfactorily address the practical needs. Due to the presence of weak forces like van der Waals interaction between layer materials such as graphite, hexagonal boron nitride (h-BN), graphitic-carbon nitride and layered double hydroxides (LDHs) [22-26], exfoliated into the related ultrathin 2D NSs.

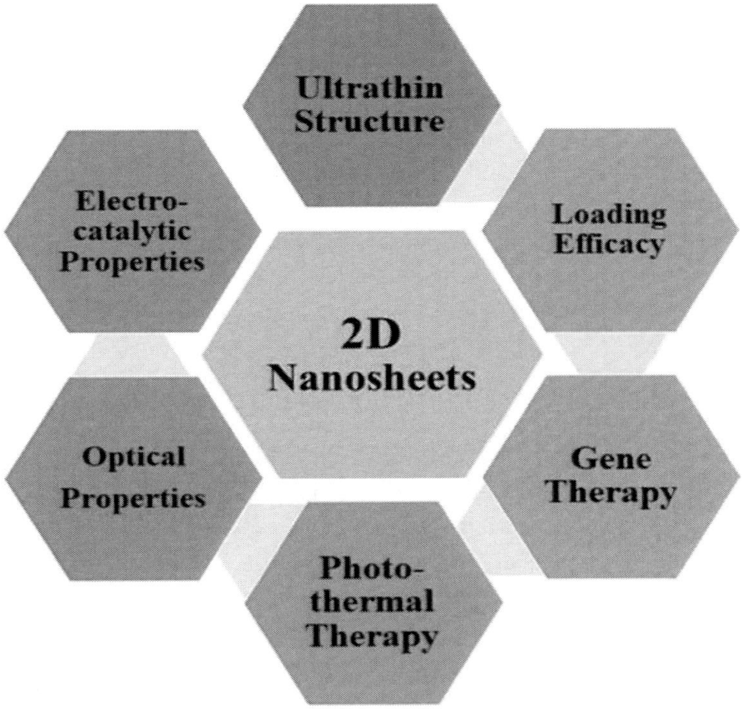

Figure 1. Different Properties of 2D nanosheets.

However, it is challenging to prepare 2D NSs with non-layered architectures because there is no inherent 2D anisotropic growth driving force due to strong chemical interactions in planes [27]. Even though ultrathin 2D NSs have numerous advantages over the corresponding bulk samples in practical applications, it is still unknown how to obtain structure and properties relationship due to the lack of an efficient characterization technique to ascertain the precise structural information of 2D NSs. In this book chapter, we intend to provide an overview about the properties of 2D NSs and their recent advances in 2D nanosheet fabrication (Figure 1).

2. Synthesis Strategies of Two-Dimensional Nanosheets

Different approaches have been reported for the synthesis of various kinds of 2D NSs. Broadly, all the synthesis methods are divided into following categories.

2.1. Mechanical Cleavage Strategy

This is very common and primitive technique to obtain 2D NSs by the exfoliation of layered precursor materials. When Novoselov and Geim isolated the 2D single layer graphene sheets from graphite exfoliation, this method came into existence. This method is extremely helpful, affordable, and capable of producing high-quality 2D structures, but it is only applicable in the laboratory due to the scale issue, which is the main difficulty. Furthermore, this procedure applicable to precursors of layered structures having van der Waals between the layers. Therefore, it is impossible to create 2D NSs of ionic solids and nonlayered materials using this approach. Layered materials are attached to the Scotch tape during this process, and the tape must then be folded and peeled numerous times until a single layer structure is not achieved. During process of making single layer graphene from graphite, graphite was applied over scotch tape, and then the forces between the layers of graphite was broken by repeatedly folding and peeling the tape. In addition to producing graphene from graphite, this method was also used to generate the 2D layered structure of transition metal chalcogenides (TMDs), h-BN, metal oxides, and other materials [28, 29].

2.2. Liquid Exfoliation Strategy

Liquid exfoliation of very useful and extensively used method for the preparation of 2D NSs. During this, bulk phase materials exfoliated in the appropriate solvent by using the ultrasonication. Ultrasonication help to delaminate the existing van der Waals force between the layers as results 2D NSs are obtained [30-31]. In this strategy selection of solvent is important because it determine quality of product. Generally, the surface energy of precursor and solvent should be match which reduced the driving energy and increase the productivity. Also, this action helps to make the stable dispersion of 2D NSs in a particular solvent. If surface energy is mismatch then agglomeration is occurred, as a result sedimentation can be seen in the bottom of the container which reduced the potential of 2D NSs. Till date, ultrasonic exfoliation is used for the synthesis of ultrathin 2D NSs by using solvents, such as water, N-methylpyrrolidone (NMP), isopropanol, dimethylformamide (DMF), and so on [32] Li et al., [33] reported the synthesis of few layer boron sheets by liquid phase exfoliation. During this process, boron powder was dispersed in 100 mL DMF or IPA (1mg/mL) then this mixture was ultrasonicated for 4hr to obtain desired boron NSs. After this process resultant mixture was centrifuged for 30min at 500 rpm to remove un-exfoliated boron as according to Figure 2. The Figure 2 (b), (c) and (d) are showing the SEM images of pure boron powder, DMF-exfoliated and IPA-exfoliated boron sheets. The SEM image of pure boron powder showed the thick or bulky phase which was converted into thin sheets of boron.

Mittal et al., [35] reported the synthesis of hexagonal boron nitride NSs (h-BNNSs) from bulk hexagonal boron nitride (hBN) via liquid phase exfoliation in the mixture of isopropyl alcohol (IPA) and de-ionized (DI) water (in 3:7 ratio). According to AFM studied they have reported 3-4 layers of NSs with thickness of the order of ~1.2 nm.

The liquid phase exfoliation method was used for synthesis of different NSs such as tellurium NSs,[36] WS_2, [37] MoS_2, [38] and graphene, [39] SnS NSs, BiOBr NSs [40] etc. Many modified liquid phase exfoliation methods are reported such as lithium-intercalated exfoliation [41] and ion-change exfoliation [42] etc.

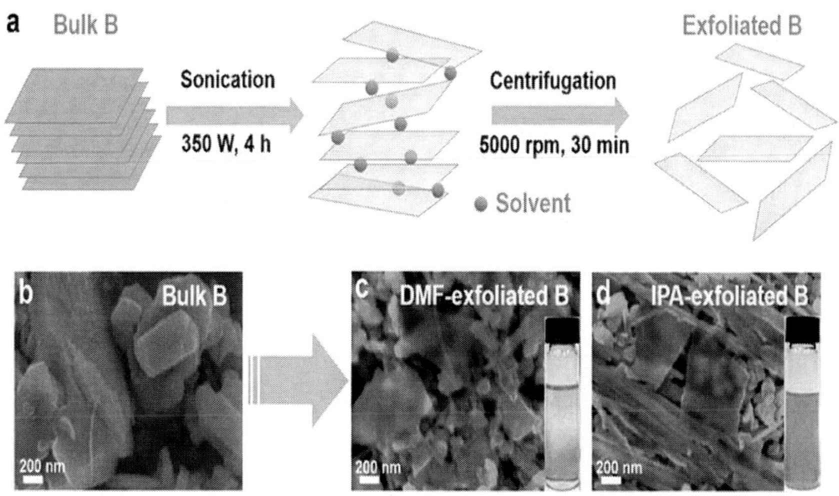

Figure 2. (a) Synthetic procedure of boron sheet (b) exfoliated sheets of boron, (b), (c) and (d) are the SEM images of pure boron powder, DMF-exfoliated and IPA-exfoliated boron sheets, respectively (Reprinted from [34], Copyright 2018, with the permission from American Chemical Society).

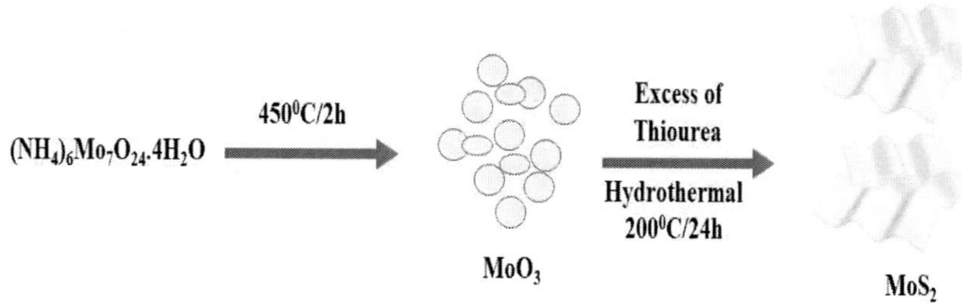

Figure 3. Synthesis of MoS_2 NSs via hydrothermal method.

2.3. Hydrothermal/Solvothermal Synthesis

This method is a very easy which has been used for synthesis of different kinds of nanostructure. During this process, if water used as the solvent, then the process called hydrothermal and it solvent other than water then it is called solvothermal method. In both conditions reaction is performed at higher temperature. The quality of product depends upon reaction time, temperature, and pressure. Previously, many 2D materials have been synthesized by this method [43]. In this regard, Muralikrishna et al., [44] reported 2D MoS_2 NSs from molybdenum trioxide (MoO_3) by a facile hydrothermal method. Figure 3 is showed the synthesis process of MoS_2 NSs, during this process specific amount of MoO_3 and urea were dispersed in water and stirrer for 30 minutes. Then, this solution was transferred into 50 mL of Teflon lined autoclave and placed at 200°C for 24 h. Finally, black precipitate was separated and washed obtained MoS_2 NSs.

2.4. Microwave-Assisted Method

The concept of microwave-assisted chemical synthesis has gained widespread acceptance as a means of accelerating and enhancing chemical processes. When compared to other standard tactics, this technology has a significantly shorter reaction time (often just a few minutes) and higher energy efficiency. These benefits make it convenient to create various 2D nanomaterials by using this method, including, α-$Ni(OH)_2$,SnO_2, CuSe and $K_{0.17}MnO_2$ [45]. In this regard, Modi et al., [46] reported the fast and easy microwave assisted solvothermal process for synthesis of 2D-SnS NSs. During this process, they have used $SnCl_2 \cdot 2H_2O$ and thioacetamide (TAA) as precursor for Sn and S, respectively. Initially, they have dispersed specific amount of $SnCl_2 \cdot 2H_2O$ and TAA separately in DI water and then mixed certain amount of polyvinylpyrrolidone (PVP). After that, this reaction mixture was microwave irritated for 30 min at 320 W. Finally, they obtained the black colour precipitates of SnS.

2.5. Chemical Vapour Deposition (CVD)

This method is very popular methods for the synthesis of 2D transition metal dichalcogenides (TMDs) and graphene. Due to its ease of use and compliance with industry standards, CVD produces high-quality 2D NSs with adjustable layer number with exceptional characteristics. During CVD, precursors are typically interacted with or broken down before being deposited as a mono- or multi-layer film at a relatively high temperature on the exposed substrate. To create ultrathin TMDs materials with superior electrical characteristics, scalable size, tunable thickness, and good crystal clarity, numerous fundamental studies have been conducted [47-49]. The reaction process is much faster to obtain few layer 2D structures for later research and application does not necessitate further mechanical exfoliation.

In this context, Choudhary et al., [50] reported the two-step method for synthesis of MoS2 NSs from CVD method. As according to the Figure 4, first step (Figure 4 (a)) involves synthesis of Mo thin films over the silicon substrate coated with SiO2 layer at room temperature. The deposition of various thickness of Mo film process carried out for various time intervals (4 to 180 s). Mo thin sheets were sputtered onto metal target. Following a thorough washing with acetone, the substrates were cleaned with methanol, ethanol, and DI water. After fixing the substrates to the heater, the chamber was vacuumed to a pressure of 10-7 Torr. The target was pre-sputtered for 5 min before to each sputtering run to ensure that the Mo target was in the same condition for each sample. In the second step (Figure 4 (b)), magnetron-sputtered Si/SiO2/Mo films were placed quartz tube furnace for low-pressure chemical vapor deposition (LPCVD). One gram of pure sulphur was packed in a ceramic boat and placed upstream of the quartz tube. To transport sulphur vapor species to Mo films further downstream, argon was employed as a carrier gas. The furnace heated to 600°C at the central zone in 30 mins in a flow atmosphere of 200 sccm Ar and a chamber pressure of 5 Torr. The furnace spontaneously cooled to room temperature after 60 minutes. The samples S1, S2, S3, S4, and S5 (Figure 4 (e)) are MoS2 films that have been converted for Mo films that have been formed at various sputtering times of 4, 10, 30, 60, and 180 s, respectively. Further, thicknesses of MoS2 films evaluated by AFM height profiles were found to be 0.72, 3.01, 5.40, 7.16 and 12.69 nm, corresponding to samples S1, S2, S3, S4 and S5, respectively.

2.6. Oriented Attachment Growth Strategy

Oriented attachment growth strategy is a well-known technique to fabricate non-layered 2D NSs having thickness in atomic range. In this method, a large number of microscopic 2D building blocks are produced and randomly dispersed during the initial phase of the reaction. After that, because of the orientational driving force, these initially formed tiny 2D nanocrystals would combine with one another and forming larger 2D nanoplates. Finally, these 2D nanoplates would attach to each other and produce ultrathin 2D NSs with atomic thickness. Various types of non-layered 2D NSs have been synthesized using oriented attachment growth strategy such as, SnS, SnO_2, CeO_2, Co_9Se_8, and etc [52-54].

3. General properties of Two-Dimensional Nanosheets

3.1. Ultrathin Structure

2D layered nanomaterials have paying attention of growing scientific because of their thickness and 2D shape, dimension-related chemical and physical features, such as their high intrinsic porosities, and abundance of available active sites. The ultrathin structure of 2D NSs can be simply explained on the basis of synthesis techniques.

Since the discovery of one-atomic-thick graphene in 2004 [55], a large number of ultrathin 2D materials with atomic thickness have been described [56–60]. Due to the strong quantum confinements, highly exposed active sites and huge specific surface areas provided by these ultrathin 2D sheet-like structures Figure 5, there is a great interest in exploring their activities. Ultrathin 2D Metal Organic Framework (MOF) NSs are a recently discovered component that can be freestanding NSs or nanofilms with lateral dimensions of a few microns and a thickness of a few nanometers [61–64]. Due to ordered pores, exposed active sites, MOF exhibit unique anisotropic properties that are advantageous for their use in optical properties [65–66].

The enormous surface areas, organised 1D channels, and active sites exposed on the surface of ultrathin 2D MOF NSs make them excellent modes for electronic devices.

The production of these NSs with the salient characteristics of outstanding high yield, structure stability and dispersion, is still strongly demanded for electronic devices (Figure 6). Additionally, it would be advantageous for novel techniques in effective production of ultrathin 2D MOF NSs. For instance, it is anticipated as effective approach to use ultrathin 2D inorganic hydroxides NSs and holds considerable potential for the shape-controlled crystal-structure transformation approach.

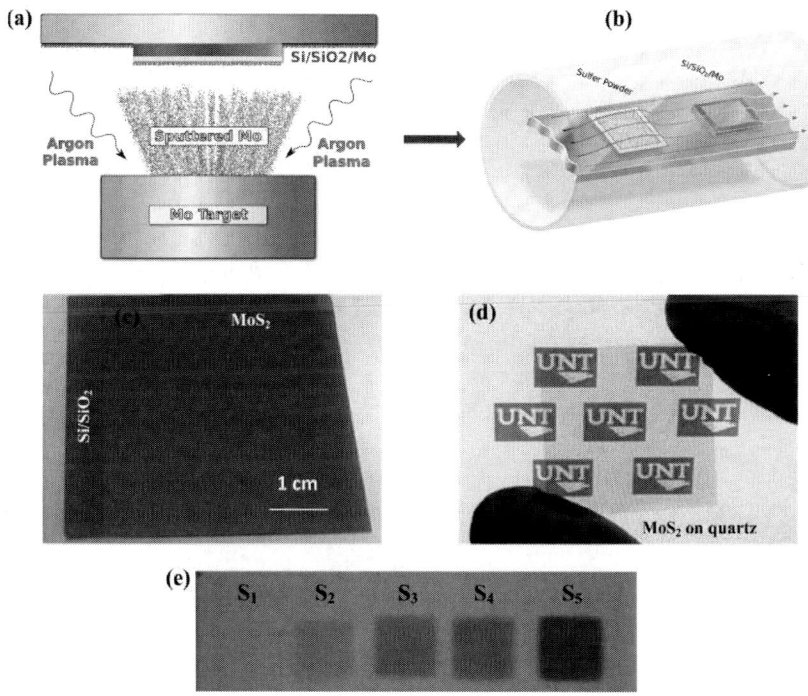

Figure 4. Representation of (a) sputtering and (b) Growth of MoS_2 thin films via CVD method (c) Si/SiO_2 and (d) transparent substrate (e) MoS_2 films samples with different thicknesses (S_1–S_5) ((Reprinted from [51], Copyright 2014, with the permission from American Chemical Society) [51].

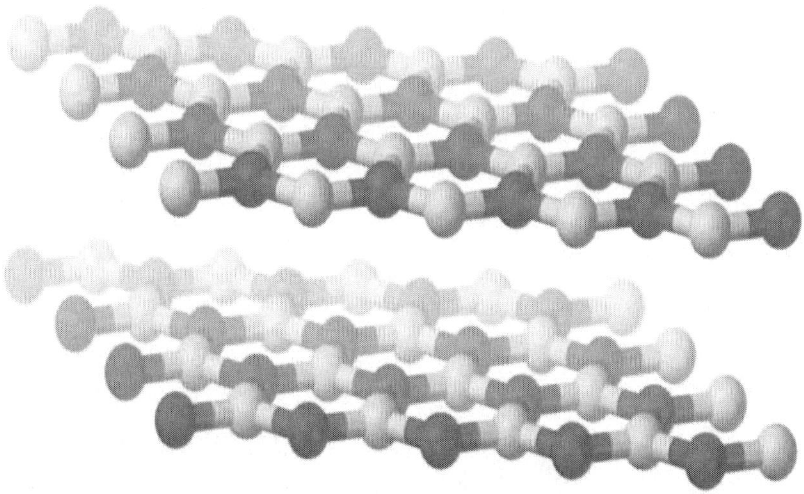

Figure 5. Schematic diagram of 2D nanosheets.

The examination of the ultrathin 2D MOF NSs' structure-dependent electrical characteristics has received a lot of attention to date. Pore volume of ultrathin 2D MOF NSs significantly affect their physical/chemical characteristics, including surface area, and charge mobility. The electrical characteristics of the incredibly thin 2D MOF NSs are still impossible to predict with any degree of accuracy, though. Therefore, theoretical calculation approach is so highly required for synthesis of innovative 2D MOF NSs [67].

3.2. Loading Efficacy

Due to their incredibly thin structure, NSs have a high surface to volume ratio, which is one of their distinguishing features. Because of this, NSs can load more drugs or genetic material more efficiently. Peng et al., [68], was demonstrated doxorubicin and indocyanine green loaded Gd3+ doped monolayered-double-hydroxide NSs, which displayed an extremely high drug loading efficiency.

Ji et al., published an intriguing statistic comparing the DNA loading effectiveness of silica nanoparticles with NSs constructed of the same substance [69].

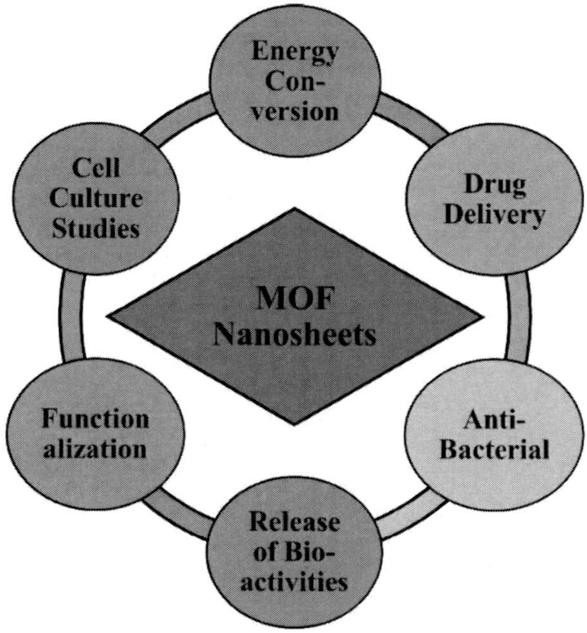

Figure 6. General applications of 2D nanosheets.

Compared to silica NSs, silica nanoparticles had a loading efficiency of 3.3% against 34%. Additional research on silica spheres is also important to note [70]. Silica NSs have a loading efficiency that is more than 10 times greater than that of nanoparticles, it is likely that the geometry of NSs contributes significantly to the higher loading capacity of genetic materials (Figure 7). NSs can also be chemically synthesized to significantly increase their ability to load genes. Polyethyleneimine is one of the most widely utilized polymers for attaching genetic material to NSs (PEI).

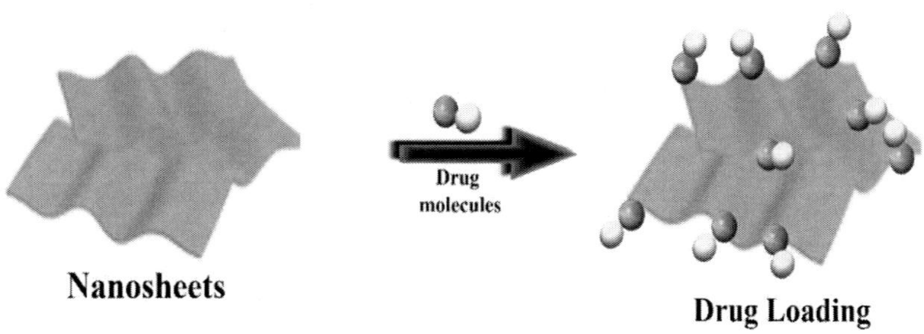

Figure 7. Drug loading efficacy of 2D nanosheets.

A branching polymer polyethyleneimine has a positive charge that enables it to electrostatically attach negatively charged genetic components like DNA and RNA. For instance, Kou et al., [71] described MoS2 NSs with modification with PEI and PEG, as a delivery vehicle.

Moreover, Teimouri et al., [72] reported that cationic polymers, such as polypropylenimine (PPI) and polyamidoamine (PAMAM), had been used to modify graphene oxide NSs However, compared to PPI or PAMAM modified sheets, PEI modified NSs had a greater transfection efficiency.

3.3. Gene Therapy

An important step in gene therapy is the successful delivery of a gene to a therapeutic target. In general, cellular absorption and endosomal escape are the identical steps in gene delivery via nanomaterials (Figure 8). Pinocytosis, a type of endocytosis, has been regarded as a significant method for cells to absorb nanomaterials [73]. Pinocytosis mediated by Caveolae, clathrin dependent endocytosis, caveolin/check whether clathrin independent endocytosis and micropinocytosis are the four forms of pinocytosis through which cells take up small molecules from the extracellular fluid [74]. The study of Huang et al., [75] offers an explanation of an entirely novel process of cellular uptake of graphene oxide-based sheets. He investigated cellular absorption of 100-200 nm graphene oxide coated with gold nanoparticles in presence of inhibitors of various endocytic pathways, including sodium azide, methyl-beta-cyclodextrin, chlorpromazine, and actin inhibitor amiloride (reduce endocytosis by ATP depletion). Further investigation revealed significant reduction in the uptake of graphene oxide nanosheets following pretreatment of cells with chlorpromazine and sodium azide, however, the uptake was only marginally reduced in the cell groups following pretreatment with methyl-beta-cyclodextrin and amiloride, when Ca Ski cells were analysed using surface-enhanced Raman scattering (SERS). According to the findings, clathrin and ATP were required for graphene oxide nanosheet uptake by cells. The study of Linares et al., showed cellular internalization of GO modified with fluorescein isothiocyanate (FITC) and polyethylene glycol (PEG) amine (FITC-PEG-GOs) by Saos-2 osteoblasts, HepG2 hepatocytes, and RAW 264.7 cells [76]. Mu et al., [77] found that cells may also phagocytose huge nanosheets.

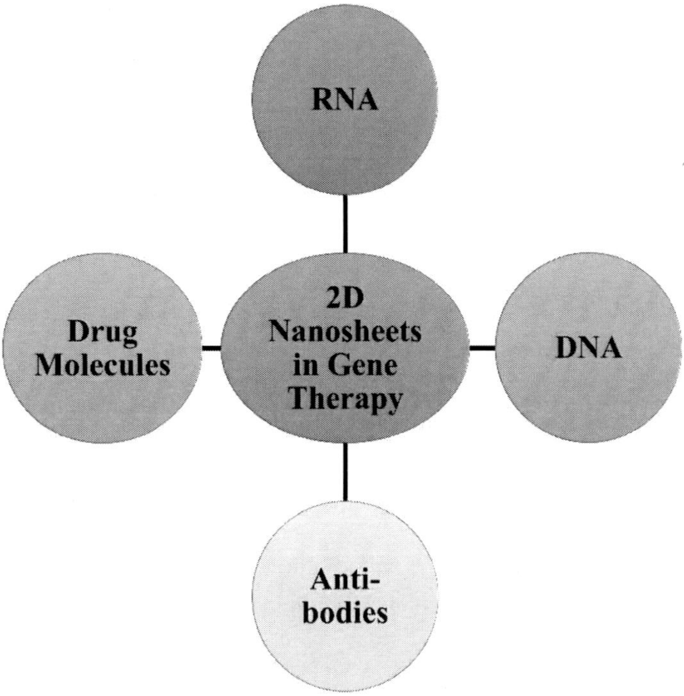

Figure 8. Gene therapy of 2D nanosheets.

They developed protein-coated GO nanosheets with an average height and diameter of 5.2 and 0.86 nm, respectively, and found that phagocytosis rather than clathrin-mediated endocytosis was more frequently utilized in the uptake of the developed GO nanosheets. The study identified a number of possible strategies for uptake of nanosheets by cells, which may be used as initial phase of gene delivery.

An important additional feature of NSs-associated gene transport is to escape endosomal degradation of desired genetic material, a phenomenon called endosomal escape. Upon entry into cells, endocytosed nanosheets are engulfed by endocytes. After successive processing in early endosomes, late endosomes, and multivesicular bodies, the Nanosheets are finally presented to lysosome, where lysosomal enzymes cause further breakdown of proteins and nucleic acids [78].

Premised on the currently understood process of cellular uptake and lysosomal degradation, a nanosheet-associated gene carrier might protect the nanosheet-associated genetic material from degradation by lysosomes by imparting ability to pass through the endosome before they enter the lysosome, thereby allowing safe delivery of the loaded genetic materials to the cytoplasmic and nuclear sites of action Several studies have shown that nano sheets may effectively perform both functions in a variety of cell types [79, 80]. (Bao et al., 2011; Kou et al., 2014). Triggering of "proton sponge effect," causing endosomal escape has been shown by loading of genes on 2D nanosheets using Cationic polymers like polyethyleneimine (PEI) [81]. The study by Behr and Varkouhi suggests that cationic polymers encourage the development of endosome proton, which causes osmotic swelling and rupture of endosomes [82]. By utilizing this method, the genetic material loaded on 2D nanosheets can cross the endosome and transfect cells in the nuclear or cytoplasmic sites.

3.4. Photothermal Therapy

A recently developed therapy called photothermal therapy (PPT) makes advantage of the heat that optical materials generate when exposed to near-infrared light (NIR, 700-1,000 nm). When administered intravenously in animal models, small nanosheets have a tendency to aggregate at tumour sites because of the improved permeability [83]. When nanosheets are exposed to NIR radiation, photons are absorbed, turning light energy into heat energy. The PPT employing nanosheets may successfully produce local temperature, which results in the selective killing of tumour cells (Figure 9) [84].

Nanosheets, including those made of graphene, black phosphorus and transition metal oxide, can be used to treat cancer cells using photothermal treatment both *in vivo* and *in vitro* [85].

3.5. Optical Properties

Another significant category that has recently attracted more attention 2D NSs. The potential of semiconductor nanostructures as components for next electro-optical technologies has attracted considerable interest. It is believed that the performance of optical systems based on nanostructures will be enhanced and less temperature sensitive [86]. ZnO nanostructures, with broad bandgap as semiconductors, are particularly appealing efficient optical and electrical nanodevices because of their high mechanical and thermal stabilities and substantial excitonic binding energies (60 meV).

In comparison to bulk materials, nanostructured materials are predicted to exhibit excellent optical characteristics. The studied photoluminescence (PL) spectra of ZnO nanostructures, which may be attributed to several structural flaws, which exist in the nanostructures [87].

Spin devices are anticipated to operate more quickly and produce less heat than conventional microelectronic devices. Future high-speed, energy-efficient electronic equipment will require this as a requirement. To find unique functional materials, researchers have made significant efforts (Figure 10). Recently, it has been extensively reported that non-magnetic layered semiconductors can acquire ferromagnetism when transition metal or rare earth atoms are used as cations. Due to their distinct structural, optical, and electrical features, layered chalcogenide semiconductors have drawn increasing attention [88].

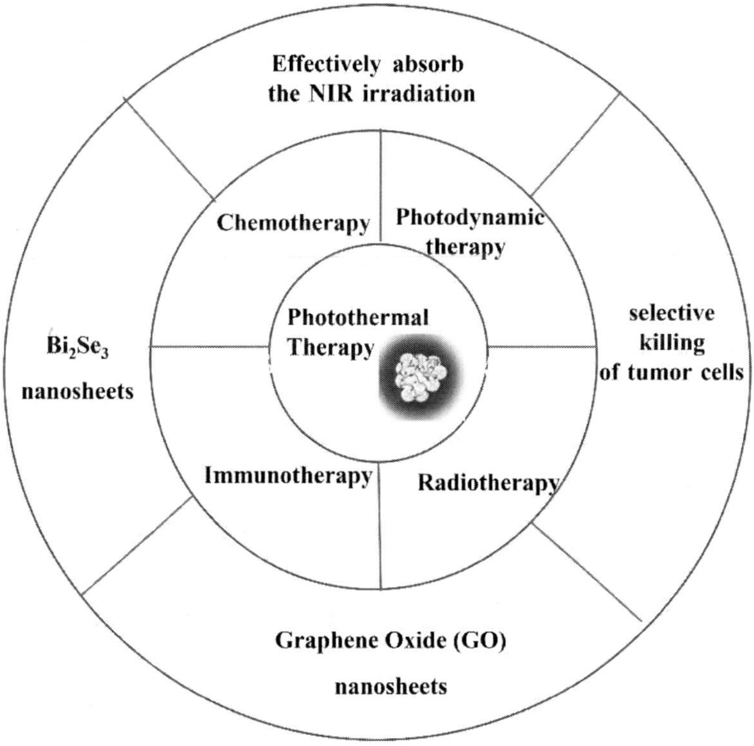

Figure 9. Photothermal therapy of 2D nanosheets.

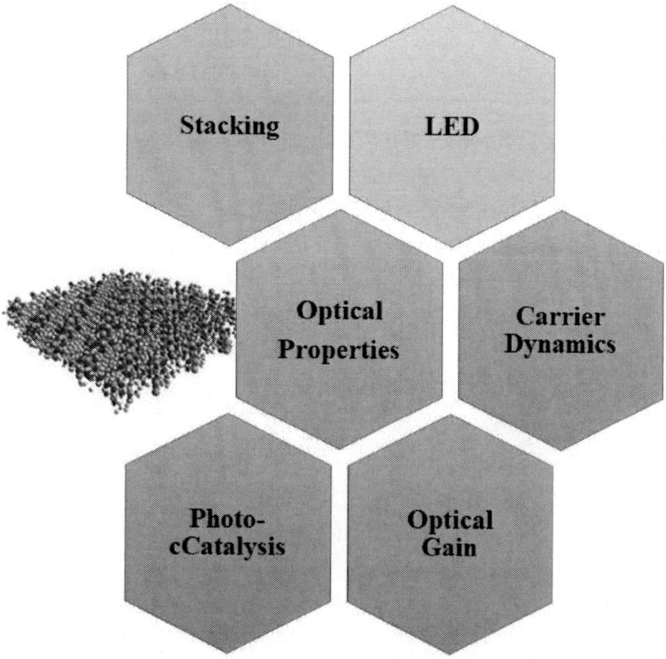

Figure 10. General Optical properties of 2D nanosheets.

The layered chalcogenide semiconductor tin (IV) sulphide (SnS2), which has a tunable optical band gap, shown to be a potential candidate in a number of fields. For instance, carbon-doped SnS2 nanostructures have been shown to have high-efficiency solar fuel catalytic properties. The SnS2 nanorods' sensing capabilities have been thoroughly investigated. The SnS2 nanosheet micro-sphere films that were prepared have been used as near-infrared photodetectors [89]. Additionally, more researchers started to take advantage of SnS2's magnetic.

3.6. Electrocatalytic Properties

Electrocatalysis focuses on the study of effect of electrode materials as well as the state of their surfaces. With significant uses in energy storage and conversion systems, it promotes the conversion of electrical and chemical energy. There are four electrochemical processes involved in the former.

In an electrolytic cell, water splitting causes the development of H_2 at the cathode and O_2 at the anode. In a hydrogen/air fuel cell, hydrogen is oxidized at the anode while oxygen is reduced at the cathode to produce electricity, with water as the primary by-product [90].

The carbon process comprises carbon dioxide released during the burning of fossil fuels to either make fuels like methanol, ethanol, and methane, or to produce fuel precursors (e.g., synthetic gas). Currently catalysts, such as Pt and some oxides of metals, exhibit great catalytic activity for electrochemical procedures, but their unavailability and high price severely restrict their use on a broad scale. Due to these costs, 2D NSs appear to be a good alternative material for charge transfer and simple electrochemical catalysis. 2D NSs have available exposed active sites on surface that open new possibilities for electrochemical catalysis [91].

Conclusion

Due to its wide-ranging uses in the fields of energy, electronic devices, and biology, among others, ultrathin 2D nanosheets have gained increasing attention from researchers in recent years as graphene's growth has accelerated. Notably, their atomically thick 2D structures provide them the ability to have unique electronic structures and also serve as excellent platforms for studying the links between structure and property.

Even though ultrathin 2D nanosheets have seen remarkable advances recently, there are number of challenges are to be overcome. These obstacles can also be seen as innovative prospects for future research areas. From a synthesis perspective, the variety of ultrathin 2D nanosheets is primarily restricted to many combination which only comprise up a slight portion of the materials that are widely prevalent in nature. Exploring and developing new ultrathin 2D nanosheets traits or electrical properties is thus one of the most crucial issues for future study. Since these required ultra-thin 2D nanosheets have naturally loose structure and few synthetic options, it is still very difficult to produce them even while these nanosheets and their derivatives have shown substantial promise in many areas. The absence of a thorough knowledge of the interaction between the nanosheets and their microscopic structures is one cause of this low performance. Despite the fact that there has been a substantial development in theoretical calculations over the past few years, there is still a distinction between theoretical simulations and realistic systems. Therefore, it is necessary to produce precise theoretical models and to get thorough knowledge of the structure-property relationship. For resolving this issue, it is important to combine the ultrathin 2D nanosheets into integrated foams or to anchor them to a highly stable matrix (like carbon fibre paper or Ni foam).

As a result, superb high-performance catalysts can be produced using these efficient procedures to perfectly meet the requirements of industrial applications. Although there are many obstacles to overcome above problems related to 2D nanosheets, we believe that with continuous efforts and sustained advancement study as well as practical applications, we will overcome them.

In conclusion, this book chapter gives detail study of synthesis and properties of 2D nanosheets, which will be helpful to give platform for opportunities in chemistry field.

References

[1] Hu, T., Mei, X., Wang, Y., Weng, X., Liang, R. and Wei, M., 2019. Two-dimensional nanomaterials: fascinating materials in biomedical field. *Science Bulletin*, *64*(22), 1707-1727.

[2] Li, X., Shan, J., Zhang, W., Su, S., Yuwen, L. and Wang, L., 2017. Recent advances in synthesis and biomedical applications of two-dimensional transition metal dichalcogenide nanosheets. *Small*, *13*(5), 1602660.

[3] Kumar, K. S., Choudhary, N., Jung, Y. and Thomas, J., 2018. Recent advances in two-dimensional nanomaterials for supercapacitor electrode applications. *ACS Energy Letters*, *3*(2), 482-495.

[4] Zhang, X., Zhang, J., Zhao, J., Pan, B., Kong, M., Chen, J. and Xie, Y., 2012. Half-metallic ferromagnetism in synthetic Co_9Se_8 nanosheets with atomic thickness. *Journal of the American Chemical Society*, *134*(29), 11908-11911.

[5] Gao, S., Sun, Y., Lei, F., Liang, L., Liu, J., Bi, W., Pan, B. and Xie, Y., 2014. Ultrahigh energy density realized by a single-layer β-$Co(OH)_2$ all-solid-state asymmetric supercapacitor. *Angewandte Chemie International Edition*, *53*(47), 12789-12793.

[6] Zhang, X., Xu, J., Wang, H., Zhang, J., Yan, H., Pan, B., Zhou, J. and Xie, Y., 2013. Ultrathin nanosheets of MAX phases with enhanced thermal and mechanical properties in polymeric compositions: Ti3Si0. 75Al0. 25C2. *Angewandte Chemie International Edition*, *52*(16), 4361-4365.

[7] Liang, L., Li, K., Xiao, C., Fan, S., Liu, J., Zhang, W., Xu, W., Tong, W., Liao, J., Zhou, Y. and Ye, B., 2015. Vacancy associates-rich ultrathin nanosheets for high performance and flexible nonvolatile memory device. *Journal of the American Chemical Society*, *137*(8), 3102-3108.

[8] Bi, W., Ye, C., Xiao, C., Tong, W., Zhang, X., Shao, W. and Xie, Y., 2014. Spatial location engineering of oxygen vacancies for optimized photocatalytic H2 evolution activity. *Small*, *10*(14), 2820-2825.

[9] Huang, J., Chen, J., Yao, T., He, J., Jiang, S., Sun, Z., Liu, Q., Cheng, W., Hu, F., Jiang, Y. and Pan, Z., 2015. CoOOH nanosheets with high mass activity for water oxidation. *Angewandte Chemie*, *127*(30), 8846-8851.

[10] Bao, J., Wang, Z., Xie, J., Xu, L., Lei, F., Guan, M., Zhao, Y., Huang, Y. and Li, H., 2019. A ternary cobalt–molybdenum–vanadium layered double hydroxide nanosheet array as an efficient bifunctional electrocatalyst for overall water splitting. *Chemical Communications*, *55*(24), 3521-3524.

[11] Guan, M., Xiao, C., Zhang, J., Fan, S., An, R., Cheng, Q., Xie, J., Zhou, M., Ye, B. and Xie, Y., 2013. Vacancy associates promoting solar-driven photocatalytic activity of ultrathin bismuth oxychloride nanosheets. *Journal of the American Chemical Society*, *135*(28), 10411-10417.

[12] Sun, Y., Sun, Z., Gao, S., Cheng, H., Liu, Q., Piao, J., Yao, T., Wu, C., Hu, S., Wei, S. and Xie, Y., 2012. Fabrication of flexible and freestanding zinc chalcogenide single layers. *Nature communications*, *3*(1), 1057.

[13] Sun, Y., Liu, Q., Gao, S., Cheng, H., Lei, F., Sun, Z., Jiang, Y., Su, H., Wei, S. and Xie, Y., 2013. Pits confined in ultrathin cerium (IV) oxide for studying catalytic centers in carbon monoxide oxidation. *Nature communications*, *4*(1), 2899.

[14] Sun, Y., Gao, S., Lei, F., Xiao, C. and Xie, Y., 2015. Ultrathin two-dimensional inorganic materials: new opportunities for solid state nanochemistry. *Accounts of chemical research*, *48*(1), 3-12.

[15] Sun, Y., Gao, S., Lei, F. and Xie, Y., 2015. Atomically-thin two-dimensional sheets for understanding active sites in catalysis. *Chemical Society Reviews*, *44*(3), 623-636.

[16] Liang, Q., Zhang, Q., Zhao, X., Liu, M. and Wee, A. T., 2021. Defect engineering of two-dimensional transition-metal dichalcogenides: applications, challenges, and opportunities. *ACS nano*, *15*(2), 2165-2181.

[17] Wang, Y., Zhao, Y., Ding, X. and Qiao, L., 2021. Recent advances in the electrochemistry of layered post-transition metal chalcogenide nanomaterials for hydrogen evolution reaction. *Journal of Energy Chemistry*, *60*, 451-479.

[18] Chia, X., Eng, A. Y. S., Ambrosi, A., Tan, S. M. and Pumera, M., 2015. Electrochemistry of nanostructured layered transition-metal dichalcogenides. *Chemical reviews*, *115*(21), 11941-11966.

[19] Lin, Y., Li, Y., Cao, Y. and Wang, X., 2021. Two-dimensional MOFs: Design & Synthesis and Applications. *Chemistry – An Asian Journal*, *16*(21), 3281-3298.

[20] Kumbhakar, P., Gowda, C. C., Mahapatra, P. L., Mukherjee, M., Malviya, K. D., Chaker, M., Chandra, A., Lahiri, B., Ajayan, P. M., Jariwala, D. and Singh, A., 2021. Emerging 2D metal oxides and their applications. *Materials Today*, *45*, 142-168.

[21] Pei, Y., Huang, L., Wang, J., Han, L., Li, S., Zhang, S. and Zhang, H., 2019. Recent progress in the synthesis and applications of 2D metal nanosheets. *Nanotechnology*, *30*(22), 222001.

[22] Lin, Y., Williams, T. V. and Connell, J. W., 2010. Soluble, exfoliated hexagonal boron nitride nanosheets. *The Journal of Physical Chemistry Letters*, *1*(1), 277-283.

[23] Wang, X., Maeda, K., Thomas, A., Takanabe, K., Xin, G., Carlsson, J. M., Domen, K. and Antonietti, M., 2009. A metal-free polymeric photocatalyst for hydrogen production from water under visible light. *Nature materials*, *8*(1), 76-80.

[24] Acerce, M., Voiry, D. and Chhowalla, M., 2015. Metallic 1T phase MoS2 nanosheets as supercapacitor electrode materials. *Nature nanotechnology*, *10*(4), 313-318.

[25] Cheng, H., Kamegawa, T., Mori, K. and Yamashita, H., 2014. Surfactant-free nonaqueous synthesis of plasmonic molybdenum oxide nanosheets with enhanced catalytic activity for hydrogen generation from ammonia borane under visible light. *Angewandte Chemie International Edition*, *53*(11), 2910-2914.

[26] Song, F. and Hu, X., 2014. Exfoliation of layered double hydroxides for enhanced oxygen evolution catalysis. *Nature communications*, *5*(1), 4477.

[27] Gao, S., Jiao, X., Sun, Z., Zhang, W., Sun, Y., Wang, C., Hu, Q., Zu, X., Yang, F., Yang, S. and Liang, L., 2016. Ultrathin Co_3O_4 layers realizing optimized CO2 electroreduction to formate. *Angewandte Chemie International Edition*, *55*(2), 698-702.

[28] Li, H., Wu, J., Yin, Z. and Zhang, H., 2014. Preparation and applications of mechanically exfoliated single-layer and multilayer MoS_2 and WSe_2 nanosheets. *Accounts of chemical research*, *47*(4), 1067-1075.

[29] Novoselov, K. S., Geim, A. K., Morozov, S. V., Jiang, D., Katsnelson, M. I., Grigorieva, I. V., Dubonos, S. and Firsov, A., 2005. Two-dimensional gas of massless Dirac fermions in graphene. *nature*, *438*(7065), 197-200.

[30] Coleman, J. N., Lotya, M., O'Neill, A., Bergin, S. D., King, P. J., Khan, U., Young, K., Gaucher, A., De, S., Smith, R. J. and Shvets, I. V., 2011. Two-dimensional nanosheets produced by liquid exfoliation of layered materials. *Science*, *331*(6017), 568-571.

[31] Finn, D. J., Lotya, M., Cunningham, G., Smith, R. J., McCloskey, D., Donegan, J. F. and Coleman, J. N., 2014. Inkjet deposition of liquid-exfoliated graphene and MoS 2 nanosheets for printed device applications. *Journal of Materials Chemistry C*, *2*(5), 925-932.

[32] Smith, R. J., King, P. J., Lotya, M., Wirtz, C., Khan, U., De, S., O'Neill, A., Duesberg, G. S., Grunlan, J. C., Moriarty, G. and Chen, J., 2011. Large-scale exfoliation of inorganic layered compounds in aqueous surfactant solutions. *Advanced materials*, *23*(34), 3944-3948.

[33] Li, H., Jing, L., Liu, W., Lin, J., Tay, R. Y., Tsang, S. H. and Teo, E. H. T., 2018. Scalable production of few-layer boron sheets by liquid-phase exfoliation and their superior supercapacitive performance. *Acs Nano*, *12*(2), 1262-1272.

[34] Li, H., Jing, L., Liu, W., Lin, J., Tay, R. Y., Tsang, S. H. and Teo, E. H. T., 2018. Scalable production of few-layer boron sheets by liquid-phase exfoliation and their superior supercapacitive performance. *ACS Nano*, *12*(2), 1262-1272.

[35] Mittal, N., Kedawat, G., Gupta, S. and Kumar Gupta, B., 2020. An Innovative Method for Large-Scale Synthesis of Hexagonal Boron Nitride Nanosheets by Liquid Phase Exfoliation. *ChemistrySelect*, *5*(40), 12564-12569.

[36] Xie, Z., Xing, C., Huang, W., Fan, T., Li, Z., Zhao, J., Xiang, Y., Guo, Z., Li, J., Yang, Z. and Dong, B., 2018. Ultrathin 2D nonlayered tellurium nanosheets: facile liquid-phase exfoliation, characterization, and photoresponse with high performance and enhanced stability. *Advanced Functional Materials*, *28*(16), 1705833.

[37] Bernal, M. M., Álvarez, L., Giovanelli, E., Arnáiz, A., Ruiz-González, L., Casado, S., Granados, D., Pizarro, A. M., Castellanos-Gomez, A. and Pérez, E. M., 2016. Luminescent transition metal dichalcogenide nanosheets through one-step liquid phase exfoliation. *2D Materials*, *3*(3), 035014.

[38] Goni, F., Chemelli, A. and Uhlig, F., 2021. High-Yield Production of Selected 2D Materials by Understanding Their Sonication-Assisted Liquid-Phase Exfoliation. *Nanomaterials*, *11*(12), 3253.

[39] Paolucci, V., D'Olimpio, G., Lozzi, L., Mio, A. M., Ottaviano, L., Nardone, M., Nicotra, G., Le-Cornec, P., Cantalini, C. and Politano, A., 2020. Sustainable liquid-phase exfoliation of layered materials with nontoxic polarclean solvent. *ACS Sustainable Chemistry & Engineering*, *8*(51), 18830-18840.

[40] Brent, J. R., Lewis, D. J., Lorenz, T., Lewis, E. A., Savjani, N., Haigh, S. J., Seifert, G., Derby, B. and O'Brien, P., 2015. Tin (II) sulfide (SnS) nanosheets by liquid-phase exfoliation of herzenbergite: IV–VI main group two-dimensional atomic crystals. *Journal of the American Chemical Society*, *137*(39), 12689-12696.

[41] Yu, H., Huang, H., Xu, K., Hao, W., Guo, Y., Wang, S., Shen, X., Pan, S. and Zhang, Y., 2017. Liquid-phase exfoliation into monolayered BiOBr nanosheets for photocatalytic oxidation and reduction. *ACS Sustainable Chemistry & Engineering*, *5*(11), 10499-10508.

[42] Han, X., Zhou, S. J., Tan, Y. Z., Wu, X., Gao, F., Liao, Z. J., Huang, R. B., Feng, Y. Q., Lu, X., Xie, S. Y. and Zheng, L. S., 2008. Crystal structures of saturn-like C50Cl10 and pineapple-shaped C64Cl4: geometric implications of double-and triple-pentagon-fused chlorofullerenes. *Angewandte Chemie International Edition*, *47*(29), 5340-5343.

[43] Ma, R. and Sasaki, T., 2015. Two-dimensional oxide and hydroxide nanosheets: controllable high-quality exfoliation, molecular assembly, and exploration of functionality. *Accounts of chemical research*, *48*(1), 136-143.

[44] Zhao, Y., Ran, W., He, J., Huang, Y., Liu, Z., Liu, W., Tang, Y., Zhang, L., Gao, D. and Gao, F., 2015. High-performance asymmetric supercapacitors based on multilayer MnO_2/graphene oxide nanoflakes and hierarchical porous carbon with enhanced cycling stability. *Small*, *11*(11), 1310-1319.

[45] Muralikrishna, S., Manjunath, K., Samrat, D., Reddy, V., Ramakrishnappa, T. and Nagaraju, D. H., 2015. Hydrothermal synthesis of 2D MoS_2 nanosheets for electrocatalytic hydrogen evolution reaction. *RSC advances*, *5*(109), 89389-89396.

[46] Zhu, J., Li, Q., Bi, W., Bai, L., Zhang, X., Zhou, J. and Xie, Y., 2013. Ultra-rapid microwave-assisted synthesis of layered ultrathin birnessite K 0.17 MnO_2 nanosheets for efficient energy storage. *Journal of Materials Chemistry A*, *1*(28), pp.8154-8159.

[47] Modi, K. H., Pataniya, P. M., Patel, V. and Sumesh, C. K., 2021. Microwave assisted synthesis of SnS nanosheets for fabrication of large area SnS/Si heterojunction. *Solar Energy*, *221*, 412-417.

[48] Lee, Y. H., Zhang, X. Q., Zhang, W., Chang, M. T., Lin, C. T., Chang, K. D., Yu, Y. C., Wang, J. T. W., Chang, C. S., Li, L. J. and Lin, T. W., 2012. Synthesis of large-area MoS2 atomic layers with chemical vapor deposition. *Advanced materials*, *24*(17), 2320-2325.

[49] Ji, Q., Zhang, Y., Zhang, Y. and Liu, Z., 2015. Chemical vapour deposition of group-VIB metal dichalcogenide monolayers: engineered substrates from amorphous to single crystalline. *Chemical Society Reviews*, *44*(9), 2587-2602.

[50] Shi, Y., Li, H. and Li, L. J., 2015. Recent advances in controlled synthesis of two-dimensional transition metal dichalcogenides via vapour deposition techniques. *Chemical Society Reviews*, *44*(9), 2744-2756.

[51] Choudhary, N., Park, J., Hwang, J. Y. and Choi, W., 2014. Growth of large-scale and thickness-modulated MoS2 nanosheets. *ACS applied materials & interfaces*, *6*(23), 21215-21222.

[52] Choudhary, N., Park, J., Hwang, J. Y. and Choi, W., 2014. Growth of large-scale and thickness-modulated MoS2 nanosheets. *ACS applied materials & interfaces*, *6*(23), 21215-21222.

[53] Zhang, X., Zhang, J., Zhao, J., Pan, B., Kong, M., Chen, J. and Xie, Y., 2012. Half-metallic ferromagnetism in synthetic Co9Se8 nanosheets with atomic thickness. *Journal of the American Chemical Society*, *134*(29), 11908-11911.

[54] Zhu, J., Li, Q., Bi, W., Bai, L., Zhang, X., Zhou, J. and Xie, Y., 2013. Ultra-rapid microwave-assisted synthesis of layered ultrathin birnessite K 0.17 MnO 2 nanosheets for efficient energy storage. *Journal of Materials Chemistry A*, *1*(28), 8154-8159.

[55] Lee, C. C., Suzuki, S., Xie, W. and Schibli, T. R., 2012. Broadband graphene electro-optic modulators with sub-wavelength thickness. *Optics express*, *20*(5), 5264-5269.

[56] Zhou, J., Lin, J., Huang, X., Zhou, Y., Chen, Y., Xia, J., Wang, H., Xie, Y., Yu, H., Lei, J. and Wu, D., 2018. A library of atomically thin metal chalcogenides. *Nature*, *556*(7701), 355-359.

[57] Hu, Z., Xiao, X., Jin, H., Li, T., Chen, M., Liang, Z., Guo, Z., Li, J., Wan, J., Huang, L. and Zhang, Y., 2017. Rapid mass production of two-dimensional metal oxides and hydroxides via the molten salts method. *Nature communications*, *8*(1), 15630.

[58] Dong, J., Zhang, K., Li, X., Qian, Y., Zhu, H., Yuan, D., Xu, Q. H., Jiang, J. and Zhao, D., 2017. Ultrathin two-dimensional porous organic nanosheets with molecular rotors for chemical sensing. *Nature Communications*, *8*(1), 1142.

[59] Basu, S. and Bhattacharyya, P., 2012. Recent developments on graphene and graphene oxide based solid state gas sensors. *Sensors and Actuators B: Chemical*, *173*, 1-21.

[60] Zhao, W., Peng, J., Wang, W., Liu, S., Zhao, Q. and Huang, W., 2018. Ultrathin two-dimensional metal-organic framework nanosheets for functional electronic devices. *Coordination Chemistry Reviews*, *377*, 44-63.

[61] Liu, J. and Wöll, C., 2017. Surface-supported metal–organic framework thin films: fabrication methods, applications, and challenges. *Chemical Society Reviews*, *46*(19), 5730-5770.

[62] Wang, Z. and Cohen, S. M., 2009. Postsynthetic modification of metal–organic frameworks. *Chemical Society Reviews*, *38*(5), 1315-1329.

[63] Sakata, Y., Furukawa, S., Kondo, M., Hirai, K., Horike, N., Takashima, Y., Uehara, H., Louvain, N., Meilikhov, M., Tsuruoka, T. and Isoda, S., 2013. Shape-memory nanopores induced in coordination frameworks by crystal downsizing. *Science*, *339*(6116), 193-196.

[64] Zhao, S., Wang, Y., Dong, J., He, C. T., Yin, H., An, P., Zhao, K., Zhang, X., Gao, C., Zhang, L. and Lv, J., 2016. Ultrathin metal-organic framework nanosheets for electrocatalytic oxygen evolution. *Nature Energy*, *1*(12), 1-10.

[65] Wang, W., Zhai, W., Chen, Y., He, Q. and Zhang, H., 2022. Two-dimensional material-based virus detection. *Science China Chemistry*, *65*(3), 497-513.

[66] Tan, C., Cao, X., Wu, X. J., He, Q., Yang, J., Zhang, X., Chen, J., Zhao, W., Han, S., Nam, G. H. and Sindoro, M., 2017. Recent advances in ultrathin two-dimensional nanomaterials. *Chemical reviews*, *117*(9), 6225-6331.

[67] Sun, L., Campbell, M. G. and Dinca, M., 2016. Electrically conductive porous metal–organic frameworks. *Angewandte Chemie International Edition*, *55*(11), 3566-3579.

[68] Peng, L., Mei, X., He, J., Xu, J., Zhang, W., Liang, R., Wei M., Evans D. G. and Duan X., 2018. Monolayer nanosheets with an extremely high drug loading toward controlled delivery and cancer theranostics. *Adv. Mater.* 30 (16), 1707389.

[69] Ji, Q., Yamazaki, T., Sun, J., Gorecka, Z, Huang, N. C., Hsu, S., Shrestha, L. K., Hill, J. P., and Ariga, K., 2017. Spongelike porous silica nanosheets: from 'soft' molecular trapping to DNA delivery. *ACS Appl. Mater. Interfaces* 9 (5), 4509–4518.

[70] Li, C., Hu, J., Li, W., Song, G., and Shen, J. 2017. Combined bortezomib-based chemotherapy and P53 gene therapy using hollow mesoporous silica nanospheres for P53 mutant non-small cell lung cancer treatment. *Biomater. Sci.* 5 (1), 77–88.

[71] Kou, Z., Wang, X., Yuan, R., Chen, H., Zhi, Q., Gao, L., Wang, B., Guo, Z., Xue, X., Cao, W., and Guo, L. 2014. A promising gene delivery system developed from PEGylated MoS_2 nanosheets for gene therapy. *Nanoscale Res. Lett.* 9 (1), 587.

[72] Teimouri, M., Nia, A. H., Abnous, K., Eshghi, H. and Ramezani, M., 2016. Graphene oxide–cationic polymer conjugates: synthesis and application as gene delivery vectors. *Plasmid*, *84*, 51-60.

[73] Sahay, G., Alakhova, D. Y. and Kabanov, A. V., 2010. Endocytosis of nanomedicines. *Journal of controlled release*, *145*(3), pp.182-195.

[74] Conner, S. D. and Schmid, S. L., 2003. Regulated portals of entry into the cell. *Nature*, *422*(6927), 37-44.

[75] Huang, J., Zong, C., Shen, H., Liu, M., Chen, B., Ren, B., and Zhang Z., 2012. Mechanism of cellular uptake of graphene oxide studied by surface-enhanced raman spectroscopy. *Small* 8 (16), 2577–2584.

[76] Linares, J., Matesanz, M. C., Vila, M., Feito, M. J., Gonçalves, G., Vallet-Regi, M., Marques, P. A. and Portolees, M. T., 2014. Endocytic mechanisms of graphene oxide nanosheets in osteoblasts, hepatocytes and macrophages. *ACS Appl. Mater. Interfaces* 6 (16), 13697–13706.

[77] Mu, Q., Su, G., Li, L., Gilbertson, B. O., Yu, L. H. and Zhang, Q., 2012. Size-dependent cell uptake of protein-coated graphene oxide nanosheets. *ACS Appl. Mater. Interfaces* 4 (4), 2259–2266.

[78] Luzio, J. P., Hackmann, Y., Dieckmann, N. M. and Griffiths, G. M., 2014. The biogenesis of lysosomes and lysosome-related organelles. *Cold Spring Harbor perspectives in biology*, *6*(9), 016840.

[79] Bao, H., Pan, Y., Ping, Y., Sahoo, N. G., Wu, T., Li, L., Li, J. and Gan, L. H., 2011. Chitosan- functionalized graphene oxide as a nanocarrier for drug and gene delivery. *Small* 7 (11), 1569–1578.

[80] Kou, Z., Wang, X., Yuan, R., Chen, H., Zhi, Q., Gao, L., Wang, B., Guo, Z., Xue, X., Cao, W. and Guo, L., 2014. A promising gene delivery system developed from PEGylated MoS2 nanosheets for gene therapy. *Nanoscale Res. Lett.* 9 (1), 587.

[81] Boussif, O., Lezoualc'h, F., Zanta, M. A., Mergny, M. D., Scherman, D., Demeneix, B., and Behr, J. P., 1995. A versatile vector for gene and oligonucleotide transfer into cells in culture and *in vivo*: polyethylenimine. *Proc. Natl. Acad. Sci.* 92 (16), 7297–7301.

[82] Varkouhi, A. K., Scholte, M., Storm, G., and Haisma, H. J., 2011. Endosomal escape pathways for delivery of biologicals. *J. Controlled Release* 151 (3), 220–228.

[83] Fang, J., Nakamura, H., and Maeda, H., 2011. The EPR effect: unique features of tumor blood vessels for drug delivery, factors involved, and limitations and augmentation of the effect. *Adv. Drug Deliv. Rev.* 63 (3), 136–151.

[84] Zhang, H., Chen, G., Yu, B., and Cong, H., 2019. Emerging advanced nanomaterials for cancer photothermal therapy. *Rev. Adv. Mater. Sci.* 53 (2), 131–146.

[85] Zhang, C., Yong, Y., Song, L., Dong, X., Zhang, X., Liu, X., Gu, Z., Zhao, Y. and Hu, Z. 2016. Multifunctional WS_2 @Poly(Ethylene Imine) nanoplatforms for imaging guided gene-photothermal synergistic therapy of cancer. *Adv. Healthc. Mater.* 5 (21), 2776–2787.

[86] Zeng, Z., Yin, Z., Huang, X., Li, H., He, Q., Lu, G., Boey, F. and Zhang, H., 2011. Single-layer semiconducting nanosheets: high-yield preparation and device fabrication. *Angewandte Chemie*, *123*(47), 11289-11293.

[87] Chen, S. J., Liu, Y. C., Shao, C. L., Mu, R., Lu, Y. M., Zhang, J. Y., Shen, D. Z. and Fan, X. W., 2005. Structural and optical properties of uniform ZnO nanosheets. *Advanced Materials*, *17*(5), 586-590.

[88] Priour Jr, D. J., Hwang, E. H. and Sarma, S. D., 2005. Quasi-two-dimensional diluted magnetic semiconductor systems. *Physical review letters*, *95*(3), 037201.

[89] Zhao, Q., Zhang, H., Liu, Y., Zhu, M. and Zhang, M., 2019. Magnetic and optical properties of two-dimensional SnS2 nanosheets doped with Ho ions. *Applied Surface Science*, *481*, 1370-1376.

[90] Yi, H., Liu, S., Lai, C., Zeng, G., Li, M., Liu, X., Li, B., Huo, X., Qin, L., Li, L. and Zhang, M., 2021. Recent advance of transition-metal-based layered double hydroxide nanosheets: synthesis, properties, modification, and electrocatalytic applications. *Advanced Energy Materials*, *11*(14), 2002863.

[91] Tao, H., Gao, Y., Talreja, N., Guo, F., Texter, J., Yan, C. and Sun, Z., 2017. Two-dimensional nanosheets for electrocatalysis in energy generation and conversion. *Journal of Materials Chemistry A*, *5*(16), 7257-7284.

Chapter 2

Boron Nitride Nanosheets

Young Nam Kim
Sooyeon Ryu
Hyunsung Jeong
and Yong Chae Jung*

Institute of Advanced Composite Materials, Korea Institute of Science and Technology (KIST), 92, Chudong-ro, Bongdong-eup, Wanju-gun, Jeonbuk, Republic of Korea

Abstract

Boron nitride (BN) refers to substances composed of boron and nitrogen atoms. Based on their structural form at the nanoparticle level, they can be categorized into three types: BN nanotubes (BNNT, 1D), BN nanosheets (BNNS, 2D), and BN spherical nanoparticles (BNNP, 3D). BN exhibits high thermal conductivity similar to carbon-based materials while also possessing remarkable insulation properties. In response to the increasing demand for miniaturized and highly integrated electronic devices in modern society, there is a growing need for materials that combine excellent thermal conductivity with insulation properties. Consequently, BN has garnered significant attention in technologically advanced areas such as electric vehicles and semiconductor fields, leading to extensive research efforts. This chapter aims to explore the fundamental characteristics of BN and showcase its potential utility to readers.

Keywords: hexagonal boron nitride (h-BN), Boron nitride sheets (BNNs), Boron nitride nanotube (BNNT)

---

* Corresponding Author's Email: ycjung@kist.re.kr.

In: Properties and Applications of Nanosheets
Editors: Nanda Gopal Sahoo and Anirban Dandapat
ISBN: 979-8-89113-381-5

1. Introduction

Recently, the fabrication of lightweight, thin, miniaturized, highly integrated, and multifunctional products have been pursued for applications in the electric and electronic fields, such as in cutting-edge electronic devices for automobiles. As electronic devices become more highly integrated, a large amount of heat is generated. The heat emitted in this way not only degrades the device's function but also causes the malfunction of peripheral devices and substrate deterioration. Therefore, significant research is being conducted on technologies for controlling the heat released. High-heat-dissipation circuit boards based on metal substrates with excellent thermal conductivity are applied to dissipate heat from parts that consume high power and generate a large amount of heat, such as power devices and light-emitting diode (LED) modules. However, owing to the limitation of the thermal conductivity of the insulating layer substrate material, their performance cannot be improved beyond a certain limit. Further research is needed to improve the heat dissipation performance of components required for heat dissipation measures, such as flip-chip semiconductor packaging substrate materials for high integration of next-generation packages, system integrated circuits (ICs), and multi-stacking package materials. (*Figure 1*) Looking at the material components of the heat dissipation material, a composite material comprising a filler material with high thermal conductivity and a polymer material is used. Representative fillers with high thermal conductivity currently include nanomaterials with a 2-dimensional (2D) structure.

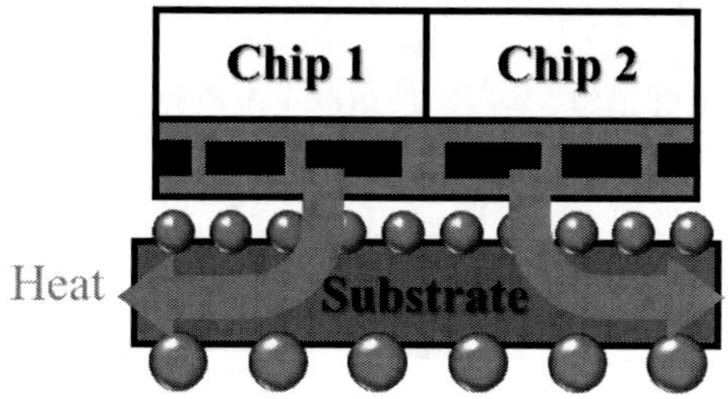

Figure 1. Schematic diagram of heat flow in a highly integrated circuit.

Since the discovery of graphene in 2004, the family of 2D materials has grown. These materials have unique physical, chemical, and electronic properties compared to their 3-dimensional (3D) counterparts. Such properties can be attributed to their band structure's dimensionality effect and modulation [1]. Graphene, commonly deployed as a thermally conductive filler, has a thin film structure comprising carbon atoms connected in a honeycomb shape. Although graphene is only 0.35 nm thick, its electrical conductivity is 100 times higher than copper; its thermal conductivity is two times higher than diamond; and it is stronger than steel. While graphene has excellent electrical conductivity due to sp2 hybridization, its application to electrical and electronic materials that require insulation is challenging due to high integration [2]. A material actively researched along with graphene is hexagonal boron nitride nanosheet (h-BNN). Graphene and h-BNN share a similar chemical structure, thermal conductivity, and dielectric constant of 3–4 [3].

Boron nitride (BN) is a nanomaterial composed of B and N with a carbon structure. It exists in various forms. The hexagonal form (h-BN) is similar to graphite; cubic form (c-BN) is isostructural to diamond; and amorphous form (a-BN) is similar to amorphous carbon. Depending on the shape, 2D BN sheets (BNNS) [4-6] 1-dimensional (1D) BN tubes (BNNT) [7, 8] or 3D BN spherical nanoparticles (BNNP) can be used. [9, 10] Among them, hexagonal BN (h-BN), a 2D network of hexagonally structured materials, has attracted considerable research interest. [11, 12] Both h-BN and graphite have a hexagonal structure, but graphite has excellent conductivity due to sp^2 hybridization. However, in h-BN, electrons are localized close to N in the electronegativity difference (B = 2.04 and N = 3.04) of other atoms and σ bonds. However, in a π bond containing an empty *p*-orbital of B and full orbital of N, the N electrons are much less delocalized than the π electrons of C–C bonds. Consequently, h-BN is called "white graphene" or "non-carbon graphene" due to its analogous 2D atomic crystal structure and van der Waals heterojunction. It shows distinctly different properties from graphite in terms of optical, electrical, and chemical properties [11, 13-16]. With these properties, h-BN appears white or translucent, has a high thermal conductivity (600 W/mK), high insulation, wide band gap (5.2–5.5 eV), high specific surface area (3300 m2/g), and thermal and chemical stability [17].

2. Preparation of Boron Nanosheets (BNNS)

2.1. Mechanical Exfoliation

BNNs were first obtained by mechanical exfoliation or fragmentation, a technique developed in 2004 [18]. Adhesive tape was used to peel off the BN layer that continuously adhered to substrates such as Si/SiO_2. Additional shear treatment can aid layer separation by disrupting the van der Waals interactions between BN layers. In addition, a ball-milling technique for the separation of the BN layer was first attempted in 2011 by Li et al., [19]. *Figure 2* is a schematic representation of BNNs obtained through the ball milling technique and the action of the ball.

Figure 2. Corresponding diagrams illustrating two observed exfoliating mechanisms under the shear force created by milling balls.

2.2. Chemical Exfoliation (Sonication)

The fabrication of single- and few-layered nanosheets from single-crystal h-BN using a chemical solution induction method was first performed by Han et al., [20]. h-BN crystals were sonicated in an organic solution to break them down into few-layered h-BN sheets. Subsequently, nanosheet exfoliation of the BN particles dispersed in a strongly polar solvent was performed by vigorous sonication and centrifugation [21]. Solvent molecules with strong polarity were exfoliated owing to the strong interactions between the BN surfaces. BNNs with a thickness of 2–10 nm were obtained through chemical treatment, and they can be obtained by chemical treatment in various solvents to obtain effective BNNs [22, 23].

2.3. Chemical Vapor Deposition (CVD)

BNNs can be synthesized by chemical vapor deposition (CVD) techniques from various precursors such as BF3-NH3, BCl3-NH3, borazine (B3N3H6), trichloroborazine (B3N3H6), or hexachloroborazine (B3N3Cl6) [24-26]. Borazine and its compounds are easier to use than BF3-NH3 or BCl3-NH3 because these precursors release multiple gases. Borazine (B3N3H6) is a compound similar to benzene, except that the atoms are B and N in equal proportions. Consequently, the two parent atoms in the final product are only present in the precursor. This avoids the need for more complex systems that use two or more gaseous precursors.

In 2009, the Gao et al., group synthesized a relatively large number of thick h-BN sheets using a catalyst-free CVD process at 1100–1300 °C [27]. B2O3 and melamine powders were mechanically mixed and placed in an induction furnace. The temperature was raised to 1000–1300 °C under N2 flow, and h-BN sheets of 25–50 nm thickness were obtained, depending on the synthesis temperature. Subsequently, a multistep thermal catalytic CVD method was used to fabricate h-BN films composed of 2–5 atomic layers under ammonia-borane (NH3–BH3) flow using Cu foil as the substrate [28]. Few-layered h-BN films were also synthesized by atmospheric-pressure CVD on polycrystalline Ni films with 5–50 nm thicknesses, depending on the growth conditions [29]. h-BN grew continuously on the entire Ni surface, with areas of uniform thickness up to 20 μm in lateral size. In another series of experiments, the polyborane decaborane (B10H14) reaction with NH3 formed BNNS on polycrystalline Ni or Cu foils at 1000 °C [30]. Recently, single-layer h-BN was grown on Cu foil using NH3-BH3 via low-pressure CVD with two heating zones [31].

3. Application of Boron Nitride Nanosheets (BNNs)

BNNs manufactured using various methods have been used in several studies. Typically, it is compounded with a polymer as a reinforcing material and applied in various fields, such as heat conduction, energy, insulation, and purification.

3.1. Thermal Conductivity

Wang et al., [32] used BNNs to increase the thermal conductivity of polyimide (PI). PIs are widely applied as electronic packaging materials owing to their excellent thermal and mechanical properties. However, PI has low thermal conductivity, whereas higher thermal conductivity is required for modern electronic devices. Such devices generate a higher amount of heat owing to the increased integration of circuits for the miniaturization and performance improvement of electronic materials. Therefore, using BNNs with excellent insulation performance and thermal conductivity, effective heat dissipation can be achieved, which can ensure the longevity and reliability of electronic devices [33-36]. First, bundled BN powder was obtained as exfoliated BNNs using NaOH and KOH solutions, as shown in Figure 3.

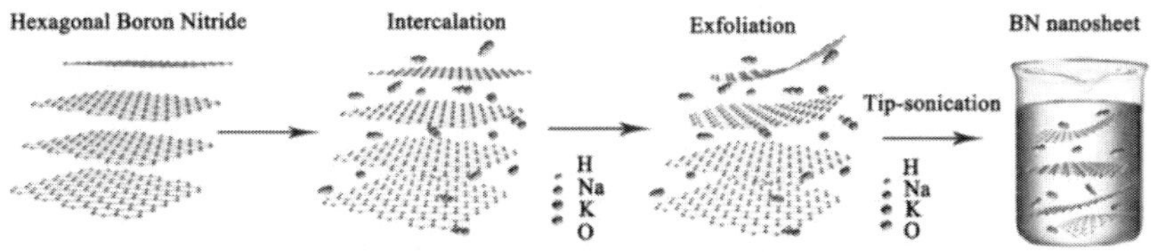

Reprinted with permission from [32].

Figure 3. Schematic diagram of the exfoliation process.

Thermal conduction was evaluated by complexing the unbundled BNNs with PI. BNNs are 2D materials in the form of sheets, and when they are composited with PI, their thermal conductivities exhibit the same tendency as when they are aligned in the in-plane direction.

Figure 4 shows the thermal conductivity and diffusivity data measured when BNNs were added according to the PI content. Because the BNNs were aligned to PI in the in-plane direction, the thermal conductivity and thermal diffusivity in the in-plane direction were high, as shown in Figures 4(a) and (b). In addition, as the content increased, the thermal conductivity increased and showed a value of 2.95 W/mK at 7 wt%. When observing the thermal diffusivity through a thermal imaging camera, the thermal diffusivity of the 7 wt% samples of PI/BNNs was much higher than that of neat PI (Figure 4(d)).

Zhou et al., [37] used fillers with various structures for effective heat conduction in polyamide-imide (PAI). BN quantum dots (BNQDs) were added to make heat conduction more effective than single BNNs. By filling the gap between the PAI and BNNs interfaces and between BNNs, BNQDs complexed with PAI can provide a path for continuous heat conduction.

As shown in Figure 5(a), BNQDs were prepared from BN powder using an ultrasonic solvothermal method. Because of the DMF solvent's polarity, BN can be easily separated into BNNs and formed into quantum dots by additional heat. The PAI/BNNs/BNQD films were prepared by combining the prepared BNQDs and BNNs with PAI. (Figure 5(b)) In PAI/BNNs/BNQDs, as the filler content increased, the thermal conductivity continuously increased in the in-plane direction. The thermal conductivity showed a value of 7.69 W/mK at a maximum of 9 wt%. This is a 5% increase compared to a BNNs-only application.

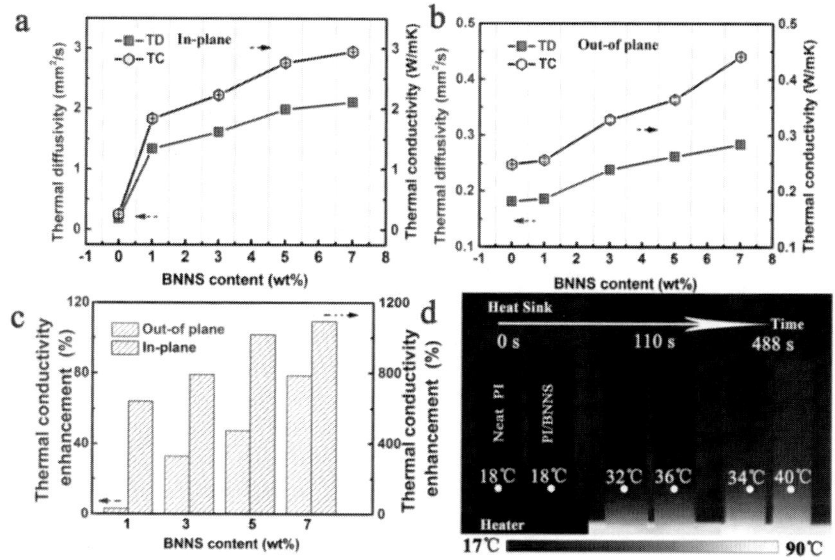

Figure 4. (a) Thermal diffusivity and thermal conductivity of neat PI and PI/BNNSs composite along in-plane direction. (b) Thermal diffusivity and thermal conductivity of neat PI and PI/BNNSs composite along out-of plane direction. (c) Thermal conductivity enhancement of PI/BNNSs composites compared to neat PI. (d) The temperature profile evolutions in time of neat PI and 7 wt% PI/BNNSs composite were captured using a calibrated infrared camera as a function of heating time. Reprinted with permission from [32].

To determine whether the PAI/BNNS-9/BNQD composite film provides a new heat conduction path when BNNs, a two-dimensional material in PAI, and BNQDs, a zero-dimensional material, are used simultaneously, its fluorescence was measured. Figures 6(a)–(c) show the BNQDs, and Figure 6(d) shows the thermal conduction paths of the BNNs and BNQDs/BNNs. The red line represents the heat conduction path when BNNs are single, and the yellow dotted line represents the heat conduction path of BNN/BNQDs. In short, the PAI/BNN/BNQDs composite film can exhibit improved thermal conductivity by providing various thermal conduction paths.

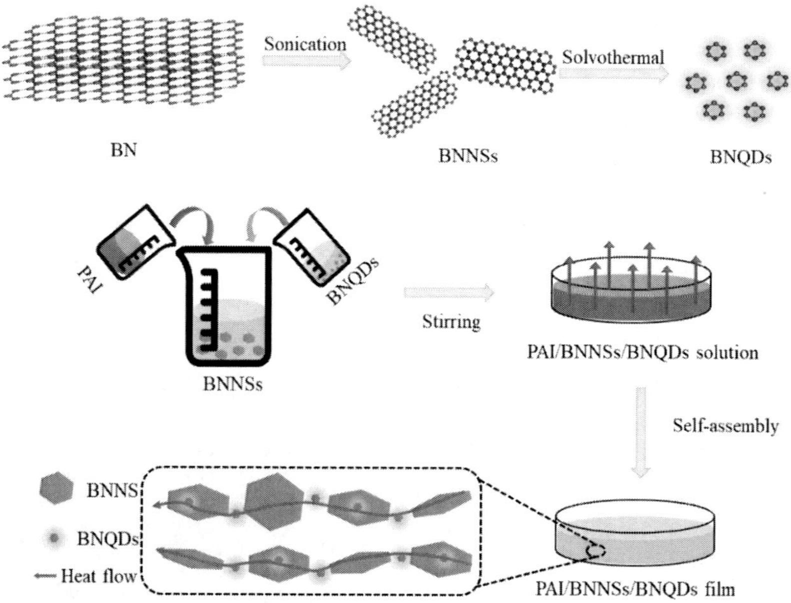

Figure 5. Schematic illustration of the preparation procedure of (a) BNQDs and (b) PAI/BNNS/BNQDs film.

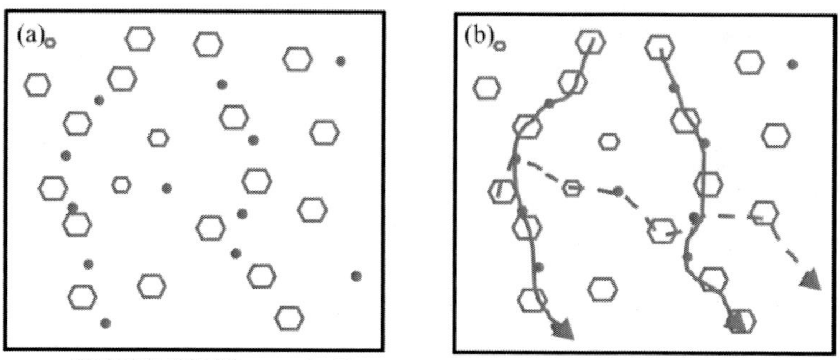

Figure 6. (a) Schematic of the PAI/BNNS-9/BNQD composite film. (b) Schematic diagram of heat flow in the PAI/BNNS-9/BNQD composite film.

3.2. Water Purification

BN-based materials and composites have been studied in various water purification fields owing to their unique structure and properties. A large specific surface area, structural defects, functional groups, and bulk π-π systems are desirable properties that enhance the interaction of contaminant molecules with the surface of BN materials. The porous structure of BN with numerous structural defects and ionic B–N bonds with "lop-sided" density characteristics make it a suitable candidate for contaminant adsorption. In addition, coexisting basic sites (N atoms) and Lewis acidic sites (B atoms) play an essential role in the uptake of contaminants [38, 39].

Weiwei Lei et al., [40] synthesized a white porous BN nanosheet using only a heat treatment without an ex-situ template. Boron trioxide and guanidine hydroxide were used as precursors. In forming the BN nanosheet, the guanidine hydroxide was decomposed by emitting many gases, including N_2, HCl, and NH_3, at high temperatures. The condensation by-product of the solid eventually acted as a nanobubble in the porous structure [41].

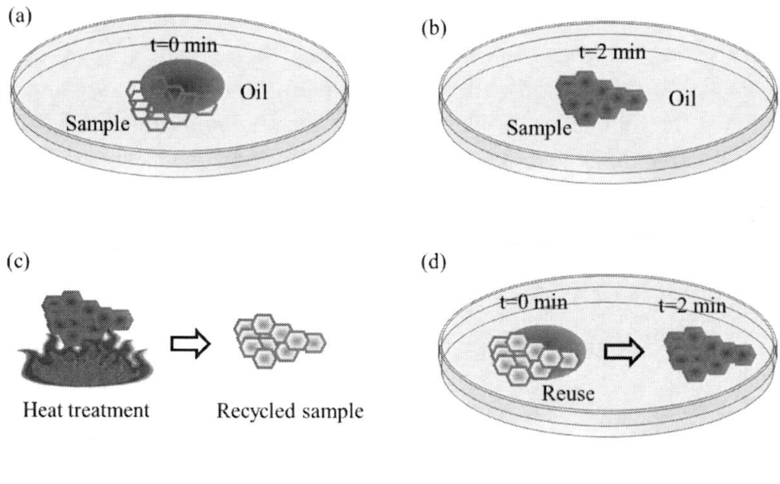

Figure 7. (a) Photograph of the set-up for oil absorption tests with white porous BN nanosheets. (b) Photograph of porous BN nanosheets saturated with oil after 2 min of absorption, inset showing the absorption process after 20 s. (c) Photograph of burning oil-saturated porous BN nanosheets in air for cleaning purpose, inset showing the color change after burning. (d) Photograph of the cleaned nanosheets for second oil absorption test, inset showing the absorption result after 2 min.

Porous BN nanosheet shows the absorption of ethanol, toluene, pump oil, engine oil, and ethylene glycol in the 2,000-3,300 wt% range. The porous BN nanosheet absorbed 29 times the weight of ethylene glycol and engine oil, which was 33 times its weight. This absorption is much higher than that of commercial activated carbon and BN particles and is much higher than that of most general absorbers and other refined materials such as graphene capsules, collagen nanocomposites, polyurethane sponges, polyurethane and iron oxide composites, MnO2 nanowires and graphene hydrogels.

Figures 8 (a-d) show how the stages of engine oil absorption. As soon as the white nanosheets were dropped onto the oil-water surface, they immediately absorbed the brown oil and turned dark brown. After 2 min, all the oil had been absorbed into the nanosheets. In addition, the initial absorbency of the porous BN nanosheets could be restored by burning them in the air (Figure 8(c)) or by washing them with ethanol or petroleum ether to remove the absorbed organic matter. (Figure 8(d)) The dye absorbance of the porous BN nanosheets is additionally confirmed. The UV-vis absorption spectrum of the aqueous solution using the anionic dye Congo red (CR) and cationic dye methylene blue (MB) showed high dye removal efficiency, as the characteristic band became very weak over time and could not be observed. Therefore, the adsorbed dye could be easily removed from the porous BN nanosheets.

Pendse et al., [1] investigated the ion transport characteristics. They evaluated its application in water purification by depositing h-BNF with a peeled nm thickness without surface functionalization on an anodized alumina oxide (AAO) membrane support. When h-BN with a thickness of 20 nm is peeled off through ultrasonic treatment, h-BN with a thickness of 4 nm can be obtained. To prepare a high-efficiency ion-selective two-dimensional membrane, h-BN was deposited on an anodized alumina oxide (AAO) membrane support. To evaluate whether h-BN can be used for water purification, its ion transport characteristics were examined.

The ion separation performance of the h-BN membrane was evaluated using ions with various hydration radii and valances. This indicates that the ion filtration reaction increased as the hydration radius increased. Although the filtration response to NaCl was lower than 40%, the ion filtration performance of the membrane started to increase for ions with a hydration radius close to 4 Å. The filtration response for ions larger than 4.5Å rose further, and the membrane showed high efficiency for FCN3- (R = 95%) and PTS4- (R = 98%). However, the h-BN membrane has a larger hydration radius than PTS4-, but shows a low ion filtration response of 20% (z−/z+ = 0.5) to Rubipy. These results suggest that the electrostatic interaction between the negative ions and negatively charged ions on the surface of the membrane plays vital role in the ion filtration properties of h-BN membranes.

In addition, to elucidate the effect of the surface charge on the ion removal performance, aadiya pendse et al., group investigated the pH-dependent separation performance of the h-BNF membrane because the solution pH is known to have a significant effect on the surface charge of monolayer BN sheets. At a relatively low pH (pH = 3), the membrane exhibited an ion filtration response of more than 10%. The filtration response increased rapidly as the pH increased above 3 and stabilized at pH > 9, and the maximum ion filtration response for the potassium ferricyanide (PFCN) solution was approximately 97.5%. At a higher pH, the solution contained more hydroxyl ions, which increased their adsorption onto the h-BN surface and consequently increased the surface charge density. Therefore, there is a stronger electrostatic repulsion between the membrane's fixed charge and the mobile ions, making the ion filtration reaction at high pH [42]. Considering the role of electrostatic interactions, the valence of anions and cations in a solution is also known to significantly affect the filtration reaction through a charged membrane. Ion exchange chromatography analysis revealed a decrease in anion and cation removal according to the z-/z+ ratio.

Therefore, the 2D h-BN membrane showed a very high filtration performance for all anionic dyes and salts while maintaining stable water permeability. The filtration performance of the salt solutions was found to increase with the hydration radius of the salt ions, and the pH filtration (pH vs. rejection % curve) of the solution tended to be similar to the zeta potential curve, suggesting that electrostatic interactions also contribute to filtration through the h-BNF membrane. Therefore, the high separation performance of the h-BNF membrane resulted from electrostatic repulsion effects due to densely filled structures, high surface charge densities, and molecular sieves.

3.3. Energy Storage and Dielectric Materials

The need for high-efficiency, high-reliability electrical energy storage devices is increasing owing to the rapid demand for 5G communications, electric vehicles, high-speed rail, smart power grids, and advanced aircraft [43, 44]. Dielectric capacitors exhibit higher operating voltages and power densities, longer lifetimes, and greater cycling stabilities, making them suitable for renewable energy storage and distributed power systems. Polymer dielectric capacitors have made significant progress owing to their intrinsic advantages, including lightweight, low cost, ease of processing, mechanical flexibility, excellent electrical breakdown strength, and relatively low dielectric loss [45-

47]. However, because most polymer dielectrics easily suffer significant energy loss due to the heat generated in a high magnetic field, both the energy density and charge/discharge efficiency of polymer dielectrics for advanced microelectronics and power systems drop rapidly [48-50].

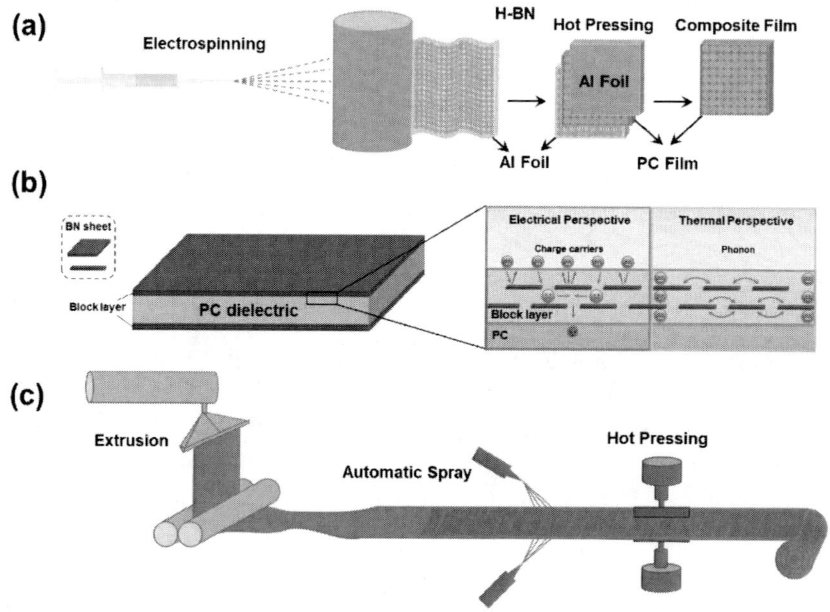

Figure 8. Flow chart of preparation of sandwich structure film (a); working principle diagram of surface h-BN functional layer (b); predicted flow chart of industrial production (c).

Liu et al., [51] fabricated BNNs in a polycarbonate polymer as a composite insulating material with a sandwich structure (h-BN/PC/h-BN) through electrospinning and hot pressing, depending on the BN thickness. This method is straightforward because the h-BN layer is coated on the surface, and heat is efficiently conducted along the surface, wherein the heat concentration of the film can be dispersed. h-BN has a large bandgap, and when an h-BN layer is introduced, a potential barrier is formed owing to the difference in the bandgap and electron affinity. Therefore, the h-BN layer can increase the barrier between the electrode and dielectric film and effectively suppress charge injection from the electrode/dielectric interface, thereby reducing the film's electrical conductivity, leakage current density, and leakage current (Figure 8).

In the same temperature range, the dielectric constant of the BN composite with a sandwich structure increased as thickness of the BN layer increased. There are two main reasons for this finding. Firstly, the dielectric constant of h-BN (4–5) is approximately higher than that of PC (3). Secondly, different interfaces are formed between the h-BN coating on the surface of the PC film and the PC film. As the thickness of h-BN increases, the h-BN/PC/h-BN composite interface increases, resulting in greater polarization and an increase in the dielectric constant [52, 53].

The energy storage properties of dielectric films and the stability of their energy storage density with temperature changes are also crucial for practical applications. According to the research team's results, energy storage densities of PC, BN-1 and BN-2 have better temperature stability and the relative rate of change of their energy storage densities with respect to room temperature within 100 °C remains within 20%. In addition, the charge and discharge efficiency of BN-1 had the best temperature stability, and the relative rate of change with respect to room temperature was maintained within 10%.

The thermal conductivity of the sandwich-structured composite film was found to greatly improve owing to the presence of the h-BN thermally conductive layer. In addition, the sandwich structure composite film with an h-BN thickness of 1 μm (BN-1) at 100 °C has maximum thermal conductivity, low leakage current density, high applied electric field, and excellent high-temperature energy storage properties. The discharge energy density (energy storage density) of the BN-1 composite film is 5.52 J/cm at an electric field of 500 MV/m at 100 °C, which is 15.10% higher than pure PC. It also exhibits excellent charging and discharging efficiencies (87.25%).

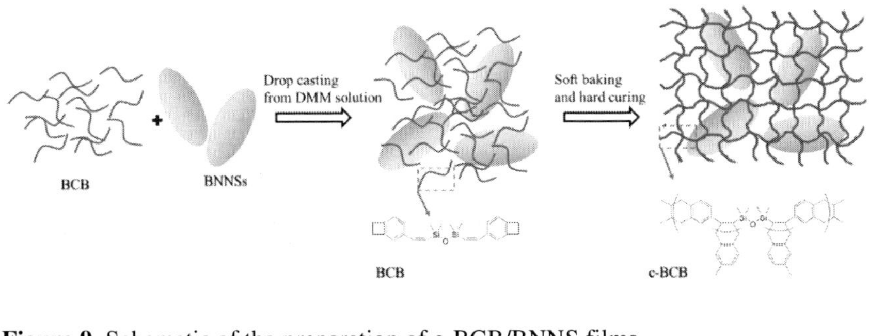

Figure 9. Schematic of the preparation of c-BCB/BNNS films.

Li et al., [48] prepared a c-BCB/BNNS composite film by compounding and curing BNNs with divinyltetramethyldisiloxane-bis(benzocyclobutene) (BCB) to replace biaxially oriented polypropylene (BOPP), which is widely used as an insulating polymer. (Figure 9)

When comparing the dielectric constants of polymers that have recently been widely used as insulating polymers (polycarbonate (PC), polyether ether ketone (PEEK), polyetherimide (PEI), fluorene polyester (FPE) and polyimide (Kapton PI)), c-BNB/BNNS maintained stable dielectric constant values and dissipation factors even at high temperatures. Also, temperature coefficient of dielectric constant of c-BCB/BNNs is approximately 65 parts per million (ppm)/°C. In contrast, for FPE and Kapton, it is 308 ppm/°C and 498 ppm/°C, respectively.

The research team compared the charge/discharge efficiency and discharge energy density (Ue) of the c-BCB/BNNS, FPE and Kapton polymers. c-BCB/BNNS outperformed other high glass transition polymer dielectrics (FPE and Kapton) in terms of discharge energy density (Ue) and charge-discharge efficiency (η) at temperatures ranging from 150°C to 250°C. For example, c-BCB/BNNS can discharge Ue at over 2.2 J cm^{-3} at 400 MVm^{-1} with an η of more than 90% at 150 °C. At 200 °C, c-BCB/BNNS gives a Ue of 2 J cm^{-3} at 400 MVm^{-1}, which is two times higher than that of PEI, and the η value is more than 1.5 times higher than that of PEI. When the temperature rises to 250 °C, the high Tg polymer dielectric cannot operate above 150 MVm^{-1} and the c-BCB/BNNS operates at up to 400 MVm^{-1} with a Ue of ~1.8 J cm^{-3}. Surprisingly, 97% of the η value of c-BCB/BNNS at 150 °C under the operating condition of 200 MVm^{-1}, which is the operating condition of the BOPP film capacitor in electric vehicles, is the same as that of BOPP at 70 °C. Also, under these conditions, the Ue of c-BCB/BNNS was more than 40% higher than that of BOPP, owing to its higher dielectric constant.

Zhai et al., [54] attempted to apply it as a sensor by combining BNNs with polyimide (PI) to form a sponge.

As shown in Figure 10, the compression-driven triboelectric properties of the BNN/PI sponges were prepared according to the BNN content. As shown in Figure 10(a), as the BNN content increases, the generated voltage value increases. However, when the content exceeds 21 wt%, the regenerated voltage decreases. This is because excessive BNNs tend to stick together owing to van der Waals forces.

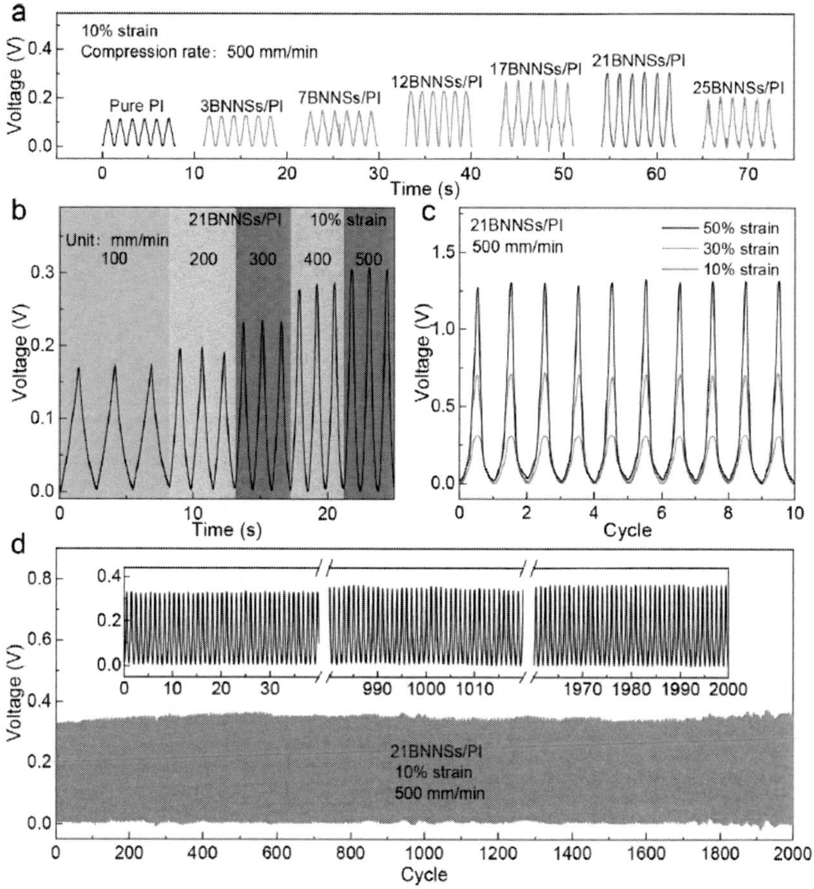

Figure 10. Compression-driven triboelectric properties of BNNSs/PI and blank PI foams. (a) Compression-driven triboelectric voltages of BNNSs/PI foams filled with different BNNS fractions under loading−unloading cycles for a compressive strain of 10% with a compression rate of 500 mm/min. (b) Compression-driven triboelectric voltages of 21BNNSs/PI foam under different compression rates of 100, 200, 300, 400, and 500 mm/min using loading−unloading cycles in a compressive strain of 10%. (c) Compression-driven triboelectric voltages of 21BNNSs/PI foam under different compression strains of 10%, 30%, and 50% with a compression rate of 500 mm/min. (d) Stability test of the compression-driven triboelectric voltages of 21BNNSs/PI foam during 2000 cycles under a compression strain of 10% using a compression rate of 500 mm/min. Reprinted with permission from [54].

Figure 11 shows the mechanism by which the voltage is generated in the PI/BNN sponge with 21 wt% BNNs added. In the 50–80% strain range, where compression occurs, a voltage of up to 1.10 V is generated because of friction between PI and BNNs.

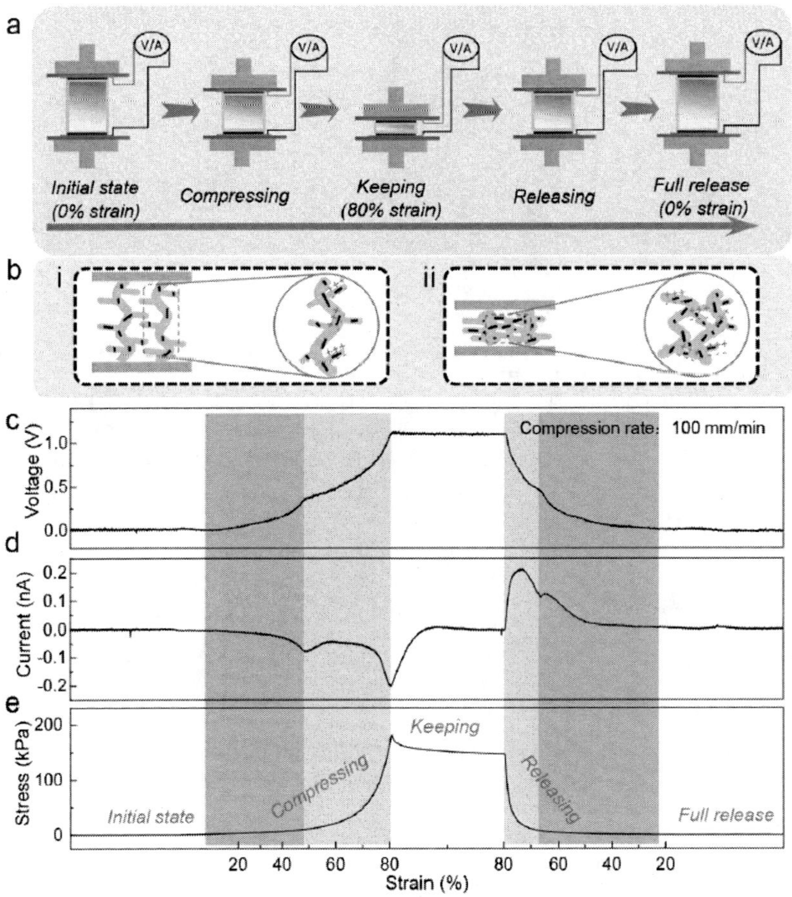

Figure 11. (a) Schematic diagram of a compression- and release-driven triboelectric electrification test of 21BNNSs/PI foams during loading–unloading processes. (b) Schematic diagram of the microscopic deformation of BNNSs/PI foams relative to their compression- and release-driventriboelectric electrification: (i) a low compression strain case and (ii) the case with a high compression strain. During compressive loading–unloading cycles of 21BNNSs/PI foam, the relationships of the transient signals of foam's generated (c) voltages and (d) currents vs (e) their compressive σ–ε cycle process. Reprinted with permission from [54].

Figure 12 shows that BNN/PI foam can be applied to footwear. Electrodes were attached to the BNN/PI foam with 21 wt% of BNNs in the insole to continuously generate electricity when a person walks. Consequently, it was able to generate up to 1 V of electricity. Second, applying the BNN/PI foam to a keyboard could also generate electricity when the keyboard is tapped.

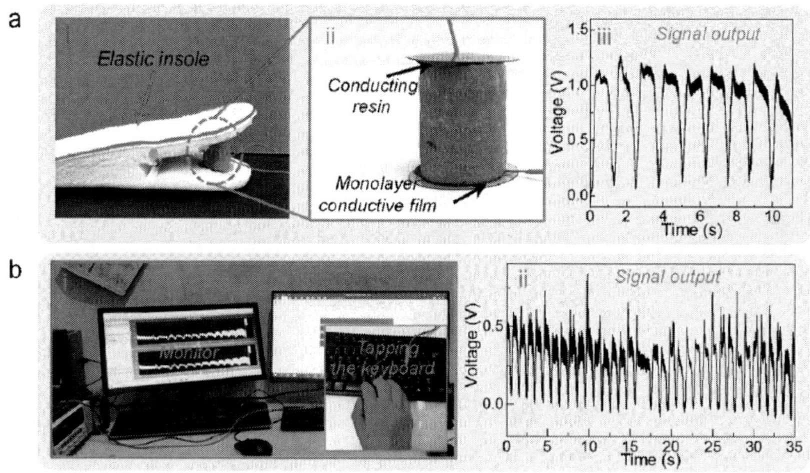

Figure 12. Application of BNNSs/PI foam for mechanical-compression sensors. (a) Application in an insole to generate energy: photos (i) showing the designed triboelectric electrification device integrated in the elastic insole, (ii) displaying the triboelectric electrification device based on 21BNNSs/PI foam, (iii) exhibiting the voltage generated by continuously and circularly squeezing the elastic insole with the foot. (b) The triboelectric electrification device based on our superelastic 21BNNSs/PI foam was built into the keys of a keyboard: (i) The voltage output was realized by tapping the keyboard (see the inset). (ii) Voltage generated by tapping the keyboard. Reprinted with permission from [54].

Conclusion

In this chapter, we introduce BN nanosheets, a representative 2D nanomaterial, as an effective solution to address the performance and lifespan limitations of electronic devices caused by the increasing heat generation due to their miniaturization and high integration. BN nanosheets possess exceptional heat conduction, a large specific surface area, and excellent insulation properties, among other advantages. These characteristics have prompted active research in areas such as energy and water treatment applications, as well as heat conduction materials. With the growing demand for portable electronic devices requiring more complex and diverse functions, effective heat management technology for materials is seen as a promising opportunity for researchers and future consumers. Furthermore, the application of such technology to eco-friendly materials holds great potential.

Acknowledgement

Authors acknowledge Korea Institute of Science and Technology, South Korea for financial support.

References

[1] Pendse, A.; Cetindag, S.; Lin, M. H.; Rackovic, A.; Debbarma, R.; Almassi, S.; Chaplin, B. P.; Berry, V.; Shan, J. W.; Kim, S., Charged layered boron nitride-nanoflake membranes for efficient ion separation and water purification. *Small* 2019, 15, (49), 1904590. https://doi.org/10.1002/smll.201904590.

[2] Zhu, H.; Li, Y.; Fang, Z.; Xu, J.; Cao, F.; Wan, J.; Preston, C.; Yang, B.; Hu, L., Highly thermally conductive papers with percolative layered boron nitride nanosheets. *ACS nano* 2014, 8, (4), 3606-3613. https://doi.org/10.1021/nn500134m.

[3] Dean, C. R.; Young, A. F.; Meric, I.; Lee, C.; Wang, L.; Sorgenfrei, S.; Watanabe, K.; Taniguchi, T.; Kim, P.; Shepard, K. L., Boron nitride substrates for high-quality graphene electronics. *Nature nanotechnology* 2010, 5, (10), 722-726. https://doi.org/10.1038/nnano.2010.172.

[4] Lin, Y.; Williams, T. V.; Connell, J. W., Soluble, exfoliated hexagonal boron nitride nanosheets. *The Journal of Physical Chemistry Letters* 2010, 1, (1), 277-283. https://doi.org/10.1021/jz9002108.

[5] Merlo, A.; Mokkapati, V. R. S. S.; Pandit, S.; Mijakovic, I., Boron nitride nanomaterials: biocompatibility and bio-applications. *Biomater Sci* 2018, 6, (9), 2298-2311. https://doi.org/10.1039/C8BM00516H.

[6] Yu, J.; Qin, L.; Hao, Y.; Kuang, S.; Bai, X.; Chong, Y.-M.; Zhang, W.; Wang, E., Vertically aligned boron nitride nanosheets: chemical vapor synthesis, ultraviolet light emission, and superhydrophobicity. *ACS nano* 2010, 4, (1), 414-422. https://doi.org/10.1021/nn901204c.

[7] Chopra, N. G.; Luyken, R.; Cherrey, K.; Crespi, V. H.; Cohen, M. L.; Louie, S. G.; Zettl, A., Boron nitride nanotubes. *Science* 1995, 269, (5226), 966-967. https://doi.org/10.1126/science.269.5226.966.

[8] Golberg, D.; Bando, Y.; Eremets, M.; Takemura, K.; Kurashima, K.; Yusa, H., Nanotubes in boron nitride laser heated at high pressure. *Appl. Phys. Lett.* 1996, 69, (14), 2045-2047. https://doi.org/10.1063/1.116874.

[9] Zhi, C.; Meng, W.; Yamazaki, T.; Bando, Y.; Golberg, D.; Tang, C.; Hanagata, N., BN nanospheres as CpG ODN carriers for activation of toll-like receptor 9. *J. Mater. Chem.* 2011, 21, (14), 5219-5222. https://doi.org/10.1039/C1JM10199D.

[10] Tang, C.; Bando, Y.; Huang, Y.; Zhi, C.; Golberg, D., Synthetic routes and formation mechanisms of spherical boron nitride nanoparticles. *Adv. Funct. Mater.* 2008, 18, (22), 3653-3661. https://doi.org/10.1002/adfm.200800493.

[11] Jiang, X.-F.; Weng, Q.; Wang, X.-B.; Li, X.; Zhang, J.; Golberg, D.; Bando, Y., Recent progress on fabrications and applications of boron nitride nanomaterials: a review. *Journal of Materials Science & Technology* 2015, 31, (6), 589-598. https://doi.org/10.1016/j.jmst.2014.12.008.

[12] Wang, J.; Wu, Y.; Xue, Y.; Liu, D.; Wang, X.; Hu, X.; Bando, Y.; Lei, W., Super-compatible functional boron nitride nanosheets/polymer films with excellent mechanical properties and ultra-high thermal conductivity for thermal management. *Journal of Materials Chemistry C* 2018, 6, (6), 1363-1369. https://doi.org/10.1039/C7TC04860B.

[13] Bryning, M.; Milkie, D.; Islam, M.; Kikkawa, J.; Yodh, A., Thermal conductivity and interfacial resistance in single-wall carbon nanotube epoxy composites. *Appl. Phys. Lett.* 2005, 87, (16), 161909. https://doi.org/10.1063/1.2103398.

[14] Yum, K.; Yu, M.-F., Measurement of wetting properties of individual boron nitride nanotubes with the Wilhelmy method using a nanotube-based force sensor. *Nano Lett.* 2006, 6, (2), 329-333. https://doi.org/10.1021/nl052084l.

[15] Kiran, M.; Raidongia, K.; Ramamurty, U.; Rao, C., Improved mechanical properties of polymer nanocomposites incorporating graphene-like BN: Dependence on the number of BN layers. *Scripta Mater.* 2011, 64, (6), 592-595. https://doi.org/10.1016/j.scriptamat.2010.12.007.

[16] Kim, K.; Kim, J., Fabrication of thermally conductive composite with surface modified boron nitride by epoxy wetting method. *Ceram. Int.* 2014, 40, (4), 5181-5189. https://doi.org/10.1016/j.ceramint.2013.10.076.

[17] Zhou, C.; Lai, C.; Zhang, C.; Zeng, G.; Huang, D.; Cheng, M.; Hu, L.; Xiong, W.; Chen, M.; Wang, J., Semiconductor/boron nitride composites: synthesis, properties, and photocatalysis applications. *Applied Catalysis B: Environmental* 2018, 238, 6-18. https://doi.org/10.1016/j.apcatb.2018.07.011.

[18] Novoselov, K. S.; Geim, A. K.; Morozov, S. V.; Jiang, D.-e.; Zhang, Y.; Dubonos, S. V.; Grigorieva, I. V.; Firsov, A. A., Electric field effect in atomically thin carbon films. *Science* 2004, 306, (5696), 666-669. https://doi.org/10.1126/science.1102896.

[19] Li, L. H.; Chen, Y.; Behan, G.; Zhang, H.; Petravic, M.; Glushenkov, A. M., Large-scale mechanical peeling of boron nitride nanosheets by low-energy ball milling. *J. Mater. Chem.* 2011, 21, (32), 11862-11866. https://doi.org/10.1039/C1JM11192B.

[20] Han, W.-Q.; Wu, L.; Zhu, Y.; Watanabe, K.; Taniguchi, T., Structure of chemically derived mono-and few-atomic-layer boron nitride sheets. *Appl. Phys. Lett.* 2008, 93, (22), 223103. https://doi.org/10.1063/1.3041639.

[21] Zhi, C.; Bando, Y.; Tang, C.; Kuwahara, H.; Golberg, D., Large-scale fabrication of boron nitride nanosheets and their utilization in polymeric composites with

improved thermal and mechanical properties. *Adv. Mater.* **2009,** 21, (28), 2889-2893. https://doi.org/10.1002/adma.200900323.

[22] Warner, J. H.; Rummeli, M. H.; Bachmatiuk, A.; Buchner, B., Atomic resolution imaging and topography of boron nitride sheets produced by chemical exfoliation. *ACS nano* 2010, 4, (3), 1299-1304. https://doi.org/10.1021/nn901648q.

[23] Wang, Y.; Shi, Z.; Yin, J., Boron nitride nanosheets: large-scale exfoliation in methanesulfonic acid and their composites with polybenzimidazole. *J. Mater. Chem.* 2011, 21, (30), 11371-11377. https://doi.org/10.1039/C1JM10342C.

[24] Pierson, H. O., Boron nitride composites by chemical vapor deposition. *J. Compos. Mater.* 1975, 9, (3), 228-240. https://doi.org/10.1177/002199837500900302.

[25] Auwärter, W.; Suter, H. U.; Sachdev, H.; Greber, T., Synthesis of one monolayer of hexagonal boron nitride on Ni (111) from B-trichloroborazine (ClBNH) 3. *Chem. Mater.* 2004, 16, (2), 343-345. https://doi.org/10.1021/cm034805s.

[26] Müller, F.; Stöwe, K.; Sachdev, H., Symmetry versus commensurability: epitaxial growth of hexagonal boron nitride on Pt (111) from B-trichloroborazine (ClBNH) 3. *Chem. Mater.* 2005, 17, (13), 3464-3467. https://doi.org/10.1021/cm048629e.

[27] Gao, R.; Yin, L.; Wang, C.; Qi, Y.; Lun, N.; Zhang, L.; Liu, Y.-X.; Kang, L.; Wang, X., High-yield synthesis of boron nitride nanosheets with strong ultraviolet cathodoluminescence emission. *The Journal of Physical Chemistry C* 2009, 113, (34), 15160-15165. https://doi.org/10.1021/jp904246j.

[28] Song, L.; Ci, L.; Lu, H.; Sorokin, P. B.; Jin, C.; Ni, J.; Kvashnin, A. G.; Kvashnin, D. G.; Lou, J.; Yakobson, B. I., Large scale growth and characterization of atomic hexagonal boron nitride layers. *Nano Lett.* 2010, 10, (8), 3209-3215. https://doi.org/10.1021/nl1022139.

[29] Shi, Y.; Hamsen, C.; Jia, X.; Kim, K. K.; Reina, A.; Hofmann, M.; Hsu, A. L.; Zhang, K.; Li, H.; Juang, Z.-Y., Synthesis of few-layer hexagonal boron nitride thin film by chemical vapor deposition. *Nano Lett.* 2010, 10, (10), 4134-4139. https://doi.org/10.1021/nl1023707.

[30] Chatterjee, S.; Luo, Z.; Acerce, M.; Yates, D. M.; Johnson, A. C.; Sneddon, L. G., Chemical vapor deposition of boron nitride nanosheets on metallic substrates via decaborane/ammonia reactions. *Chem. Mater.* 2011, 23, (20), 4414-4416. https://doi.org/10.1021/cm201955v.

[31] Kim, K. K.; Hsu, A.; Jia, X.; Kim, S. M.; Shi, Y.; Hofmann, M.; Nezich, D.; Rodriguez-Nieva, J. F.; Dresselhaus, M.; Palacios, T., Synthesis of monolayer hexagonal boron nitride on Cu foil using chemical vapor deposition. *Nano Lett.* 2012, 12, (1), 161-166. https://doi.org/10.1021/nl203249a.

[32] Wang, T.; Wang, M.; Fu, L.; Duan, Z.; Chen, Y.; Hou, X.; Wu, Y.; Li, S.; Guo, L.; Kang, R., Enhanced thermal conductivity of polyimide composites with boron nitride nanosheets. *Scientific reports* 2018, 8, (1), 1-8. https://doi.org/10.1038/s41598-018-19945-3.

[33] Chen, H.; Ginzburg, V. V.; Yang, J.; Yang, Y.; Liu, W.; Huang, Y.; Du, L.; Chen, B., Thermal conductivity of polymer-based composites: Fundamentals and applications. *Prog. Polym. Sci.* 2016, 59, 41-85. https://doi.org/10.1016/j.progpolymsci.2016.03.001.

[34] Li, Q.; Guo, Y.; Li, W.; Qiu, S.; Zhu, C.; Wei, X.; Chen, M.; Liu, C.; Liao, S.; Gong, Y., Ultrahigh thermal conductivity of assembled aligned multilayer graphene/epoxy composite. *Chem. Mater.* 2014, 26, (15), 4459-4465. https://doi.org/10.1021/cm501473t.

[35] Hu, J.; Huang, Y.; Zeng, X.; Li, Q.; Ren, L.; Sun, R.; Xu, J.-B.; Wong, C.-P., Polymer composite with enhanced thermal conductivity and mechanical strength through orientation manipulating of BN. *Compos. Sci. Technol.* 2018, 160, 127-137. https://doi.org/10.1016/j.compscitech.2018.01.045.

[36] Dai, W.; Lv, L.; Lu, J.; Hou, H.; Yan, Q.; Alam, F. E.; Li, Y.; Zeng, X.; Yu, J.; Wei, Q., A paper-like inorganic thermal interface material composed of hierarchically structured graphene/silicon carbide nanorods. *ACS nano* 2019, 13, (2), 1547-1554. https://doi.org/10.1021/acsnano.8b07337.

[37] Zhou, S.; Xu, T.; Jiang, F.; Song, N.; Shi, L.; Ding, P., High thermal conductivity property of polyamide-imide/boron nitride composite films by doping boron nitride quantum dots. *Journal of Materials Chemistry C* 2019, 7, (44), 13896-13903. https://doi.org/10.1039/C9TC04381K.

[38] Li, J.; Jin, P.; Tang, C., Cr (III) adsorption by fluorinated activated boron nitride: a combined experimental and theoretical investigation. *RSC Adv.* 2014, 4, (29), 14815-14821. https://doi.org/10.1039/C4RA01684J.

[39] Ihsanullah, I., Boron nitride-based materials for water purification: Progress and outlook. *Chemosphere* 2021, 263, 127970. https://doi.org/10.1016/j.chemosphere.2020.127970.

[40] Lei, W.; Portehault, D.; Liu, D.; Qin, S.; Chen, Y., Porous boron nitride nanosheets for effective water cleaning. *Nature communications* 2013, 4, (1), 1777. https://doi.org/10.1038/ncomms2818.

[41] Glasner, A.; Makovky, A., 36. The thermal decomposition of guanidine perchlorate. Part I. Reaction products and mechanism. *Journal of the Chemical Society (Resumed)* 1953, 182-184. https://doi.org/10.1039/jr9530000182.

[42] Huang, H.; Song, Z.; Wei, N.; Shi, L.; Mao, Y.; Ying, Y.; Sun, L.; Xu, Z.; Peng, X., Ultrafast viscous water flow through nanostrand-channelled graphene oxide membranes. *Nature communications* 2013, 4, (1), 1-9. https://doi.org/10.1038/ncomms3979.

[43] Hu, J.; Zhang, S.; Tang, B., Rational design of nanomaterials for high energy density dielectric capacitors via electrospinning. *Energy Storage Materials* 2021, 37, 530-555. https://doi.org/10.1016/j.ensm.2021.02.035.

[44] Li, Q.; Yao, F.-Z.; Liu, Y.; Zhang, G.; Wang, H.; Wang, Q., High-temperature dielectric materials for electrical energy storage. *Annual Review of Materials Research* 2018, 48, 219-243. https://doi.org/10.1146/annurev-matsci-070317-124435

[45] Tan, D. Q., Review of polymer-based nanodielectric exploration and film scale-up for advanced capacitors. *Adv. Funct. Mater.* 2020, 30, (18), 1808567. https://doi.org/10.1002/adfm.201808567.

[46] Wang, Q.; Zhu, L., Polymer nanocomposites for electrical energy storage. *J. Polym. Sci., Part B: Polym. Phys.* 2011, 49, (20), 1421-1429. https://doi.org/10.1002/polb.22337.

[47] Wei, J.; Zhu, L., Intrinsic polymer dielectrics for high energy density and low loss electric energy storage. *Prog. Polym. Sci.* 2020, 106, 101254. https://doi.org/10.1016/j.progpolymsci.2020.101254.

[48] Li, Q.; Chen, L.; Gadinski, M. R.; Zhang, S.; Zhang, G.; Li, H. U.; Iagodkine, E.; Haque, A.; Chen, L.-Q.; Jackson, T. N., Flexible high-temperature dielectric materials from polymer nanocomposites. *Nature* 2015, 523, (7562), 576-579. https://doi.org/10.1038/nature14647.

[49] Song, Q.; Zhu, W.; Deng, Y.; Zhu, M.; Zhang, Q., Synergetic optimization of thermal conductivity and breakdown strength of boron nitride/poly (vinylidene fluoride) composite film with sandwich intercalated structure for heat management in flexible electronics. *Compos. Part A Appl. Sci. Manuf.* 2020, 135, 105933. https://doi.org/10.1016/j.compositesa.2020.105933.

[50] Fan, L.; Yang, D.; Huang, L.; Fan, M.; Lei, C.; Fu, Q., Polymer nanocomposite with enhanced energy storage capacity by introducing hierarchically-designed 1-dimension hybrid nanofiller. *Polymer* 2020, 201, 122608. https://doi.org/10.1016/j.polymer.2020.122608.

[51] Liu, G.; Zhang, T.; Feng, Y.; Zhang, Y.; Zhang, C.; Zhang, Y.; Wang, X.; Chi, Q.; Chen, Q.; Lei, Q., Sandwich-structured polymers with electrospun boron nitrides layers as high-temperature energy storage dielectrics. *Chem. Eng. J.* 2020, 389, 124443. https://doi.org/10.1016/j.cej.2020.124443.

[52] Luo, H.; Ma, C.; Zhou, X.; Chen, S.; Zhang, D., Interfacial design in dielectric nanocomposite using liquid-crystalline polymers. *Macromolecules* 2017, 50, (13), 5132-5137. https://doi.org/10.1021/acs.macromol.7b00792.

[53] He, D.; Wang, Y.; Song, S.; Liu, S.; Deng, Y., Significantly enhanced dielectric performances and high thermal conductivity in poly (vinylidene fluoride)-based composites enabled by SiC@ SiO2 core–shell whiskers alignment. *ACS Appl. Mater. Interfaces* 2017, 9, (51), 44839-44846. https://doi.org/10.1021/acsami.7b14751

[54] Zhai, Q.; Yang, J.; Qiao, W.; Qiao, J.; Gao, H.; Li, Z.; Wang, P.; Tang, C.; Xue, Y., Ultralight and Highly Resilient Boron Nitride Nanosheet/Polyimide Foams for Energy Harvesting and Sensing. *ACS Appl. Polym. Mater.* 2022. https://doi.org/10.1021/acsapm.1c01795.

Chapter 3

Cerium Oxide Nanosheets: Synthesis, Characterization, and Challenges

Divya Mahar[1]
Nitish Semwal[1]
Manjunath Chatti[2]
and Mahesh Chandra Arya[1,*]
[1]Department of Chemistry, DSB Campus, Nainital, Kumaun University, Uttarakhand, India
[2]Australian Centre for Electromaterials Science, School of Chemistry, Monash University, Clayton, Victoria, Australia

Abstract

Cerium oxide is a rare-earth oxide that has been used in various fields for its diverse applications. The diversified morphology of cerium oxide nanomaterials includes nanorods, nanoparticles, nanocubes, nanotubes, etc. Each of these morphologies varies in surface-to-volume ratio and hence differs in catalytic potentials. The cerium oxide-based two-dimensional (2D) materials have been reported to possess high surface area, potential quantum size effect, and a high degree of anisotropic effect apart from having an ultrafine thickness. Owing to their upgraded properties, various methods have been employed for the synthesis of cerium oxide nanosheets which include aqueous phase synthesis, bio-template method, hydrothermal method, complexing co-precipitation method, exfoliation method, and carbonization precipitation method. Each of these methods has its limitations which have been discussed in the book chapter. The formation of ceria nanosheets via aqueous phase

* Corresponding Author Email: mcarya181@gmail.com.

In: Properties and Applications of Nanosheets
Editors: Nanda Gopal Sahoo and Anirban Dandapat
ISBN: 979-8-89113-381-5

synthesis was found to be difficult due to the lack of the driving force for the anisotropic growth. Whereas in the hydrothermal method and exfoliation method the produced quantity of ceria nanosheets was reported to have low yield. This literature review exhaustively explores various methods, characterization, and potential applications of cerium oxide nanosheets available in the literature. It also highlights the challenges before the scientific community for the synthesis of 2D-ceria nanosheets. The work ends with the opinion of the authors about the potential applications of the 2D-ceria nanosheets, the challenges in their synthesis, and the various ways to meet them.

Keywords: Ceria (CeO_2) nanosheet, hydrothermal method, exfoliation method, co-precipitation method, solvothermal method, aqueous phase synthesis, bio templates

1. Introduction

Cerium oxide is a rare-earth oxide that has been used in various fields for its diverse applications. It is used as a ceramic material, polishing agent, sunscreen, photocatalyst, electrolytic material in solid oxide fuel cells, and an ingredient in the catalytic convertors of vehicles [1]. Among various metal oxides, cerium oxide or ceria has emerged as one of the main compounds that have attracted the attention of the researchers and scientific community in search of a substitute for PGM catalysts. Its reversible change in the oxidation state Ce^{3+}/Ce^{4+} makes it an excellent catalyst due to the presence of a high density of oxygen vacancy. It is a proven fact that catalysts show multifold efficiency when used in nano forms. The nanoceria has become one of the foremost used catalysts for various reactions. Among several reactions catalyzed by nanoceria, its efficiency in catalyzing hydrogen evolution reactions has made it a promising catalyst for an economical and eco-friendly energy source [2]. The band gap of cerium oxide nanoparticles depends on the method of synthesis and it can also be achieved by using suitable dopants. This has added one more dimension to its utility as a photocatalyst. Cerium oxide nanomaterials can be synthesized in various morphologies which include nanorods, nanoparticles, nanocubes nanotubes, etc. Each of these morphologies is unique in the sense that these vary in surface-to-volume ratio and hence different catalytic potentials [3]. In recent years, researchers are engaged in exploring the potential of cerium oxide-based two-dimensional (2D) materials. These materials have been reported to possess high surface area, potential quantum size effect, and a high degree of anisotropic effect apart from having an ultrafine thickness [4]. Nanosheets of cerium oxide are one of these 2D-cerium oxide materials. These have been found to show higher catalytic potential than the CeO_2 nanotubes and nanorods. Literature review reveals their higher capacity as catalysts as well as photocatalysts [5]. Owing to their upgraded properties, various methods have been employed for the synthesis of cerium oxide nanosheets. This work exhaustively explores various methods, characterization, and potential applications of cerium oxide nanosheets available in the literature. These methods include an aqueous phase synthesis, bio-template method, hydrothermal method, complexing co-precipitation method, exfoliation method, and carbonization precipitation method. Each of these methods has its limitations which have been discussed in the book chapter.

This literature review also highlights the challenges before the scientific community for the synthesis of 2D-ceria nanosheets as it lacks the driving force for the anisotropic growth of nanoparticles, due to the cubic crystal structure of metal oxides. The book chapter ends with the opinion of the authors about the potential applications of the 2D-ceria nanosheets, the challenges in their synthesis, and the various ways to meet them.

2. Methods of Synthesis of 2D-Ceria Nanosheets

As the researchers turned their attention toward the synthesis of 2D-ceria nanosheets, a number of methods were developed for the same. These methods vary in their precursor, medium, and reaction conditions. However, there cannot be a clear-cut demarcation as all of them have common aqueous phase reactions and the formation of the products as precipitates.

These methods can be classified as:

1. Synthesis of ceria nanosheets via hydrothermal method
2. Synthesis of ceria nanosheets via solvothermal method
3. Synthesis of ceria nanosheets by exfoliation method
4. Synthesis of ceria nanosheets using bio-templates
5. Synthesis of ceria nanosheets by coprecipitation or assisted coprecipitation method

The result of the literature review has been reported under these headings with a detailed characterization of the nanosheets prepared by each of these methods.

Let's discuss these methods one by one.

2.1. Synthesis of Ceria Nanosheets via Hydrothermal Methods

Hydrothermal synthesis is one of the most common methods used for the synthesis of nanomaterials due to its economical nature, less requirement of apparatus, and compatibility with most lab conditions. It is the reaction of precursors in the aqueous medium under suitable reaction conditions followed by moderate heating.

Plenty of research papers are available on the synthesis of 2D-ceria nanosheets via the hydrothermal route. Some of these have been reviewed and reported here.

2.1.1. Aqueous Phase Synthesis of Ultrathin Ceria Nanosheets

Aqueous phase synthesis of ultrathin ceria nanosheets was primarily reported by Yu et al., (2010) [6], In this synthesis, the 2D-ultrathin ceria nanosheets were formed by the aqueous route process, by using a solution of 6-aminohexanoic acid (AHA) and anhydrous cerous nitrate [$Ce(NO_3)_3$]. In a solution of 1 mmol of 6-aminohexanoic acid (AHA) in 6 mL of deionized water, the aqueous solution of 0.25 mmol cerium nitrate was added continuously through an injector at 95°C for 75 min and further, it was cooled down, during the process 5.5 pH was maintained. after that, the nanocrystals of ceria formed and the formation of ultrathin ceria nanosheets takes place via a self-organization procedure, accompanied by an in-situ recrystallization method. The obtained ultrathin ceria nanosheet has a thickness of 2.2 nm approx. The process of formation of 2D-ceria nanosheet was examined by a modified reaction mixture in which NaCl (0.25 mmol) was added in the reaction mixture, and the 6-aminohexanoic acid (AHA) ceria nanocrystals show electrostatic repulsion, on the increasing ionic strength, by the result of it irregular 3-D crystal accumulation was observed rather than the 2D-ceria nanosheets. After that, it was concluded that the electrostatic repulsion force shows an important effect on the self-arrangement of 2D-ceria nanosheets because on decreasing electrostatic repulsion force the crystals of ceria aggregates in the form of irregular 3D-spheres of ceria nanocrystals.

2.1.2. Characterization

Ultrathin ceria nanosheets synthesized by aqueous phase synthesis were characterized by TEM, AFM, and XRD techniques. The thickness of the ultrathin ceria nanosheet was about 2.2 nm thick having an edgewise dimension of up to 4 μm as confirmed by atomic force microscopy (AFM). It has been seen that the produced self-folded sheets of ceria have a lateral-dimensions of 4 μm, and it was confirmed by the TEM images. And the cubic fluorite cell structure was illustrated by the XRD (X-ray diffraction) images. To illustrate the formation of nanocrystals via using electron microscopy one part of the prepared solution was taken out before time (t = 15 min) it has been seen that the formed ceria nanocrystals were of 2-2.5 nm size. Whereas the HRTEM image shows that the formation of a 2D polycrystalline complex acts as an intermediate state, in the formation of a ceria nanosheet.

2.1.3. Synthesis of Ceria Nanosheets via Hydrothermal Process

Li et al., (2011), used a traditional method for the synthesis of ceria nanosheets, known as a hydrothermal method (Figure 1) [7]. This method synthesized ceria nanosheets without using a bio template or any surfactant. this process involves aqueous ammonia ($NH_3.H_2O$) which plays a very important role in the synthesis process, it acts as a precipitant and was mainly used for the customization of ceria nanosheets. Aqueous ammonia shows concentration depended on activity in the hydrothermal process when no aqueous ammonia was used then there was no formation of ceria nanosheets and upon the addition of 0.5 mL concentration of aqueous ammonia solution, the crystallinity of ceria crystals was not good, after increasing the concentration of $NH_3.H_2O$ up to 0.7 mL the hexangular crystals starts merging, at 1 mL it shows perfect hexangular morphology of ceria nanosheets having a size of 15 nm. It concludes that on the further increase of $NH_3.H_2O$ concentration, the qualities of crystal also get increased. This process was environment friendly and the major reason was the precursor and organic reagents were not used in this method. This method further can be used for the manufacturing of nanomaterials and devices.

In 40 mL of distilled water, 1 mmol cerous nitrate [$Ce(NO_3)_3.6H_2O$] was dissolved, and then in this solution, 1 mL of aqueous ammonia ($NH_3.H_2O$) was transferred and after that, it was placed into the 50 mL Teflon-lined stainless-steel chamber. Then this sealed autoclave was placed into an oven for 24 h at 220°C and left for the cooling process to room temperature. After that, using a centrifuge process, the yellow precipitate was separated and washed with distilled water and ethanol several times. Then the obtained product was air-dried for 12 h at 60°C.

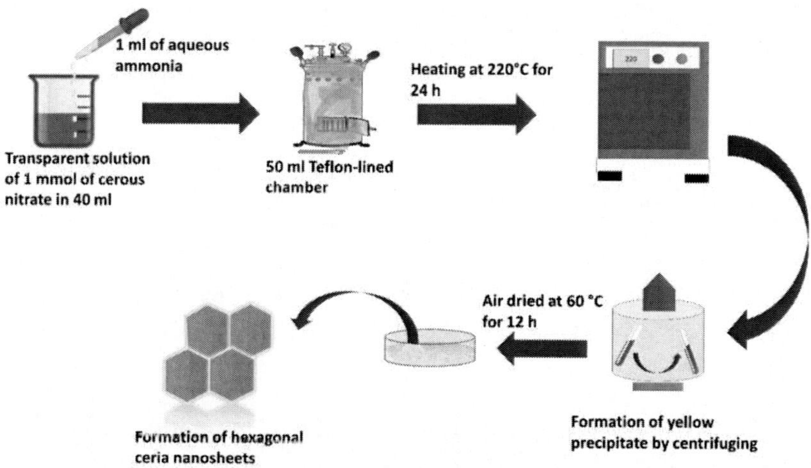

Figure 1. Synthesis of ceria nanosheets via hydrothermal process [7].

2.1.4. Characterization

The hexagonal morphological structure of ceria nanosheets was illustrated by the TEM and ED (Electron diffraction) methods. The electron diffraction patterns show the cubic fluorite structure of the ceria nanosheet. The surface morphology of the ceria nanosheet was displayed by high-resolution transmission electron microscopy (HRTEM). Whereas the XRD images illustrate the crystalline, hexangular structure and the high purity of ceria nanosheets. For the chemical composition of ceria nanosheets, EDAX spectra (energy-dispersive X-ray spectroscopy) were carried out, and it confirms the presence of cerium (Ce) and oxygen (O). On the contrary Raman spectra show only first-order Raman peak at ~463.5 cm^{-1} having F_{2g} symmetry and shows the space group O^5_h of the ceria nanosheets afterward it further confirms the cubic fluorite structure of produced ceria nanosheets.

2.1.5. Synthesis of Hexagon Ceria Nanosheets via Hydrothermal Method

Various methods have been used in the formation of ceria nano-structure such as nanosheets, nanowires, nanotubes, nanocubes, etc. They all exhibit several morphological special properties. The ceria nanostructures are formed by using cerous nitrate [$Ce(NO_3)_3.6H_2O$], $Ce_2(SO_4)_3.9H_2O$, $CeCl_3.7H_2O$ as a cerium source, and NaOH, $NH_3.H_2O$ was used as a mineralizer. Meng *et al.*, (2015), introduced the hydrothermal method for the synthesis of hexagonal ceria nanosheets (Figure 2), in which $CeCO_3OH$ was used as a precursor, $CeCl_3.7H_2O$ as a cerium source, NH_4HCO_3 (ammonium hydrogen bicarbonate) was used as a mineralizer and ethylenediamine used as complexant, which increases the viscosity of the reaction mixture by the coordination reaction and affect the diffusion coefficient which helps in the formation of hexagon ceria nanosheets [8]. Ceria nanosheets show size-dependent properties such as catalytic properties, oxygen sensors, fuel cells, room temperature ferromagnetism, etc.

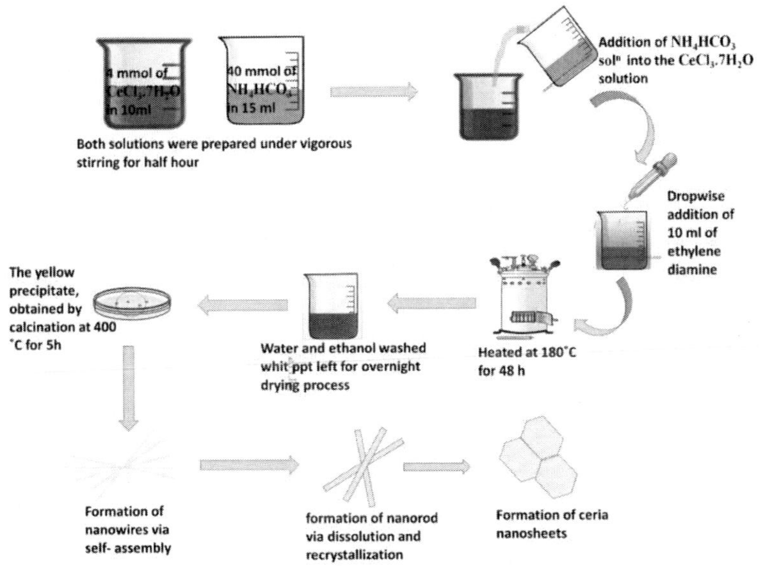

Figure 2. Schematic diagram of the synthesis of hexagonal ceria nanosheets via hydrothermal method [8].

In this method, colorless precipitation was formed with the gradual mixing of 10 mL of an aqueous solution of $CeCl_3.7H_2O$ (used as cerium source) and 15 mL of an aqueous solution of NH_4HCO_3 (ammonium bicarbonate) (used as a mineralizer). Afterward, 10 mL of ethylenediamine (used as a complexant) was added drop-by-drop to the reaction mixture after continuously stirring for a half an hour. Now this solution was placed into the 50 mL Teflon-lined stainless-steel chamber. Then this sealed autoclave was placed into an oven for 48 h at 180°C and left for the cooling process to room temperature. After that distilled water and the ethanol-washed white precipitate were kept for the overnight drying process at 70ºC. the yellow-colored ceria nanosheets were obtained after the calcination process at 400ºC for 5 h.

2.1.6. Characterization

The morphology and structure of ceria nanosheets was examined by XRD and (HR)TEM. It displays the FCC crystalline structure of both Ce_3CO_3OH nanosheets and ceria (CeO_2) nanosheets. The thickness of Ce_3CO_3OH nanosheets and ceria (CeO_2) nanosheets (after the calcination process) was 40-50 nm and 300-400 nm edgewise.

The XRD images also illustrate the different phases of the formed product which depend on the reaction time, the product formed at a reaction time of 1 hour was pure orthorhombic and the product obtained at the reaction time of 6-9 h displays a mixed crystal orthorhombic nature, while at reaction time 48 h the orthorhombic nature of product gets disappeared and it shows the hexagonal crystal nature. Whereas HRTEM images illustrate the porous and loose hexagon structure of ceria (CeO_2) nanosheets. The monocrystalline hexagon structure of ceria nanosheets was further confirmed by SAED (select area electron diffraction). Whereas the mechanism of the formation of nanosheets was confirmed by SEM (scanning electron microscopy). In the initiation reaction time, the nanowires were formed having a diameter of 10 nm, after that these nanowire gets assembled into the nanorods with a length of 100-300 nm and a diameter of 20 nm. As the reaction time increases, the nanorods get aligned in the same direction and at time = 48 h the nanorods get converted into nanosheets. The composition and the oxidation state of produced hexagonal ceria nanosheets were confirmed by XPS (X-ray photoelectron spectra). it displays that the product shows +4 (major) and +3 (minor) oxidation states [8].

2.1.7. Synthesis of Porous Ceria (CeO_2) Nanosheets on Optical Fiber via Hydrothermal Method

A novel and facile method was used to produce porous ceria nanosheets on optical fiber, by using the hydrothermal method [9]. The controlled synthesis of ceria (CeO_2) nanosheets depends on the temperature adjusted in the hydrothermal equipment. The coating of ceria (CeO_2) on the surface of optical fiber increases the optical properties and it can be used in the fiber laser and sensor.

The single-mode fiber (diameter = 125μm) has been used as an optical fibre, these fibres were further sterilized by using deionized water and acetone solution for approx. 10 min. a reaction mixture was prepared by mixing PVP (2 mmol) in distilled water (15 mL) with half hour stirring and afterward 5 mmol of Cerous nitrate ($Ce(NO_3)_3.6H_2O$) was dissolved in this solution. A homogeneous solution was prepared with Urea (15 mmol) and PVP (2 mmol) in 15 mL of distilled water with a half-hour strenuous stirring, accompanied by the half-hour stirring of the 2 aqueous solutions. Afterward, the nitric acid (1mL) was added drop by drop to this solution. Then the single-mode fibres were placed into the solution for 1 hour with continuous stirring. Then these fibres were rapidly placed into a Teflon-lined stainless-steel chamber and heated for 48 h at 120, 160, and 200°C, and cooled down to room temperature, later on, washed fibres were dried out for 12 h at 80°C. then the produced pale-yellow ceria (CeO_2) was air calcinated at 80°C for 12 h.

2.1.8. Characterization

The FCC cubic fluorite structure, purity, and crystallinity of porous ceria nanosheet were confirmed by XRD (X-ray diffractometer). The SEM (Scanning electron microscopy) illustrates the size and morphology of produced porous ceria nanosheets. It displays that the ceria nanosheets' surface was rough and porous and the optimum thickness of ceria nanosheets was 50 nm at 120°C, after that on increasing the temperature the thickness of the produced product also increased ranging from 100 to 200 nm. The XPS spectra illustrate the composition and oxidation states of porous ceria (CeO_2) nanosheets. It shows the presence of Ce^{3+} on the surface of ceria nanosheets, therefore the internal oxygen vacancies were created as a result of it, maintaining neutrality. Raman spectroscopy is a very facile technique to illustrate the optical properties of produced porous ceria nanosheets. It shows the defects produced due to the oxygen vacancies in the ceria nanocrystal, and the photoluminescence spectra display that as the oxygen vacancies increase, so does the intensity of PL emission increase.

2.1.9. Synthesis of Fe-Doped-Ceria Nanosheets

The hydrothermal method for the formation of Fe-doped-ceria nanosheets has been illustrated in literature and it shows the combustion catalytic activity of Fe-doped ceria nanosheets toward chlorinated organic volatile compounds (1,2-dichloroethane) [10]. A reaction mixture was prepared by mixing hexamethylenetetramine (3.0 g) and acetic acid (0.5 ml) in the solution of cerium (III) nitrate hexahydrate [$Ce(NO_3)_2.6H_2O$] and Ferric (III) nitrate nonahydrate [$Fe(NO_3)_2.9H_2O$) in 60 ml deionized water. After 30 min strenuously stirring, the reaction mixture was crystallized with the help of a crystallization kettle, then the sealed vessel was placed into an oven at 170ºC for 15 h for the drying process. The ICP-AES technique was used for the detection of the produced 5Fe-CeO_2-HT, and the molar ratio of Fe/(Fe +Ce) was 4.3%.

2.1.10. Synthesis of Ceria Nanosheets via Using the Low-Temperature Hydrothermal Method

A low-temperature hydrothermal method for the synthesis of ultrathin-ceria (CeO_2) nanosheets was used by the researchers for its benefits such as the requirement of low temperature and being comparatively less expensive, than the other methods [11]. The produced ultrathin ceria nanosheets were successfully used for the detection of formaldehyde and humidity and worked as bifunctional substances. Ceria nanosheets show these properties due to their specific morphology and high surface area, which makes them the perfect material for the sensor detector.

A reaction mixture was prepared by dissolving 2.5 mmol cerium nitrate ($Ce(NO_3)_3.6H_2O$) in the NaOH solution (50 mL), afterward the prepared solution was placed into a Teflon-lined stainless-steel chamber and heated for 24 h at 100ºC, and cooled down to room temperature, later on, washed white precipitate dried out overnight at 70ºC and then the produced yellow ceria (CeO_2) was air annealed at 400ºC for 2 h.

2.1.11. Characterization

The crystallinity and purity of produced ceria nanosheets were confirmed by the XRD (X-Ray diffractometer). Whereas the phase morphology was illustrated by the FESEM (field emission scanning electron microscopy), it displays the universal distribution of ceria (CeO_2) particles on nanosheets. The high-resolution image of FESEM shows that the thickness of produced ceria (CeO_2) nanosheets was 10 nm approx., and the length and width were approximately 500 nm and 150 nm respectively. A sinusoidal loop (IV isotherm) was shown via the N_2 adsorption-desorption isotherms graph, it illustrates the internal morphology of ultrathin ceria nanosheets, and it shows that the adsorption takes place by the capillary condensation.

2.2. Synthesis of Ceria Nanosheets via Hydrothermal Methods

The hydrothermal method has been used for the synthesis of Fe-doped ceria nanosheets by Wang *et al.*, (2017) [10.]. As Figure 3 shows a homogenous reaction mixture was prepared by mixing Ferric (III) nitrate nonahydrate [$Fe(NO_3)_2.9H_2O$) and cerium nitrate hexahydrate [$Ce(NO_3)_2.6H_2O$] in ethylene glycol (50 mL) on the 1 h of strenuously stirring. Afterward, the reaction mixture was poured into a crystallization kettle and placed into an oven for 28 h at 180ºC for the drying process, after the filtration the water/ethanol-washed product dried at 80ºC for 12 h. The Fe-doped ceria (CeO_2) nanosheet was obtained by the air-calcination process for 4 h at 450ºC. The oxides of ruthenium (Ru) and vanadium (V) were used as supporting catalysts and prepared by the impregnation method. The ICP-AES technique was used for the detection of the produced 5Fe-CeO_2-ST, and the molar ratio of Fe/(Fe +Ce) was 4.4%.

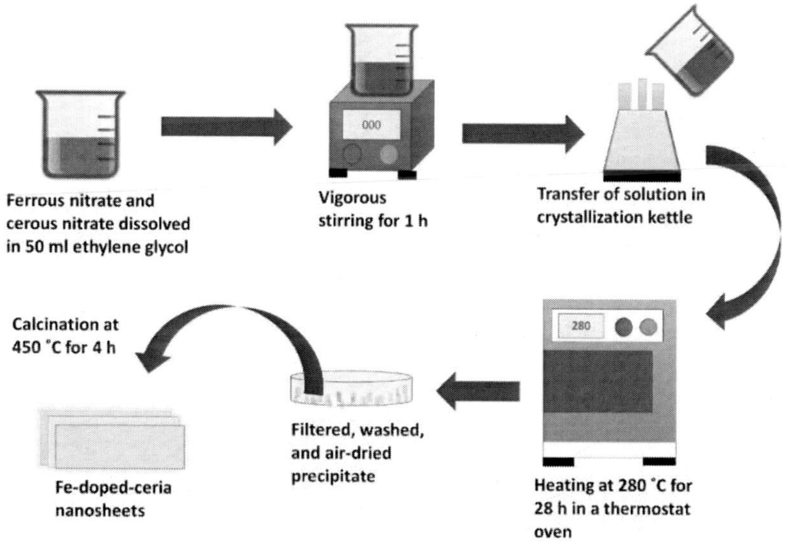

Figure 3. Synthesis of Fe-doped-ceria nanosheets via solvothermal process [10].

2.3. Synthesis of Ceria Nanosheets via Exfoliation Method

This is a method based on peeling off the ceria nanosheets from the 3D nanostructure of the ceria.

2.3.1. Synthesis of Co-Doped Ceria Nanosheets via Exfoliation Method

Jiang *et al.*, (2020) introduced a method named thermal decomposition exfoliation method for the synthesis of Co-doped ceria nanosheets [12]. Doping of cobalt on ceria nanosheets enhances the catalytic activity of 2D-ceria nanosheets by increasing the oxygen vacancies. The Co-doped ceria nanosheets show excellent catalytic properties in hydrogen evolution reaction (HER) in an alkaline medium by providing a larger surface area, and larger oxygen vacancies for the adsorption of hydrogen and dissociation of water. The produced Co-doped ceria nanosheets consist of a thickness of 4.3 nm.

A multilayer Co-doped ceria nanosheets were produced by the solvothermal treatment, in which a reaction mixture was prepared by the mixing of the cerium (III) nitrate, cobalt nitrate (metal source), hexamethylenetetramine (HMTA) (ligands), and ethanol and ethylene glycol (solvent) at 170°C for 8 h. Furthermore, $CeCO_3OH$ was used as a precursor, which tends to shed the Co-doped ceria nanosheets via the calcination process at 400°C for 2 h, which helps to remove CO_3^{2-} and OH^-.

2.3.2. Characterization

The XRD (X-ray diffractometer) displays that the Co-doped ceria nanosheets exhibit cubic fluorite structure as ceria, and the morphology of the produced product was the same as the precursor. Whereas SEM (scanning electron microscopy) images illustrate the flaky morphology of Co-doped ceria nanosheets. While Raman spectra confirm the increase in the number of oxygen vacancies on the surface area of Co-doped ceria nanosheets, cobalt doping is the reason behind the adsorption of activated oxygen and charge transfer. Furthermore, the AFM (atomic force microscopy) displays that the thickness of produced layered Co-doped-ceria nanosheets was 4.3 nm, while the single layer ceria shows a thickness of 2.71 Å. The uniform arrangement of O, Ce, and Co in the ceria nanosheets was confirmed by the EDS (Energy-dispersive X-ray spectrometry). The XPS displays the chemical composition and the oxidation states (Ce^{3+}/Ce^{4+}) of the Co-doped ceria nanosheets, and the doping % of the cobalt was ~14 a.t.%. The EIS (Electrochemical impedance spectroscopy) illustrates that the doping of cobalt provides the transportation of charge in the ceria (CeO_2) sheets so that the Co-doped ceria nanosheets exhibit tiny charge transfer resistance. The doping of cobalt also increases the surface area of ceria (CeO_2) nanosheets. The hydrogen-free energy (ΔG_{H+}) of CeO_2 was ΔG_{H+}= -0.31eV, for the production of hydrogen gas (H_2), it displays the strong interactivity between hydrogen and ceria (CeO_2) whereas for CeO_2

(having one oxygen vacancy) the value of hydrogen-free energy was 0.11eV, which displays the poor coordination between the hydrogen and ceria, while in Co-doped-CeO_2 (having one oxygen vacancy) the value of ΔG_{H+} was 0.07eV, displays the position of a hydrogen atom was on the 2-coordinated hydrogen atom. It illustrates that the Co (doping metal) helps to activate the nearby oxygen atom in CeO_2, which shows good catalytic performance for hydrogen evolution reaction (HER) [12].

2.4. Synthesis of Ceria Nanosheets Using Bio-Templates

A lot of flat bio-templates have been used for the direct synthesis of 2D-ceria nanosheets. This method has an advantage over other methods as it gives a bulk synthesis of ceria nanosheets.

2.4.1. Synthesis of 2D-ceria (CeO_2) Nanosheet via Using China Rose Petal Bio Template

A conventional and nature-friendly method was introduced by Qian *et al.*, (2011) for the synthesis of ceria nanosheets having a thickness of around 7 nm via using China rose petals as a bio template. China rose petals exhibit a strong and facile route for the synthesis of ceria nanosheets because they have secretory cone-shaped cells [13].

In this method, ceria nanosheets were prepared by the biologically induced mineralization process. Sterilized and dust-free 4 g of China rose petals were pre-treated for about 1 hour at pH 4 with ethanol/water solution for the modification of the cell wall of the petals. Then in this mixture, 2.5 g of cerous nitrate was added with strenuous stirring. After the soaking process, the petals of the China rose were removed with the help of forceps, afterward, these petals were washed several times with distilled water and left over for the overnight drying process. At 550ºC the bio template was removed by the calcination process, during this process hierarchical ceria nanosheet is produced by the conversion of the biomembrane of the cell wall.

2.4.2. Characterization

The XRD confirms the cubic face-centered structure of the ceria nanosheet after the calcination process, whereas during the calcination process, TGA (thermogravimetric analysis) shows the development of the ceria nanosheet. Field-emission scanning electron microscopy (FESEM) displays the morphological structure of petals and TEM images show the formation of ceria nanosheet on the petal cell wall, these hierarchy petals were proved as a good copy of the cell wall. It also confirms that the lamella of the petals contain a single-layered nanosheet of ceria of thickness around 7 nm. The XPS spectra illustrate that the oxygen vacancies and the proportion of Ce^{3+} ions are more in ceria nanosheets.

2.4.3. Nitrogen-Doped Cerium Oxide (CeO_2) Nanosheets Synthesis

A novel method was reported for the synthesis of N-doped ceria nanosheets, in which the synthesis was carried out on the template of natural plant material poppy flower petals (Figure 4) [14]. The biomembrane of the petal cell acts as the accurate template and the nitrogen from the corpus (plant part) is self-doped in the developing nanosheets. In the actual process of the above-reported method, 2 g of cerous nitrate was dissolved in 120 mL of distilled water.

Sterilized and dust-free fresh poppy petals were activated by dipping them in an ethanol/water mixture for 3 hrs. These pre-treated petals were soaked in a 2 g cerous nitrate solution in 100 mL of water maintained at a pH of 3. After the soaking process, the petals are removed by forceps and after that, these petals are washed with distilled water and then left for the overnight drying process. After that, at 550°C the bio template was removed by the calcination process for 120 min. And gained products were stored for further analysis.

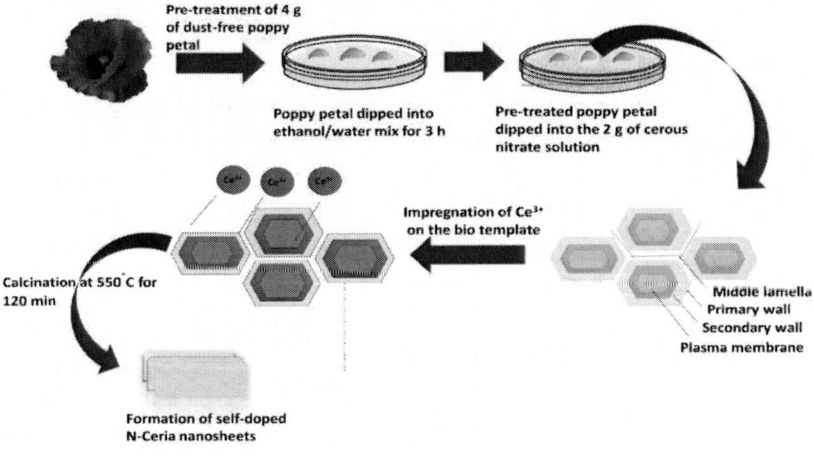

Figure 4. Schematic diagram of the synthesis of N-doped-ceria nanosheets [14].

2.4.4. Characterization

Characterization of petal template for surface structure and morphology was performed by field emission scanning electron microscopy (FESEM) and imaged via environment scanning electron microscope (ESEM) whereas the N-doped ceria nanosheets were characterized by SEM, TEM, AFM, XRD, and EDS techniques [13]. The ESEM image showed that the poppy petal is mainly made up of conical epidermal cells whereas SEM images show the biomorphic microstructure of N- doped ceria nanosheets after the calcination process. TEM images display that the structure of the nanosheet looks like a replica of gauze. The XRD illustrates the crystal structure of the ceria nanosheet. Afterward, the components of the ceria nanosheet are confirmed by EDS which displays that Ce, N, and O elements are the main components of the ceria nanosheet. Atomic force microscopy (AFM) displays the formation of a 2D-ceria nanosheet having a thickness of 4.978 nm approx. Raman spectra show two peaks at 464 cm^{-1} of bulk ceria oxide and ceria nanosheet, in which bulk ceria shows a sharp peak of first-order vibrational mode having F_{2g} symmetry and ceria nanosheet shows comparatively broader & weak peaks. TPR (Temperature programmed reduction) process is used for the analysis of the catalytic activity of cerium oxide regarding H_2. It shows a peak at 586 cm^{-1} which could be allotted to oxygen vacancies in the ceria nanosheet. The broad peak of surface oxygen vacancies of bulk ceria was observed at lower temperatures from 316 to 576°C whereas the shoulder peak at 764 shows the reduction of the lattice oxygen.

2.5. Synthesis of Ceria Nanosheets by Coprecipitation or Assisted Coprecipitation Method

2.5.1. Synthesis of Ceria Nanosheets Using Aqueous-Phase Precipitation Method

A novel method for the synthesis of petal and belt-like ceria nanosheets, named the aqueous phase precipitation method has been reported in the literature (Figure 5) [15]. This method was a surfactant or template-free, non-hydrothermal method, and there was no organic solvent used. The precipitant used in this method was ammonium bi-carbonate ($NaHCO_3$). The reaction is carried out at 2 different temperatures (0°C-25°C to 50°C-150°C). At lower temperatures synthesis of sheet-like structure of ceria, nanosheets takes place whereas at higher temperatures polyhedral ceria nanosheets were obtained via dissolution and recrystallization of precursor (known as the Ostwald ripening process). In this method, the morphology of ceria nanosheets majorly depends on the reaction temperature, and the precursor that defines the different morphology (petal and belt-like) of ceria nanosheets was $Ce_2(CO_3)_3$. To study the effect of the synthesized precursor at 0ºC, different precipitants were used named sodium bicarbonate ($NaHCO_3$) formed petal-like phase morphology ($Ce(CO_3)_3.xH_2O$), aqueous ammonia (AA), and ammonium carbonate (AC) formed plate-like phase morphology ($Ce(CO_3)_3.6H_2O$). While using sodium bicarbonate as a precipitant at different temperatures displays different phase morphology [such as petal-like (0ºC), belt-like (25ºC), irregular sheet (50ºC), olive-like (75-100ºC), and irregular particles (150ºC)] (Figure 5).

A solution of 1.39 g of cerium (III) nitrate (dissolved in 200 mL of deionized water at 0ºC) and 0.75 g of sodium bicarbonate (dissolved in 200 mL of deionized water at 0ºC) was prepared by the instant mixing, followed by stirring at 0ºC for a half an hour and placed for 15 h after that the washed, filtered, and dried residue calcinated at 450ºC for 4 h. By this process the morphology of produced ceria nanosheet was petal-like. Whereas at a reaction temperature of 70ºC-100ºC, this reaction was carried out through the reflux, and a hydrothermal environment was used at 150ºC to 170ºC.

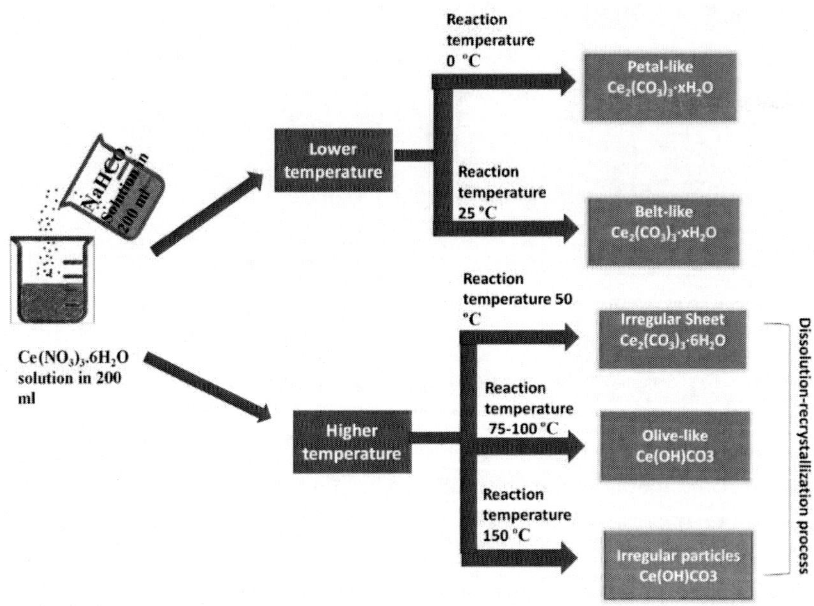

Figure 5. Schematic diagram of the synthesis of ceria nanosheets via an aqueous-phase precipitation process [15].

2.5.2. Characterization

SEM images illustrate the morphology of different synthesized ceria nanosheets at different temperatures. Below 75°C, the morphology of ceria nanosheets was sheet-like, and its thickness was reported about 20-50 nm, on the contrary at 0°C the self-assembled flower-like morphology was shown by ceria nanosheets, whereas at 25°C, the belt-like morphology was obtained having the length and width of 5-10 μm and 0.5-1.2 μm respectively. At higher temperatures (75°C-100°C) the olive-like morphology was shown by ceria nanosheets, and it was assumed that the reason behind it was probably the dissolution-recrystallization method. The olive-like structure consists of multilayer nanosheets. While at 150°C irregular-shaped particles were found. The TEM images show that the polyhedral nanoparticles (200-300 nm) were formed by the thermal decomposition of sheet-like precursors at high temperatures (150°C-170°C), which confirms the dissolution-recrystallization and it is favorable at high temperatures for the formation of polyhedral nanoparticles, while due to the less solubility and slow rate of crystallization sheet like precursors were favorable at a lower temperature. The XRD technique illustrates the crystal cubic FCC fluorite structure, and the purity of ceria nanosheets, it also shows that the size of grain size nanoparticle of ceria nanosheets increases with increasing temperature, and the calculated average size of it was 15-20 nm. At lower temperatures (0°C-25°C) the uncalcined precipitates show poor crystallinity, and it increases at rising temperatures (50°C-150°C). The orthorhombic phase was shown more stable than that of the hexagonal phase of the synthesized precursor ($CeOHCO_3$). The results of XRD were further confirmed by the FT-IR spectroscopy, which displays the characteristic peaks of the OH group, at 3456 cm^{-1} (stretching vibration) and 3620 (Bending vibration), and the intense peaks of CO_3^{2-} were shown at 1510 and 1443 cm^{-1}, and the peaks of CH group was shown at 2000-3000 cm^{-1}. Which confirms the presence of the hexagonal phase of $CeOHCO_3$ precursors.

2.5.3. Synthesis of Cu^{2+}/La^{3+} -doped ceria (CeO_2) Nanosheets via Complexing Co-Precipitation Method

The catalytic activity of the ceria (CeO_2) for the oxidation of CO can be enhanced by doping transition metal oxides. A novel and facile complexing co-precipitation method for the synthesis of transition metal doped ceria nanosheets is available in literature in which the coordination polymers were used as a precursor in the synthesis of Cu^{2+}/La^{3+} doped ceria nanosheets [16]. The homogeneous dispersion of doping metal and the morphology of Cu^{2+}/La^{3+} doped ceria was majorly affected by the 1,2,4,5-benzene tetracarboxylic acid (BTA). The synthesized Cu-doped-CeO_2 nanosheets exhibit higher catalytic activity toward CO oxidation rather than the La-doped-CeO_2 nanosheets.

In this method, a solution mixture of 1.25 mmol of 1,2,4,5-benzene tetracarboxylic acid (BTA) in 25 mL of ethanol and 1.25 mmol of cerium (III) acetate hydrate ($Ce(CH_3COO)_3.1.5H_2O$) in 25 mL of water was prepared and kept for 1 hour at 80°C. After that, the ethanol/water washed white precipitate left for the overnight drying process in a vacuum at 45°C. then the resultant metal oxide was obtained by the calcination process at 450°C and 650°C for 1 h and 2 h simultaneously. Afterward, to obtain the metal-doped ceria nanosheets the ethanol solution of 1,2,4,5-benzene tetracarboxylic acid (BTA) was mixed with the aqueous solution of cerium (III) acetate hydrate ($Ce(CH_3COO)_3.1.5H_2O$). Later on, the obtained product was followed by the calcination process at 450°C and 650°C for 1 h and 2 h respectively. For the determination of the concentration of doped metal ICP-MS (inductively coupled plasma spectrometer) technique was used.

2.5.4. Characterization

SEM (scanning electron microscopy) and TEM (transmission electron microscopy) techniques were used for the illustration of the morphology of coordination polymers (CP) precursors. SEM images display that the thickness of BTA-Ce was less than 100 nm, and they exhibit sheet-like morphology with up to 10% content of doping cations. At 10%, they appeared in a folded sheet-like structure. While TEM images illustrate the scaly structure of ceria nanosheets before and after the doping process. Whereas EDS (energy dispersive spectrometer) displays the uniform distributions of metal cations in ceria nanosheets, which help in the catalysis process. The chemical composition and the oxidation states of Cu^{2+}/La^{3+} doped ceria nanosheets were illustrated by the XPS (X-ray diffraction spectroscopy). It shows that more oxygen vacancies were produced when Ce^{4+} gets converted into Ce^{3+} and the reduction of Ce^{4+} increases by increasing the content of Cu^{2+}. The crystallinity and purity of the product were investigated by the XRD technique, which also confirms the homogeneous doping of metal cations in the ceria lattice.

2.5.5. Synthesis of Ceria (CeO_2) Nanosheets via Carbonization Precipitation Method

Lai *et al*., (2020) introduced a method (i.e., simple, low cost, suitable under mild conditions) named the precipitation carbonization method for the preparation of uniformed Ceria (CeO_2) nanosheets, during the reaction procedure the growth mechanisms include the oriented attachment mechanisms and Ostwald ripping process respectively [17]. The thickness of the uniformly prepared ceria nanosheets was approx. 180nm. The 2D-ceria nanosheets exhibit excellent catalytic activity for CO oxidation rather than the ceria nanotubes, ceria nanorods, etc. The several methods that have been used such as the hydrothermal method, sol-gel method, etc. The preparation of ceria nanomaterials since past few years has some advantages with some drawbacks such as environmental pollution, excessive cost, etc. This is because of the use of organic additivities, and the templates, complexant, etc. In this method, the $Ce_2(CO_3)_3.6H_2O$ was used as a precursor and was produced by the raw-material cerium chloride ($CeCl_3$) reaction at pH 5 maintained by using NaOH.

In this method, a solution of 0.9 mol/L cerium chloride solution (in 200 mL) was prepared and for the further reaction the pH of 9.2 (initial pH) was maintained by the 1.0 mol/L NaOH solution for the precipitation of cerium ions, furthermore, the flow of CO_2 was carried out with the flow rate of 1.0 L/min for the carbonization process, followed by stirring with 400 r/min stirring speed. During the carbonization reaction, the reaction pH also gets decreased to 4.93 pH, this process takes a time of 60 min., afterward, at different-2 reaction times the product was collected, followed by filtration and washing. Then the washed and filtered product cerium carbonate (precursor) was obtained via the carbonization process. Then the product was dried, and the calcination process was carried out at 400°C for 2 h. The 2nd experiment was performed under the same conditions but the reaction pH 5 was maintained during the reaction. Whereas, in the 3rd experiment, the concentration of cerium chloride solution was taken at 0.3 mol/L and the concentration of NaOH solution was 0.3 mol/L and all other reaction conditions were the same as the experiment 2nd. These 3 experiments were carried out at 25°C temperature.

2.5.6. Characterization

The FT-IR spectra illustrate the phase composition of three obtained precursors produced by the carbonization reaction., It shows the broad peak at 3334 $cm^{-1}$and 1630 cm^{-1} irrespective of stretching and bending vibrations of -OH bond, and the symmetric stretching band of CO_3^{2-} observed at 1477 $cm^{-1}$and 1375 cm^{-1}. These peaks in FT-IR confirm the presence of -OH and - CO_3^{2-} groups in three obtained precursors. The phase composition and phase purity were confirmed by XRD. It displays that the first precursor obtained in the basic conditions (pH = 9.2) was $Ce_2(CO_3)_3.8H_2O$ and the second and third precursors that were obtained in the acidic conditions (pH = 5) were $Ce_2(CO_3)_3.6H_2O$. These 3 precursors show the orthorhombic crystal and structures.

The SEM (scanning electron microscopy) illustrates the formation of ceria nanosheets during the carbonization reactions. In the first experiment, the gas-liquid-solid reaction system was used after the initiation of the carbonization reaction. At very first, nanograins of $Ce(OH)_3$ were formed in alkaline conditions, afterward when the carbonization reaction proceeds these nanograins get associated to reduce the high surface energy of nanograins, and form a slurry of $Ce(OH)_3$. Due to the dissolution of $Ce(OH)_3$ it gets converted into $Ce_2(CO_3)_3.8H_2O$, and at the reaction time of 50 min, the aggregated nanograins form an irregular-shaped flaky structure of $Ce_2(CO_3)_3.8H_2O$ via the oriented attachment and Ostwald ripping process.

In the second experiment, the reaction was initiated by the slow addition of NaOH, and in the presence of a high concentration of cerium chloride, at pH 5 (acidic environment), the reaction opted in the gas-liquid system. The driving force of the system was the formation of carbonic acid (H_2CO_3), after the carbonization reaction started the nanograins of cerium (III) hydroxide ($Ce(OH)_3$) get aggregated and formed flaky crystals, in this method the crosslinking phenomenon was observed between the crystals, as the reaction time proceeds the flaky crystals of cerium carbonate get dissolved and produced Ce^{3+} and CO_3^{2-}, these ions grow as a sheet-like structure on the surface of cerium carbonate sheets, at reaction time was reached at termination point (50 min), the larger and thicker nanosheets were obtained by the crosslinking process.

While in the third experiment, a low concentration of cerium chloride was used in an acidic environment (pH 5). As the reaction time proceeds the regular, uniform, and thin product was produced via the aggregation and rearrangement process and it has been observed via SEM that the cross-linked phenomenon doesn't occur in this experiment. And the precursor formed was $Ce(OH)_3$, which has a small, and regular size.

The flaky crystal structure of cerium carbonate was further confirmed by the TEM (transmission electron microscopy), which displays that the nanograins formed the flaky structure of cerium carbonate.

2.5.7. Synthesis of Pd/Al_2O_3-Ceria Nanosheets via Coprecipitation Method

A novel and facile method were introduced by Lv *et al.*, (2022), named as coprecipitation process, for the synthesis of Pd/Al_2O_3-ceria nanosheets [18].

The Pd/Al_2O_3-ceria nanosheets show excellent catalytic activity and selectivity rather than the Pd/CeO_2 because together the ceria (CeO_2) and alumina (Al_2O_3) show a magnificent synergistic effect. The obtained Pd-Al_2O_3-CeO_2 was effectively used in the one-pot synthesis of methyl isobutyl ketone (MIBK) from ketone. As alumina provides high surface area and strength whereas ceria provides redox properties. Methyl isobutyl ketone is used as an industrial chemical, for the synthesis of surfactants, and is widely used in the medicine and the coating industry.

A reaction mixture was formed by the mixing of aluminium nitrate ($Al(NO_3)_3.9H_2O$) and Cerous nitrate ($Ce(NO_3)_3.6H_2O$). To maintain the pH (9-10) of the solution, the aqueous ammonia ($NH_3.H_2O$) was added steadily. afterward left for the 12 h at 25°C. Then the washed solid precipitate was dried out at 120°C for 5 h. The Al_2O_3-CeO_2 was obtained by heating at 550°C for 4 h. The 5, 10, 15, 20, 25, and 30 mass fractions of ceria were maintained in Al_2O_3-CeO_2. Furthermore, the ultrasonic waves passed through the solution of Al_2O_3-CeO_2 containing $PdCl_2$ for 30 min and were left for 5 h. followed by the drying process at 120°C for 2 h. Then the addition of potassium borohydride (KBH_4) with strenuous stirring (for 5 h) was carried out in the catalyst-containing ice bath. After that, the washed solid was dried out at 50°C for 5 h.

2.5.8. Characterization

The XRD (X-Ray diffractometer) displays the crystallinity and purity of Pd/Al_2O_3-CeO_2, whereas the SEM (scanning electron microscopy) displays the surface morphology of produced Pd/Al_2O_3-CeO_2, it displays the flaky structure and confirms the presence of palladium particles on the surface of the catalyst (Pd/Al_2O_3-CeO_2). While the even distribution of the palladium on the surface of the Al_2O_3-CeO_2 was confirmed by the HRTEM (High-resolution transmission electron microscopy).

2.5.9. Cold-Co-Precipitation Method (CP)

In this method, the transparent solutions of cerium nitrate hexahydrate [$Ce(NO_3)_2.6H_2O$] and Ferric (III) nitrate nonahydrate [$Fe(NO_3)_2.9H_2O$) were obtained, when dissolved in 200 mL of deionized water and 0.75 g of ammonium hydrogen carbonate (NH_4HCO_3) was dissolved into 200 mL (deionized water) under stirring at 0°C. After that, the rapid mixing of ammonium hydrogen carbonate solution with the mixture of metal solutions was done with strenuous stirring for a half an hour. The ICP-AES technique was used for the detection of the produced 5Fe-CeO_2-CP, and the molar ratio of Fe/(Fe +Ce) was 4.7%.

2.5.10. Characterization

The morphology of Fe-doped ceria nanosheets was characterized by SEM (scanning electron microscopy). The produced Fe-doped ceria nanosheets exhibit different morphological characteristics depending on the synthesis method. In the solvothermal method (ST) on increasing the amount of iron (Fe) the size of produced uniform Fe-doped sheets decreased. At a 15% concentration of iron for 20Fe-ceria-ST the Fe-doped ceria sheets get disappeared. Whereas in the hydrothermal method (HT) plate-like Fe-doped-CeO_2 was formed and was larger in size than the pure ceria nanosheets. While the petal-like Fe-doped-ceria nanosheets were obtained in the cold-co-precipitation method, which was independent of the amount of iron. The crystallinity of the Fe-doped ceria nanosheets was illustrated by XRD (X-ray diffraction) technique. It shows that the crystallinity of Fe-doped-CeO_2 nanosheets depends on the growth mechanisms of different synthesis methods. Ostwald ripening procedure was followed in the solvothermal method, the concentration of precursor (CeO_2 or Ce_2O_3) increases on the increasing iron (Fe) quantity as a result of it the distortion in perfect cubic fluorite happened by which the crystallinity of Fe-doped-CeO_2 (ST) nanosheets decreases. In the hydrothermal method, the Fe-doped-CeO_2 (HT) was synthesized by the dissolution- re-crystallization mechanism, in which the grain size is decreased (due to the reducing effect of Fe^{3+} ions on Gibbs free energy), and the crystallinity of the Fe-doped-CeO_2 (HT) was improved. On the contrary, in the co-precipitation method, Fe-doped-CeO_2 nanosheets were formed by self-assembly at a low temperature (0ºC) into the flower-like structure. As a result, it was shown that there is no significant role of Iron (Fe).

The Raman spectra show that the doping of iron in the ceria (CeO_2) lattice cold-co-precipitation method was better than the other two methods. Fe-doped-CeO_2 nanosheets having larger oxygen vacancies and surface oxygen with better stability were produced by the solvothermal synthesis, it is the most facile method. The FT-IR spectroscopy displays that in the presence of Fe-doped-CeO_2 catalyst, the catalysis combustion of 1,2 dichloroethane (DCE) takes place, in which firstly it is converted into volatile organic compounds (VOC), after that followed by aldehyde, acetate and at last acetates gets oxidized into CO_2 [10].

2.5.11. Synthesis of Phosphate-Modified Ceria Nanosheets via an Organophosphate Mediated-Route/Impregnation Route

Catalytic combustion has been used for the oxidation of several chlorinated volatile organic compounds e.g., 1,2-dichloroethane, dichloromethane (DCM), chlorobenzene, trichloroethylene (TCE), etc. Catalytic combustion converts these chlorinated volatile organic compounds into water, HCl, and CO_2. For the oxidation process, several metal oxides have been used as a catalyst, among all of the other metal oxides cerium oxide (CeO_2) shows excellent catalytic properties because of its high activity and high abundance. However, due to the strong bonding of the chlorine with the active site of ceria, it decreases the catalytic activity of ceria. To rectify this, the doping of transition metal has been used as a good alternate pathway, but via this method, the by-product formed was organic chlorinated compounds. To resolve this problem, an organophosphate route was introduced for the synthesis of phosphate-modified ceria (CeO_2) nanosheets [19]. It helps to retain the catalytic activity of ceria nanosheets as it is, by eliminating toxic organic volatile compounds. The TMP-modified ceria nanosheets exhibit higher catalytic activity than the AP, PA, and SP (lowest activity) (inorganic phosphates).

The ceria nanosheets were prepared by the aqueous phase precipitation method, A reaction mixture was prepared by mixing 2.78 g of cerium nitrate solution (in 100 mL) and 1.50 g of ammonium bicarbonate solution (in 100 mL) followed by stirring for half an hour at 30°C aged for 24 h. Furthermore, the washed and filtered products were left for the overnight drying process at 80°C, and the belt-like ceria nanosheet was obtained by air-calcination for 4 h at 500°C.

For the synthesis of phosphate-modified CeO_2, a reaction mixture was prepared by mixing trimethyl phosphate (TMP), phosphoric acid (PA), ammonium phosphate (AP), or sodium phosphate (SP) in 1.1 mL of deionized water. After that, the 0.86 g of ceria (CeO_2) nanosheets were added to the solution of phosphate and aged for 24 h, and dried at 80°C, followed by the calcination process for 4 h at 450°C with the elevation rate of 5°C/min. The produced catalyst was named Po_x-CeO_2-n, AP- CeO_2-n, PA- CeO_2-n, and SP-CeO_2-n. (where n = molar ratio of P/Ce).

2.5.12. Characterization

The SEM (scanning electron microscopy) illustrates the belt-like structure of ceria nanosheets having a thickness of 20-50 nm, with 5-10μm length. Whereas, SAED (the selected area electron diffraction) confirms the polycrystalline nature of phosphate-modified-ceria nanosheets. The XRD illustrates the cubic fluorite structure and the purity of all phosphate-modified ceria nanosheets was the same as the ceria nanosheet. While the FT-IR spectra show the 3 different peaks at 1158 cm^{-1} attributed to nonbridging $(P\text{-}O\text{-}P)_{symm}$, 1000 cm^{-1} to $P\text{-}O^-$ stretching vibrations, and 950 cm^{-1} to nonbridging $(P\text{-}O\text{-}P)_{asymm.}$ and the intensity of these peaks increases with the increasing phosphate content. The XPS spectra display the distribution of O and P on the surface of ceria nanosheets, it shows that on increasing the phosphate content the -OH content gets increases and O_2^{2-}/O species decreases. The SEM-EDS spectra show that the P/Ce atomic ratio of POx-CeO_2-0.05, POx-CeO_2- 0.1, POx-CeO_2-0.2, and POx-CeO_2-0.3 was 0.04,0.09, 0.15, and 0.28 respectively, which was somewhat less than the value of precursor.

2.5.13. Synthesis of Pt-Ni Coupled Ceria Nanosheet via One-Pot Production Method

To enhance the catalytic activity of Platinum (Pt) based catalysts, metal oxide has been used as a Co-catalyst. Ceria (CeO_x) shows excellent oxygen storage and transport properties due to the presence of oxygen vacancies in its lattice structure. Even though ceria has low conductivity and it has poor interactivity with Pt-based catalysts, Kwon *et al.*, (2020) reported a one-pot synthesis method for the synthesis of Pt-Ni coupled ceria nanosheets. The Pt-Ni coupled ceria enhance the electrocatalytic properties, which can be used in polymer electrolyte membrane fuel cells (PEMFCs) [20]. The alloying of platinum with Ni results in efficient binding activity for the reactive components and enhances its electrocatalytic activity. Whereas due to the very small size, the nanosheets of ceria exhibit high conductivity. And due to the synergic effect of Ni and ceria, the produced Pt/Ni coupled ceria nanosheets display high electrocatalytic activity in the oxidation of alcohol and reduction of oxygen.

In this method $Pt(acea)_2$ (0.05 mmol), $Ni(acea)_2$ (0.05 mmol), $Ce(acea)_3$. xH_2O (0.05 mmol) (precursor of Pt, Ni, and Ce) in 1:1:1 mixed with oleyl-amine (10 mL)/oleic acid (0.5 mL) solution, followed by heating at 80°C for 1 h in a vacuum region furthermore for air dried for 1 h, after that, it was heated for 30 min with strenuously stirring at 210°C with the elevating rate 5°C/min^{-1}, and again heated at 320°C with elevating rate 3°C/min^{-1} for 40 min. then the ethanol (10 mL) and hexane (10 mL) diluted mixture were washed and centrifuged 3 times with ethanol and hexane. Pt-Ni nano dendrites and Pt_3Ni were prepared by the addition of ketjenblack carbon (10 mg) to the solution, followed by an ultrasonication process for 2 h.

2.5.14. Characterization

The TEM images illustrate that the diameter of Pt/Ni dendrites organized on the nanosheets of ceria was approx. 63.7 ± 9.3 nm and the thickness of the ceria nanosheets was 4.6 nm. The synthesis of PT/Ni coupled ceria nanosheets was observed by TEM images, at different temperatures. It shows the formation of Pt-Ni nano dendrites at 200ºC and continuously grows till the temperature reaches 230ºC. Furthermore, the synthesis of ceria nanosheets takes place at the reaction temperature of 250ºC. While,at the reaction temperature of 320ºC the formation of welly-organized Pt/Ni coupled ceria nanosheets takes place. Whereas XRD confirms the FCC cubic fluorite structure of Pt-Ni/ceria nanosheets Pt-Ni/ceria, and Pt3Ni/ceria. The ceria nanosheets were smaller in size (8.5 nm) rather than the ceria nanoparticles (21.1 nm), which was confirmed by the border diffraction peak of ceria nanosheets on the comparison of ceria nanoparticles. Electrochemical impedance spectroscopy (EIS) displays that the smaller the size of ceria nanosheets, the more conductivity will be because of the enhancement of the number of oxygen vacancies. On the contrary, XPS (X-ray photoelectron spectroscopy) confirms the reason behind the increasing conductivity was increased oxygen vacancies in ceria nanosheets. It shows that the concentration of Ce^{3+} ions in ceria nanosheets (58.4%) was higher than the ceria nanoparticles (31.4%), and as a result of it, the oxygen vacancies in the ceria nanosheets were higher.

3. Application of 2D- ceria (CeO_2) Nanosheets

It is reported that the ceria nanosheets having a thickness of 2.2 nm, synthesized via aqueous phase synthesis have been used for many purposes due to their very small thickness, they show quantum size effect and excellent catalytic properties because of the presence of oxygen vacancies, they are also used in oxygen sensors and solid oxide fuel cells [6]. The ceria nanocrystals and metal oxides show a large band energy gap, which increases the catalytic activity and also shows an impact on the optical and electronic properties of ceria nanocrystals.

A method was introduced for the synthesis of the ceria nanosheet via using a bio template of China rose petals, the produced ceria (CeO_2) nanosheets exhibit a thickness of 7 nm approx. [13]. The specific morphology and surface area of the ceria nanosheets play an important role in the specific chemical and physical properties of ceria (CeO_2). It can be used as an oxygen storage and sensor, fuel cells, ultraviolet blockers, and in catalysis activities. The catalysis activities were increased by the enhancement in the number of oxygen vacancies.

The ceria (CeO_2) nanosheets are formed by the novel aqueous-phase precipitation method [15]. These synthesized ceria nanosheet (thickness 20-50 nm), shows the oxidative catalytic properties in the oxidation of CO and chlorinated volatile organic compounds including 1,2-dichloroethane.

In the traditional method like the hydrothermal method and sol-gel method the precursors, surfactant, or bio templates are required, but in the aqueous phase precipitation method, no such things were needed. In this method, the reaction temperature and aqueous ammonia played an important role in the modification of the morphology of surface structures [15].

The hexagon ceria nanosheet having a thickness of 40-50nm, produced by the hydrothermal synthesis shows excellent optical properties and room temperature ferromagnetism due to the oxygen vacancies and effect of Ce^{3+} ions [8].

The Fe-doped-CeO_2 nanosheets were prepared by different methods named the hydrothermal method, solvothermal method, and co-precipitation method. These Fe-doped-CeO_2 nanosheets display excellent catalytic oxidation properties in the oxidation of chlorinated volatile organic compounds including 1,2-dichloroethane (DCE) and convert them into less hazardous compounds. The Fe-doped-CeO_2 produced from the hydrothermal method shows the best catalytic activity at low temperatures (100-175ºC). While at higher temperatures (175-350ºC) the Fe-doped-CeO_2 produced from the solvothermal method shows better catalytic activity [10].

The Cu^{2+}/La^{3+}-doped ceria nanosheets were synthesized via a complexing-coprecipitation process, displaying catalytic oxidation properties in the oxidation of CO, in which Cu^{2+}-doped ceria shows better efficiency in catalytic oxidation rather than the La^{3+}-doped ceria nanosheet [16].

The N-doped ceria nanosheets having a thickness of 5 nm, were produced by using a bio template of poppy petals exhibit photocatalytic activities, and play an important role in hydrogen gas production by photocatalytic water splitting. The doping of nitrogen in ceria enhances the visible-light absorbency via substituting the oxygen present at lattice positions [14].

The porous CeO_2 nanosheets were synthesized via a hydrothermal process on an optical fiber, having a thickness of 50 nm approx., due to their very small size they exhibit optical, electrical, and magnetic properties. The Ceria nanosheets coated with optical fiber are used as fiber sensors and lasers [9].

The phosphate-modified ceria nanosheets having a thickness of 20-50 nm, show magnificent catalytic properties for the oxidation of chlorinated volatile organic compounds (CVOC) named 1,2-dichloroethane, dichloromethane, chlorobenzene, etc, and convert them into preferable products such as H_2O, HCl, and CO_2. Ceria (CeO_2) acts as an oxygen carrier for the oxidation of CVOC because it shows redox properties (Ce^{3+}/Ce^{4+}), but it has a disadvantage in that the ceria gets decayed due to the strong attraction of chlorine (Cl^-) with the active sites and reduce the catalytic efficiency. To resolve this problem the phosphorus-modified ceria nanosheet has been synthesized by Dai *et al*., (2018), and it shows high catalytic efficiency with durability [19].

The Pt-Ni coupled ceria nanosheets were prepared by a one-pot synthesis method. The synthesized Pt-Ni coupled ceria nanosheets display electrocatalytic properties for the oxidation of methanol, and oxygen reduction, also they possess excellent electrocatalytic activities for the polymer electrolyte membrane fuel cells since it show good oxygen storage and transport properties as it has high oxygen vacancy and display redox properties [20].

The Co-doped-CeO_2 nanosheets (thickness 4.3 nm) have been synthesized via exfoliation methods [12]. The doping of cobalt in the ceria nanosheet enhanced the oxygen vacancies, which makes them the most effective for the dissociation of water and the adsorption of hydrogen. The Co-doped-ceria nanosheets act as a durable cheap catalyst and possess excellent hydrogen evolution reaction activities [12].

The ultrathin ceria (CeO_2) nanosheets (thickness ~10 nm) have been synthesized via a low-temperature hydrothermal method [11]. These ultrathin ceria nanosheets can be used as a humidity sensor, for the detection of formaldehyde vapor, and act as bi-functioning material.

The Pd/Al_2O_3-CeO_2 shows the highest catalytic activity rather than the Pd-CeO_2, since there are higher synergistic effects between the alumina and ceria, and they increase the stability of Pd/Al_2O_3-CeO_2 nanosheets. It displays excellent catalytic activities in acetone hydrogenation since it reduces the reaction barriers and accelerates the rate of reaction. The Pd/Al_2O_3-CeO_2 nanosheet acts as a catalyst in the one-pot conversion of acetone into methyl isobutyl ketone [18].

4. Challenges and Limitations

The aqueous phase synthesis of ceria nanosheets faced difficulties in the synthesis of 2D-ceria nanosheets because of the lack of the driving force for the anisotropic growth of nanoparticles, due to the cubic crystal structure of metal oxides. A network of nanocrystals of ceria has been made up to form the polycrystalline ceria nanosheets, which act as an intermediate. During the synthesis reorientation of nanocrystals and Ostwald ripening takes place and formed resultant single crystalline ceria (CeO_2) nanosheets. It has been shown that in the formation of nanosheets via self-organization electrostatic repulsion force plays an important role, in increasing the electrostatic repulsion the nanocrystals get shaped into the nanosheets [6].

The hydrothermal method and exfoliation method have been successfully used traditionally for the synthesis of 2D-ceria nanosheets, but the disadvantage of these methods was the production of a small number of ceria nanosheets. And secondly, there is a lack of driving force between the nanoparticles for the anisotropic growth, and it is due to the cubic fluorite structure of metal oxides including ceria (CeO_2). Some of the researchers used a bio template of China rose petals, which make an ideal platform for inorganic assembly via hierarchical arrangements [13].

5. Opinion

The rare earth oxide ceria (CeO_2) draws an attraction in the field of catalysis activities, oxygen storage, oxygen transport, fuel cells, UV ray blockers, etc. ceria shows excellent earth abundance, it has low cost, and is highly active. The synthesis of 2D-ceria nanosheets always has been challenging in past decades, but different methods have been introduced for the synthesis of ceria nanosheets, and for the metal-doped-ceria nanosheets, such as the hydrothermal method, sol-gel method, co-precipitation method, aqueous phase precipitation method, carbonization precipitation method, complexing co-precipitation method, etc. these methods have some advantages and some of the disadvantages. The aqueous phase precipitation method has some specific advantages over the others, such that this method is a template or surfactant-free, non-hydrothermal method, there is no such organic solvent was used. And different temperature-based morphological structures of ceria nanosheets were obtained by this method (e.g., petal, sheet, olive-like structure, etc.). Another environmentally friendly and less costly process is the carbonization precipitation method, which is used for the synthesis of 2D-ceria nanosheets. This method is also free from any organic additives, complexant, and templates. It includes Ostwald ripping and oriented mechanism for the growth of 2D-Ceria nanosheets.

Conclusion

In conclusion, this book chapter provides a comprehensive and detailed analysis of cerium oxide (ceria) nanosheets. The study explores various synthesis methods, and characterization of the ceria nanosheets, and also investigates their potential applications. It also highlights the challenges faced in synthesizing ceria nanosheets, such as the lack of driving force for anisotropic growth and low yields in certain methods. These challenges deepen the need for further research and development to optimize synthesis techniques and enhance the production of high-quality nanosheets. Furthermore, ceria nanosheets show remarkable catalytic and photocatalytic properties. Their unique features, including potential quantum size effect, high surface area, and anisotropic behavior, contribute to their enhanced catalytic performance. The reversible oxidation states of cerium and the presence of oxygen vacancies make ceria nanosheets highly efficient catalysts for various reactions. The potential applications of ceria nanosheets are broad and promising. They hold great potential in fields such as catalysis, energy, and environmental remediation. With their tunable band gap through suitable dopants, ceria nanosheets also exhibit promising prospects as photocatalysts for sustainable energy conversion and environmental remediation processes.

References

[1] Masui T., Hirai H., Hamada R., Imanaka N., Adachi G., Sakata T., and Mori H. (2002) Synthesis and characterization of cerium oxide nanoparticles coated with turbostratic boron nitride. *J. Mater. Chem.* **13**, 622–627. https://doi.org/10.1039/B208109A.

[2] Jiang, Z., Guo, Z., Sun, B., Jia, Y., Li, M., & Liu, J. (2010). Highly sensitive and selective butanone sensors based on cerium-doped SnO_2 thin films. *Sensors and Actuators B:Chemical*, *145*(2), 667–673. https://doi.org/10.1016/j.snb.2010.01.014.

[3] Liang, H., Raitano, J. M., He, G., Akey, A. J., Herman, I. P., Zhang, L., & Chan, S. W. (2011). Aqueous co-precipitation of Pd-doped cerium oxide nanoparticles: chemistry, structure, and particle growth. *Journal of Materials Science*, *47*(1), 299–307. 10.1007/s10853-011-5798-8.

[4] Zhao, B., Ma, L., Shi, H., Liu, K., & Zhang, J. (2018). Calcium precursor from stirring processes at room temperature for controllable preparation of nano-structure CaO sorbents for high-temperature CO_2 adsorption. *Journal Of CO_2 Utilization*, *25*, 315-322. https://doi.org/10.1016/j.jcou.2018.04.012.

[5] Pan, C., Zhang, D., & Shi, L. (2008). CTAB assisted hydrothermal synthesis, controlled conversion and CO oxidation properties of CeO_2 nanoplates, nanotubes, and nanorods. *Journal of Solid State Chemistry*, *181*(6), 1298–1306. https://doi.org/10.1016/j.jssc.2008.02.011.

[6] Yu, T., Lim, B., & Xia, Y. (2010). Aqueous-phase synthesis of single-crystal ceria nanosheets. *Angewandte Chemie*, *122*(26), 4586-4589. https://doi.org/10.1002/ange.201001521.

[7] Liu, B., Li, Q., Du, X., Liu, B., Yao, M., Li, Z., Liu, R., Liu, D., Zou, X., Lv, H., and Li, D. (2011). Facile hydrothermal synthesis of CeO_2 nanosheets with high reactive exposure surface. *Journal of alloys and compounds*, *509*(23), 6720-6724. https://doi.org/10.1016/j.jallcom.2011.03.156.

[8] Meng, F., Zhang, C., Fan, Z., Gong, J., Li, A., Ding, Z., Tang, H., Zhang, M., and Wu, G. (2015). Hydrothermal synthesis of hexagonal CeO_2 nanosheets and their room temperature ferromagnetism. *Journal of Alloys and Compounds*, *647*, 1013-1021. https://doi.org/10.1016/j.jallcom.2015.06.186.

[9] Wang, J., Cao, Z., Feng, M., Zhang, G., & Yu, B. (2018). Controlled synthesis and optical properties of porous ceria nanosheets on optical fiber via a simple hydrothermal route. *Journal of Materials Science: Materials in Electronics*, *29*(5), 3730-3735. https://doi.org/10.1007/s10854-017-8306-3.

[10] Wang, W., Zhu, Q., Dai, Q., & Wang, X. (2017). Fe doped CeO_2 nanosheets for catalytic oxidation of 1, 2-dichloroethane: Effect of preparation method. *Chemical Engineering Journal*, *307*, 1037-1046. https://doi.org/10.1016/j.cej.2016.08.137.

[11] Zhang, P., Zhang, L. X., Xu, H., Xing, Y., Chen, J. J., & Bie, L. J. (2021). Ultrathin CeO_2 nanosheets as bifunctional sensing materials for humidity and formaldehyde detection. *Rare Metals*, *40*(6), 1614-1621. https://doi.org/10.1007/s12598-020-01619-7.

[12] Jiang, S., Zhang, R., Liu, H., Rao, Y., Yu, Y., Chen, S., Yue, Q., Zhang, Y., and Kang, Y. (2020). Promoting the formation of oxygen vacancies in two-dimensional cobalt-doped ceria nanosheets for efficient hydrogen evolution. *Journal of the American Chemical Society*, *142*(14), 6461-6466. https://doi.org/10.1021/jacs.9b13915.

[13] Qian, J., Chen, F., Zhao, X., & Chen, Z. (2011). China rose petals as biotemplate to produce two-dimensional ceria nanosheets. *Journal of Nanoparticle Research*, *13*(12), 7149-7158. doi: 10.1007/s11051-011-0626-2.

[14] Qian, J., Zhang, W., Wang, Y., Chen, Z., Chen, F., Liu, C., Lu, X., Li, P., Wang, K., and Chen, A. (2018). Visible-light-driven nitrogen-doped petal-morphological ceria

nanosheets for water splitting. *Applied Surface Science*, *444*, 118-125. https://doi.org/10.1016/j.apsusc.2018.03.052.

[15] Dai, Q., Bai, S., Li, H., Liu, W., Wang, X., & Lu, G. (2014). Template-free and non-hydrothermal synthesis of CeO_2 nanosheets via a facile aqueous-phase precipitation route with catalytic oxidation properties. *CrystEngComm*, *16*(42), 9817-9827. https://doi.org/10.1039/C4CE01436G.

[16] Zhou, X., Ling, J., Sun, W., & Shen, Z. (2017). Fabrication of homogeneously Cu^{2+}/La^{3+}-doped CeO_2 nanosheets and their application in CO oxidation. *Journal of Materials Chemistry A*, *5*(20), 9717-9722. https://doi.org/10.1039/C7TA00924K.

[17] Lai, A., He, Q., Rao, M., Gao, G., & Xiao, Y. (2020). synthesis of highly uniform ceria nanosheets by carbon dioxide carbonization and their growth mechanism. *Journal of Solid-State Chemistry*, *290*, 121593. https://doi.org/10.1016/j.jssc.2020.121593.

[18] Lv, Z., Zhang, Y., Gong, Y., Guo, Z., & Zhang, C. (2022). Novel Pd/Al_2O_3–CeO_2 Nanosheets for One-Pot Synthesis of Methyl Isobutyl Ketone from Acetone. *Catalysis Letters*, 1-12. https://doi.org/10.1007/s10562-022-03991-8.

[19] Dai, Q., Zhang, Z., Yan, J., Wu, J., Johnson, G., Sun, W., Wang, X., Zhang, S., and Zhan, W. (2018). Phosphate-functionalized CeO_2 nanosheets for efficient catalytic oxidation of dichloromethane. *Environmental science & technology*, *52*(22), 13430-13437. https://doi.org/10.1021/acs.est.8b05002.

[20] Kwon, Y., Kim, Y., Hong, J. W., Whang, Y., Kim, S., Wi, D. H., Byon, H. R., and Han, S. W. (2020). One-pot production of ceria nanosheet-supported Pt-Ni alloy nanodendrites with high catalytic performance toward methanol oxidation and oxygen reduction. *Journal of Materials Chemistry A*, *8*(48), 25842-25849. https://doi.org/10.1039/D0TA09310F.

Chapter 4

Nanosheet Structures of Pb Halide and Pb-Free Halide Perovskites

Debrina Jana*
Department of Chemistry, Gour Mohan Sachin Mandal Mahavidyalaya, West Bengal, India

Abstract

Perovskite are a class of semiconductor nanocrystals, which exhibit exclusive features that separate them from traditional quantum dots. Unlike traditional quantum dots, these nanocrystals possess unusual band-structure which renders these materials effectively defect tolerant and thereby allowing them to emit brightly without surface passivation. In addition to this, they possess high carrier mobility, long carrier lifetimes, long range carrier diffusion, and small effective carrier masses. So far, perovskite nanocrystals have been applied to numerous optoelectronic applications such as light emitting diodes, lasers, scintillators, solar cells, and photodetectors. No wonder, extensive research is going on different aspects of these nanocrystals to unveil their potential as much as possible. In this chapter, nanosheet and nanosheet-like such as nanoplatelet structure of perovskites have been discussed. In doing so, the origin of the perovskite, stability of the perovskite structure, importance of the nanosheet structure, lead halide perovskites in reduced dimensionality, various synthetic protocols, application of nanosheet structures, lead free perovskite nanosheet structure and finally the improvement scopes in this area have been addressed.

Keywords: perovskite, nanosheet, nanoplatelet, lead halide, lead free halide, anisotropic, reduced dimensionality

---

* Corresponding Author's E-mail: debrina.chem@gmail.com.

In: Properties and Applications of Nanosheets
Editors: Nanda Gopal Sahoo and Anirban Dandapat
ISBN: 979-8-89113-381-5

1. Introduction

Perovskites are a class of well-known compounds which have the similar structure to $CaTiO_3$. It was first discovered in the Ural Mountains of Russia by Gustav Rose in the year 1839 and was named after Russian mineralogist Count Lev Alekseyevich von Perovski. Later, in 1926, Victor Goldschmidt first used it as a general term for the crystal structure group. Though its existence and usage is known for quite long, the topic 'perovskite' has become very popular for the last two decades. Perovskite materials have emerged as the *most promising candidate* for various optoelectronic and photonic device applications. Colloidal perovskite NCs have been found to combine the outstanding optoelectronic properties of bulk perovskites with strong quantum confinement effects at the nanoscale. Credit especially goes to the metal halide perovskite (MHP). [1-5] Having the generic formula ABX_3, they adopt the simple and ideal cubic (or pseudocubic) crystal structure where A is the large monovalent cation occupying the cuboctahedral voids between adjacent octahedra. B is the small transition metal, usually divalent with a coordination number of 6 occupying the centre of the corner-sharing octahedra formed from the halide X ions (Figure 1a).

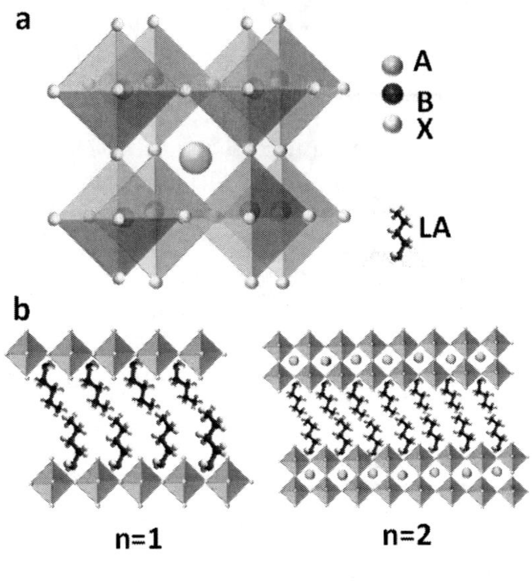

Figure 1. (a) Metal-halide perovskite have a 3D crystal structure with general formula ABX_3.(b) Crystal structures of hybrid 2D perovskite having n = 1 and n = 2.

B cations are divalent (like Pb^{2+}, or Sn^{2+}) and the A cations are large monovalent alkali metals (most commonly cesium) or small organic cations like methylammonium (MA) or formamidinium (FA). Now, based on the type of A (inorganic or organic), halide perovskites can be divided into two categories: All inorganic metal halide and organic-inorganic metal halide perovskites. All inorganic lead halide perovskites (LHPs) with $CsPbX_3$ (X = Cl^-, Br^-, and I^-) stoichiometry have been known since the end of the 19th century, although their perovskite crystal structure and semiconductive nature were not reported until the 1950s. Their hybrid organic-inorganic analogues, $MAPbX_3$ (MA = $CH_3NH_3^+$; methylammonium) and $FAPbX_3$ (FA = $CH(NH_2)^{2+}$; formamidinium) have been known since the late 1970s. Perovskites have high optical absorption coefficients and exhibit narrow-band bright photoluminescence, in addition to their halide-dependent tunable bandgaps, low exciton binding energies, and long-range carrier diffusion. In spite of having large number of defects, they exhibit fewer trap states which makes them defect tolerant. [5] These materials find extensive application in optoelectronics and photovoltaics owing to their easy synthesis and processability, near-unity Photoluminescence Quantum Yield (PLQY) and wide absorption range. The power conversion efficiency of perovskite solar cells has speeded from 3.8 to 25.7% after only 10 years of development, [6] signifying great potential for commercial consumption. In addition to photovoltaics, MHP materials are also widely investigated for use in photodetectors, [7] X-ray imaging, [8] lasers [9-11] and light emitting diodes (LEDs) [12]. In spite of significant achievements in lab-scale photovoltaics and optoelectronic devices, the large-scale commercial applications of Pb-based perovskites are currently limited by both the toxicity of Pb and the intrinsic instability of these materials. It has been revealed that the device instability is mainly due to the degradation of perovskites in the presence of moisture, heat, and light. [13] Toxic lead halides are considered as environmental hazards and potential carcinogens, [14] which could probably lead to the withdrawing of Pb-based perovskites from the future commercial market of photovoltaics. Risk of Pb release during long period usage and the long-term device instability raises serious concerns on the potential. These disadvantages of Pb-based perovskite materials are stimulating the need of replacing Pb in the perovskite community, with an intrinsically stable and environmentally friendly element. As such, a wide range of compounds comprising Sn^{2+}, Ge^{2+}, Sb^{3+}, and Bi^{3+}, and other transition metal cations are currently of interest. Substitutions at the metal cation positions, including group 14 metal elements (Sn and Ge), adjacent kin elements of Pb (Sb and Bi), and other transition metal elements have been explored as alternatives to Pb-based perovskites. These new members of the perovskite family offer rich crystal structures and similar optoelectronic properties to lead-halide perovskites with potential applications in photovoltaics and optoelectronic devices. [15-2] All these perovskite nanostructures can come in different shapes which may have impact on their properties leading to diverse applications. In this chapter, we will focus mainly on the nanosheet (NS) and nanosheet like structure such as nanoplatelets (NPLs) of Pb halide and non-Pb halide perovskite materials. In doing so, let's start with basic aspects of a perovskite structure and then proceed gradually to the advancement of nanosheet structures of perovskites.

2. Stability of Typical Perovskite Structure

The Goldschmidt tolerance factor and the octahedral factor are extensively used to predict stable crystal structures of perovskites. [21-23] Goldschmidt proposed an empirical approach to investigate the formability of ABX_3 perovskites almost a century ago. In this approach the perovskite structure is described as a collection of rigid spheres, with sizes given by the ionic radii R_A, R_B, and R_x. Goldschmidt tolerance factor is given by

$$T = \frac{(R_A + R_x)}{\sqrt{2}(R_B + R_x)}$$

The ideal perovskite has a cubic structure, with T = 1 but perovskites with orthorhombic, rhombohedral and tetragonal crystal structure having lower t values ranging between 0.75 and 1 are more common. This distortion of the structure arises from the small A site cations which causes a tilting of the octahedra to optimize A-X bonding. The Goldschmidt tolerance constraint restricts the possible combination of elements that can form stable 3D perovskites. Another factor, octahedral factor μ governs the stability of the BX_6^{2-} octahedra, thereby helps to predict the stability of the perovskite structure.

$$\mu = \frac{R_B}{R_x}$$

The structure is stable when $\mu > 0.41$. Thus, a simple calculation of both T and μ can preliminarily give idea about the formation probability and stability of a particular combination of A, B and X ions.

3. Shapes of Lead Halide Perovskites

The most well-known perovskites among Pb halide perovskites are $MAPbX_3$ and $CsPbX_3$. The electronic and optical properties of all inorganic and organometal halide perovskites are strongly dependent on the nature of the constituent ions. As for example, doping with different metal ions in $ABX_3(AB_nB'_{1-n}X_3)$ and varying the composition of halide $(ABX_nX'_{3-n})$ ions tunes their optoelectronic properties. [24] Additionally, dimensionality may be used to tune the electronic and optical properties. The most popular crystal structures of perovskites are cubic and orthorhombic. Initially, most of the synthetic procedures for the formation of ABX_3 NCs reported the formation of cubic shaped nanostructure. With advancement of the field, the fascinating 2D structures of chalcogenides also inspired researchers to think beyond the cubic or spherical shape of PNCs.

4. Importance of the Nanosheet Structure of Perovskites

Alongside the rise of perovskites, past decade has also witnessed the success of two-dimensional (2D) van der Waals materials having distinct properties than their bulk counterparts. Like the 2D van der Waals materials, where atomic layers can be isolated and manipulated, it was rightly assumed that the 2D perovskites should offer exciting opportunities to unveil new functionalities and/or for designing materials. Atomically thin 2D materials offer intriguing properties in terms of electron mobility, superconductivity, quantum Hall effect etc., thus makes it potential candidates for next-generation, high performance, ultrathin and flexible optoelectronic devices. [25] To achieve highly performing optoelectronic instruments, controlling the size, shape, crystal quality, thickness and crystallographic orientation of the semiconducting nanostructures is required. Among different optoelectronic applications, photodetectors are of particular attention because of their wide-ranging applications in biomedicine, environmental monitoring, telecommunications, imaging techniques, security checks, and industrial processing control. Two advantages offered by 2D materials for flexible photodetectors are as follows. Firstly, ultrathin planar materials having nice smooth surfaces are highly compatible with flexible substrates, which makes them ideal building blocks for the fabrication of flexible devices. Secondly, compared to their bulk counterparts, the ultrathin layered materials with lower dimensions can lead to a higher optical response, and longer photocarrier lifetime because of spatial confinement effects. [7] Nevertheless, weak photocurrent generation due to the material's limited optical absorption cross section puts an inherent limitation. For example, only 2.3% is generally found for graphene across the visible spectrum and around 10% at the band edge for 2D MoS_2, [26] which leads to a rather low photosensitivity. Definitely, significant improvement could be achieved by finding new 2D materials with a higher light absorptivity and longer carrier diffusion length. Undoubtedly, researchers started actively exploring the controllable growth of oriented optoelectronic nanostructures and thin films of metal-halide perovskites which are a class of emerging high-performance semiconductors. Halide perovskites with their high light absorptivity and carrier diffusion length undoubtedly makes them better candidates in high-performance photodetectors. As mentioned in the earlier section, most of the halide perovskite comes in cubic morphology, hence getting a 2D structure was a challenge initially.

5. Lead Halide Perovskites in Reduced Dimensionality

Halide perovskites with the ABX_3 formula have adjacent octahedra linked through all of their vertices, thereby forming a 3D network structure of connected octahedra. A reduction in the connectivity of the octahedra network leads to halide perovskites of reduced dimensionality, such as structurally 2D/quasi-2D layered perovskites. They have the general formula of $(LA)_m(A)_{n-1}B_nX_{3n+1}$, where LA is a large monovalent (m = 2) or divalent (m = 1) organic cation, and n is an integer that indicates the thickness of the perovskite layer. In the crystal structure of 2D/quasi-2D perovskites, 2D anion slabs of $[(A)_{n-1}B_nX3_{n+1}]^{2-}$ are sandwiched by a specific number of organic spacers. [27] The crystal structure units of 3D and 2D/quasi-2D perovskites are shown in Figure 1a and b. For example, n = 1, 2, and 3 correspond to layered perovskites that are one, two, and three unit cells thick respectively (Figure 1b), and n = ∞ corresponds to the bulk perovskite phase. Examples of the A site cations include butylammonium, ethylammonium, benzyl-ammonium, iodobenzylammonium etc. These 2D structures offer large compositional flexibility allowing control of the structural distortion, quantum and dielectric confinement, exciton-phonon coupling and Rashba splitting, which in turn tune their optical, electronic and spin properties. [27] The alternating layers of conducting inorganic layers and insulating organic layers generate a quantum well-like structure. Moisture stability, which was a major concern for 3D perovskite could also be avoided by the presence of the large organic spacer cations in between the inorganic halide octahedrons. [28] These properties have attracted 2D perovskites for use in optoelectronic and photovoltaic applications.

6. Formation of Nanosheet Structure of 3D Lead Halide Perovskites

As already mentioned in the earlier section, though predominantly cubic shaped NCs were found to be formed, there are synthetic reports available which could yield nanosheet like structures also. We also need to consider the structure under this category due to their similar characteristics like nanosheets. Formation of perovskite films have been also mentioned briefly as they can be considered as the long-range version of a continuous nanosheet structure. Few synthetic procedures and corresponding literature reports of getting nanosheet structures of perovskites made by different procedures are briefly discussed below.

i. Spin-Coating

Till date, there have been several reported methods of preparing organic-based and all-inorganic lead halide perovskite films, which are suitable forms to study the optical and electrical properties, as well as to fabricate opto-electronic devices. The simplest of them is spin-coating from precursor solutions of perovskite compounds. [29] Drawback of this method lies in the formation of inhomogeneous, polycrystalline structures with a large surface roughness.

ii. Thermal Evaporation and Chemical Vapour Deposition

To obtain better homogeneous and higher crystalline perovskite film, thermal evaporation technique has been applied. It can be done through either co-evaporation of lead halide and alkyl amino halide or single source (perovskite) evaporation. [30, 31] In co-evaporation technique, one requires the fine balancing of the evaporation rates of two sources which is difficult due to the much higher vapor pressure of the organic component compared to the inorganic counterpart. Single source evaporation is relatively simpler but it requires dedicated equipment. Son Tung Ha et al. reported a synthetic protocol for the formation of nanoplates which gives rise to high crystal quality and good optical properties. [31] They reported the synthesis of lead halide perovskite family NPLs with lateral dimensions from 5 – 30 μm and thicknesses from several atomic layers to several hundred nanometers. The $CH_3NH_3PbI_3$ platelets prepared by this chemical vapor deposition method have an electron diffusion length of more than 200 nm, which is two times higher relative to polycrystalline films obtained by spin-coating. The presented method involves two steps: First, the growth of lead halide NPLs on muscovite mica utilizing van der Waals epitaxy in a vapor transport chemical deposition system. Next, the as-grown platelets are converted to perovskites by a gas–solid hetero-phase reaction with methyl ammonium halide molecules. These nanoplates could be almost 20 μm long and the thickness could be varied from ~10-600 nm. Large-sized ultrathin all-inorganic $CsPbI_3$ perovskite nanosheets having lateral dimension exceeding 50 μm and smallest thickness around 3 nm can be synthesized based on the synergistic effect of an atmospheric pressure chemical vapor deposition strategy combining a space-confined method and van der Waals epitaxy. [32] Microstructure characterization techniques confirm the high crystal quality, while the corresponding performances like long carrier life (360 ns), high resistivity, and good stability were also verified, implying great promise for their applications in high-performance integrated optoelectronics. This method may pave the way for a uniform fabrication of other 2D nonlayered materials as well.

iii. Solution−Vapor-Phase Method (For Single Crystal)

One interesting study reports formation of regular arrays of perovskite microplate crystals utilizing solution-Vapor phase method in addition to lithographic technique. For integrated electronic and optoelectronic systems, lithography is required to make patterns. Now perovskite materials are not compatible with the typical lithographic process as they are soluble in various solvents and thus cannot be easily patterned into discrete device arrays. To this end, the direct growth of perovskite crystals at specific locations of prepatterned electrodes offers an alternative strategy for developing highly integrated systems. Wang et al. reported a facile two-step strategy for the scalable growth of large arrays of $MAPbI_3$ perovskite microplates with controlled physical dimensions and spatial location on diverse substrates (including silicon wafers, transparent glass, and pre-patterned gold electrodes). [33] For the production of regular arrays of perovskite crystals, lead iodide (PbI_2)) microplates were first grown on a substrate with pre-patterned (using lithography) surface functionalization to control nucleation and growth and then converted to $MAPbI_3$ perovskite microplate single crystal under methylammonium iodide (MAI) vapor. Steps involved are as follows: Functionalization of the SiO_2/Si with self-assembled monolayers of (octadecyl)trichlorosilane (OTS) to produce a hydrophobic surface, then lithographically patterned to create periodic arrays of hydrophilic areas, followed by production of PbI_2 seed particles in hydrophilic regions using a hot aqueous PbI_2 solution as seeding solution, immersed in a saturated PbI_2 solution at 80°C to further grow the seeds into larger PbI_2 microplates and finally, intercalation of the PbI_2 microplate arrays with methylammonium iodide vapor. The obtained patterned microplates showed controlled spatial location and periodic arrangement over a large area. It was also demonstrated that the directly grown microplate arrays can be used to create independently addressable photodetector arrays and well-performing field effect transistors.

iv. Two-Step Dipping Technique

Another method to prepare lead halide perovskite films is through a two-step dipping technique. In this method, a pre-deposited lead halide film is dipped into a solution of alkyl amino halide to convert the lead halides to their corresponding perovskites. [34] This method works well with a layered perovskite compound $(RNH_3)_2PbI_4$ in which R- is an alkyl group with two or more carbon atoms in the chain, which makes it easier for alkyl amino halide molecules to diffuse into the lead halide octahedron network through the van der Waals gap within the organic bilayer and, as a result, the reaction could be done in a few minutes time. However, in the case of $CH_3NH_3PbX_3$ perovskite, due to the lack of a van der Waals gap in the three dimensional structure of lead halide, the required dipping time would be much longer (i.e., 1–3 hours). Such a long immersion time in an organic solvent is not beneficial for formation of smooth film as parts of the lead halide and the formed perovskite could be dissolved and/or dispersed. [35, 36]

v. Colloidal Synthesis Route by Hot Injection Method

Colloidal synthetic procedure has been the most convenient and low cost method for preparing lead halide perovskite nanostructures. As mentioned earlier, cubic shape of the NCs was the most frequent form which are formed in the colloidal route. Protesescu et al. [5] have pioneered a simple convenient highly reproducible method to prepare lead halide perovskite nanostructures with tunable optical property across the visible region. This synthetic route takes benefit of the ionic nature of the chemical bonding in these compounds. Cs-oleate is reacted with a Pb(II)-halide in a high boiling solvent (octadecene) at 140–200°C, thus leading to controlled arrested precipitation of Cs^{+}, Pb^{2+}, and X^{-} ions into $CsPbX_3$ NCs. Mixed-halide perovskites, $CsPb(Cl/Br)_3$ and $CsPb(Br/I)_3$, can be readily produced by combining appropriate ratios of PbX_2 salts, which is widely referred as halide exchange. Results of optical characterization of these NCs have been shown in Figure 2. Note that Cl/I perovskites cannot be obtained due to the large difference in their ionic radii, which is in good agreement with the phase diagram for bulk materials.

Breaking the crystal symmetry and promoting anisotropic growth is required while seeking for a synthesis route to prepare NPLs or NSs of a nonlayered material. Few noteworthy examples in the case of $CsPbX_3$ nanosheet/nanoplate like structures are discussed as follows. Shamsi et al. have adopted an approach for the synthesis of colloidal orthorhombic single crystal $CsPbBr_3$ nanosheets with thickness in the range of 2–3 unit cells (i.e., around 3 nm) and with lateral dimensions from 300 nm up to a few micrometers (Figure 3a-d). [37] The process is an alteration of the synthesis reported by Protesescu et al. [5] and is carried out using standard air-free techniques. Their major change is to use Cs-oleate dissolved in oleic acid instead of octadecene (as done instead in the work of Protesescu et al.) and this change alone leads to NPLs, over a wide temperature range (from 50 to 150°C).

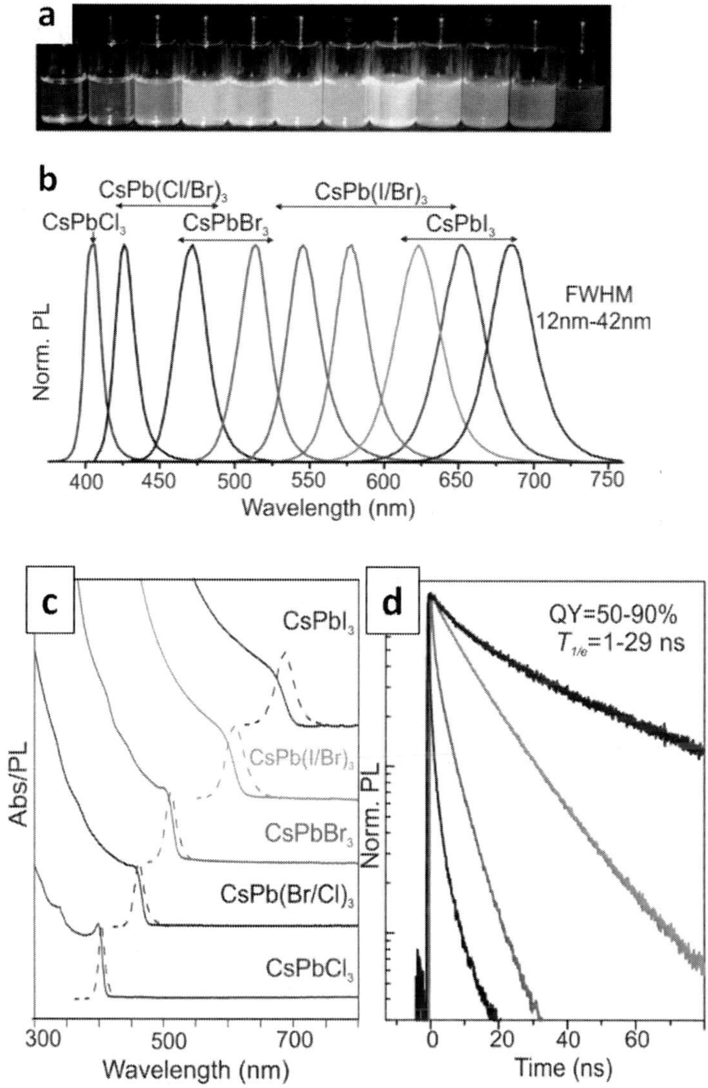

Figure 2. Colloidal $CsPbX_3$ NCs (X = Cl, Br, I) exhibit size- and composition-tunable bandgap energies covering the entire visible spectral region: (a) colloidal solutions in toluene under UV lamp (λ = 365 nm); (b) PL spectra (λ_{exc} = 400 nm for all but 350 nm for $CsPbCl_3$ samples); (c) optical absorption and PL spectra; (d) time-resolved PL decays for all samples shown in (c) except $CsPbCl_3$. Reproduced with permission. [5] Copyright 2015, American Chemical Society, further permissions related to the material excerpted should be directed to the ACS.

A generalized approach was reported by Longfei et al. enabling the direct synthesis of high-quality, phase-pure $CsPbX_3$ (X = Cl, Br, and I) perovskite and mixed perovskite nanosheets with uniform and ultrathin thicknesses down to a few monolayers. [7] The 2D perovskite nanosheets possess lateral dimensions over 100 nm to micro meter scale irrespective of the composition by controlling reaction conditions. In this work, the synthetic protocol was initially adopted by Protesescu et al., [5] but the difference was to use oleic acid instead of octadecene while preparing the Cs-oleate precursor. The underlying reason is that, at room temperature, Cs-oleate precursor is easily precipitated out in octadecene and it needs heating again at 100°C to obtain a homogeneous solution. This step can be avoided if oleic acid is used as solvent as Cs-oleate complex is well dissolved in oleic acid at room temperature. These perovskite nanosheets are highly luminescent in both colloidal and dried states. They could tune the emission wavelength of perovskite nanosheets across a wide region (380-600 nm) while maintaining narrow emission line widths (12-30 nm) by adjusting the halide composition. Notably, this is the first demonstration of photodetectors based on colloidal $CsPbBr_3$ nanosheets. The high on/off ratios and fast response time exhibited by the photodetectors are attributed to the excellent optoelectronic properties and large lateral sizes of $CsPbBr_3$ nanosheets. Thus this study demonstrates that the nanosheets could be ideal building blocks for a wide range of functional devices including photodetectors, light-emitting diodes, and photovoltaics.

So far, generation of NSs/NPLs directly were discussed by colloidal method. Next example shows a general strategy for the growth of $CsPbX_3$ (X = Cl, Br, I) perovskite nanosheets from the assembly of nanorods (NRs), done by Li et al. [38] It consists of a two-step process to grow a series of 2-D $CsPbX_3$ (X = Cl, Br, I) NCs with thickness in the range 3−6 nm and lateral dimensions from ~100 nm to ~1 μm (Figure 3e-g). In the first step, they synthesized 1-D $CsPbX_3$ (X = Cl, Br, I) NRs by reacting Cs-oleate with PbX_2 (X = Cl, Br, I) in a mixture of long-chain capping ligands (oleic acid and oleylamine), shorter ligands (octanoic acid and octylamine), and octadecene at 80−120°C for 5−10 s. Subsequently, 2-D $CsPbX_3$ NPLs and NSs were grown by the assembly of corresponding NRs in a Teflon-lined stainless-steel autoclave under solvothermal condition at 160°C for 1−5 h (Figure 3e). It should be noted that the solvothermal conditions, with elevated temperature and pressure, are critical to the formation of 2-D $CsPbCl_3$ and $CsPbBr_3$ NCs. The relatively high pressure under solvothermal condition can facilitate the spatial constraint by decreasing the viscosity of the growth medium and the free volume available for 1-D NRs, which induces 1-D to 2-D shape transition from the assembly of the NRs. In the following assembly process (Stage II), irregular NPLs can be formed by lateral crystallographic oriented attachment of small NPLs and stacking of two layers of thin NPLs. Finally, with longer reaction times, larger NSs (Stage III) can be formed from a self-limited assembly, oriented attachment, and fusion of NPLs under the solvothermal generated pressure. The solvothermal method takes advantage of the elevated pressure generated inside the sealed autoclave at moderate temperatures, which can facilitate the assembly of NRs and thin NPLs to eliminate high-energy surfaces and therefore reduce the total surface energy of the NCs. The broad lateral size range of the 2-D $CsPbX_3$ (X = Cl, Br, I) NPLs and NSs obtained by the solvothermal route can provide more opportunities to utilize 2-D all-inorganic perovskite NCs for further applications in photovoltaic cells and optical and electronic devices.

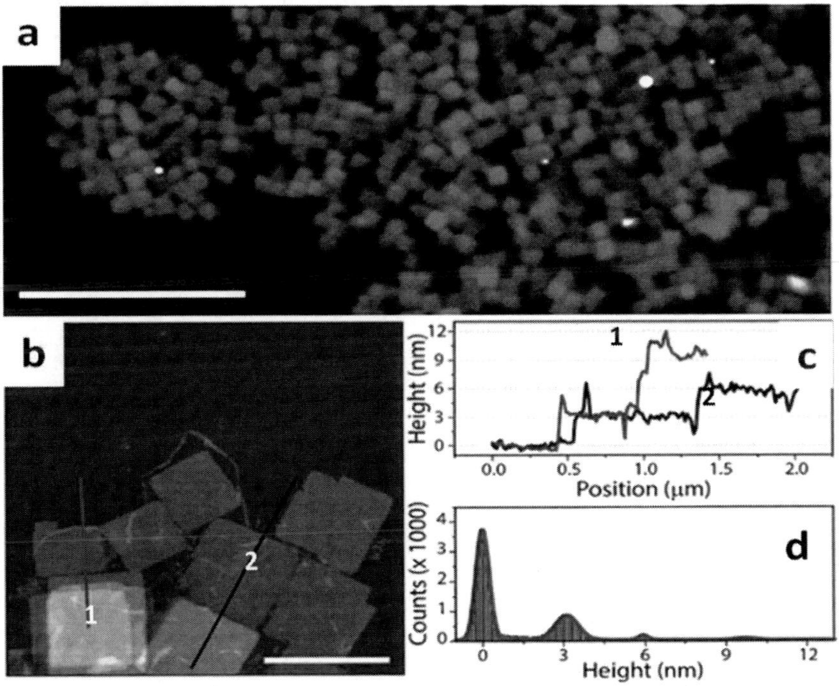

Figure 3. (Continued).

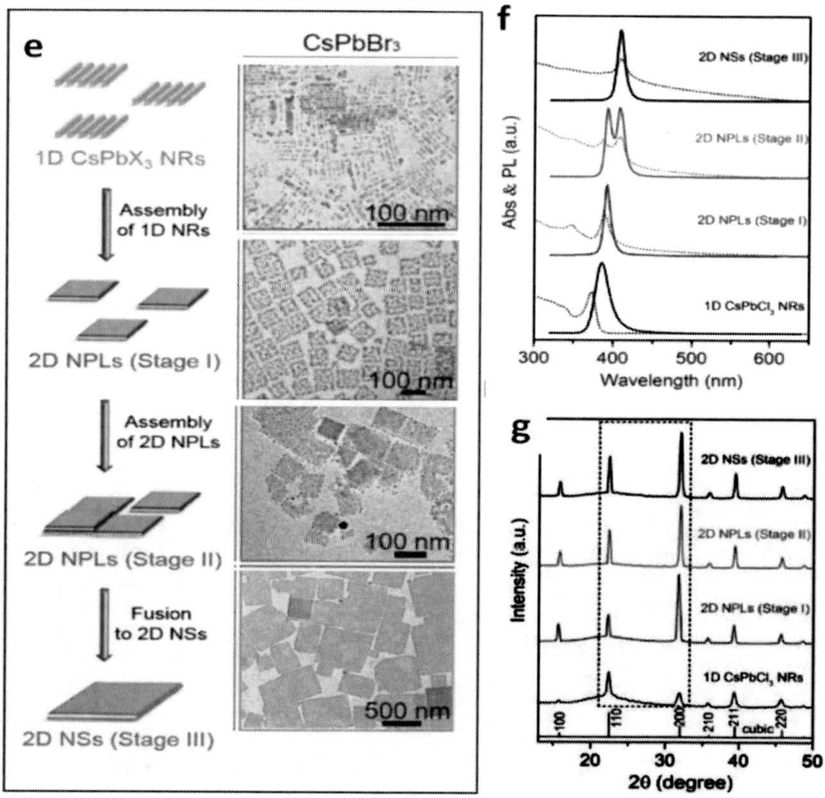

Figure 3. Confocal (a) and atomic force (b–d) microscopy analysis of the NSs. Scale bars correspond to 10 μm (panel a) and 1 μm (panel b). (c) shows the height profiles obtained from the image in (b) alongside the two lines (number 1 and number 2 line). (d) Height distribution over the whole image. The peak centered at 0 nm corresponds to the bare glass substrate (Images a-d Reproduced with permission.[37] Copyright 2016, American Chemical Society, further permissions related to the material excerpted should be directed to the ACS). (e) Schematic of the formation of 2-D $CsPbX_3$ NPLs and NSs by the oriented attachment of $CsPbX_3$ NRs. TEM images of the following are shown from top to bottom: $CsPbBr_3$ NRs synthesized in three-neck round bottom flasks under atmospheric pressure, $CsPbBr_3$ NPLs grown in Teflon-lined autoclave at 160 °C for 3.0 h, large NPLs after 4.0 h and NSs after 5.0 h. UV–vis (dotted line) and PL (solid line) (f) spectra and (g) XRD patterns of 1-D $CsPbCl_3$ NRs obtained by a hot-injection method at 120 °C, represented by the bottom curve) and 2-D $CsPbCl_3$ NCs obtained with different solvothermal treatment time at 160 °C, represented by second to the bottom up to the top curves respectively- 2D NPLs (Stage I), 2D NPLs (Stage II), 2D NSs (Stage III), (Images e-g Reproduced with permission.[38] Copyright 2018, American Chemical Society).

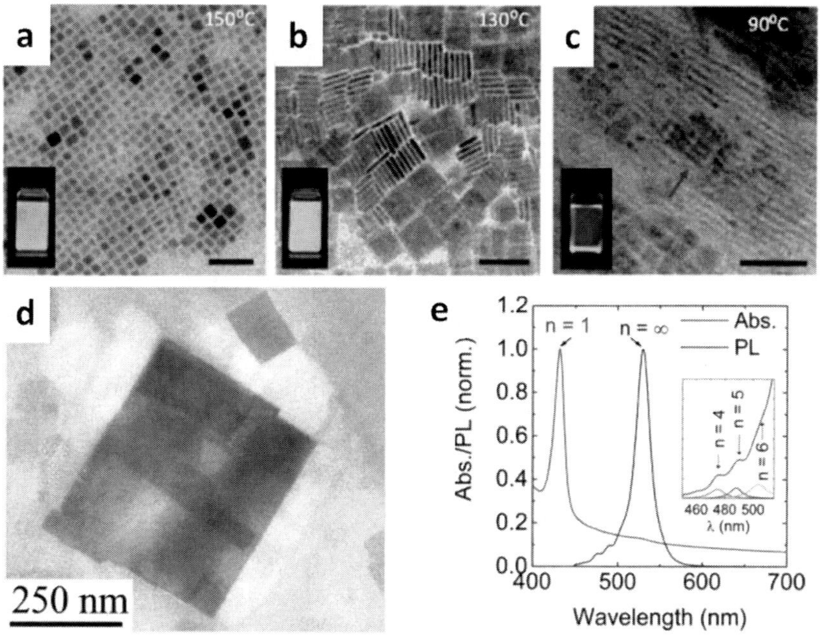

Figure 4. Study of the influence of reaction temperature in $CsPbBr_3$ colloidal synthesis. (a) At 150°C, green emitting 8–10 nm nanocubes are formed. (b) At 130°C, cyan emitting nanoplates with lateral dimensions of 20 nm and thickness of a few unit cells (~3 nm) are formed. (c) At 90°C, blue emitting thin nanoplates are observed, along with several-hundred nanometers lamellar structures. NPL growth was observed along the lamellae (red arrow) (a–c, scale bar is 50 nm), Images (a-c) Reproduced with permission. [39] Copyright 2015, American Chemical Society. (d) TEM of $MAPbBr_3$ nanoplatelet solution. (e) Absorption (red) and PL (blue) spectra of the $MAPbBr_3$ nanoplatelet solution. The inset shows an enlarged view of the blue shoulder of the emission spectrum, showing discrete peaks corresponding to nanoplatelets with different thicknesses (n: number of layered unit cells). The green, brown, and orange curves are multipeak deconvolutions (Images d,e Reproduced with permission. [40] Copyright 2015, American Chemical Society reproduced from Ref. 40).

Another notable example in colloidal method is provided by Bekenstein et al. who reported the colloidal synthesis of all-inorganic cesium lead halide perovskite nanoplates which could assemble to either form stacked columnar structures or to fuse into thin oriented sheets. [39] Colloidal synthesis of $CsPbBr_3$ perovskite NPLs was carried out by modifying the NC synthesis reported by Protesescu et al. [5] It was observed that temperatures in the lower range between 90 and 130°C tend to strongly favour asymmetric growth producing quasi 2D geometries. Here, $PbBr_2$ was solubilized in octadecene (ODE) with oleic acid and oleylamine, and then Cs-oleate was injected at elevated temperatures (90−130°C) to form NPLs. One important observation of this group is that the reaction temperature plays a critical role in determining the shape and thickness of the resulting NPLs. Reactions conducted at 130°C yield lower symmetry NPLs exhibiting cyan emission (Figures 4a and b) whereas at 90 and 100°C, very thin NPLs were detected along with lamellar structures ranging 200−300 nm in length (Figure 4a-c). As evident from the TEM images, NPLs grow along and inside the lamellar structures (Figures 4c and d), suggesting that organic meso-structures serve as growth directing soft templates that break the crystal's inherent cubic symmetry and dictate the 2D growth. Two different hierarchical structures were found to be formed by self-assembly of NPLs. First one is assembly into stacked columnar phases, which form in concentrated NPL solutions when ligands strongly interact. The second type of assembled structures are large 2D sheets formed by lateral crystallographic oriented attachment of single plates and they typically form in dilute samples where ligand passivation is destabilized.

Examples are there in organometal halide perovskite also. Tyagi et al. prepared colloidal NCs and 2D NPLs of methylammonium lead bromide ($MAPbBr_3$) perovskites and identify the optical signatures of their excitonic states using a combination of optical spectroscopy and electron microscopy. [40] Crystalline 2D perovskite NPLs of single unit cell thick (n = 1) were formed showing clear evidence of quantum confinement with a sharp excitonic absorption 0.5 eV blue-shifted from that of the 3D bulk perovskite phase (Figure 4d, e). The thickness-dependent absorption and emission spectra of these colloidal NPLs closely resemble those of solid-phase layered perovskites. $MAPbBr_3$ perovskite NCs were synthesized with octylammonium bromide as the long-chain ligand. The synthesis involved addition of lead bromide, methylammonium bromide, and octylammonium-bromide to a stirring solution of oleic acid and 1-octadecene at 80°C. As usual, the reaction was quenched with acetone. Purification was done to obtain a colloidal solution of thin 2D nanosheets of submicron lateral dimensions and varying thickness. They have the bulk $MAPbBr_3$ crystal structure but with thickness as small as a single unit cell, with octylammonium bromide ligands bound to each face.

With the purpose of directly correlate dimensionality with quantum confinement effects in sheets/platelets, control of the thickness down to a discrete number of monolayers, with differences in thickness from sample to sample at the level of a single monolayer needs to be achieved using colloidal synthesis route. Effort was made by Akkerman et al. to synthesize $CsPbBr_3$ NPLs with precise thickness tunability in the range from 3 to 5 unit cells. [41] They fabricated a room-temperature synthesis scheme where acetone was injected into a mixture of the precursors, which initiated the nucleation and growth process. Shape control was made possible due to the low temperature. At this small thickness, the $CsPbBr_3$ NPLs emitted in the blue region of the spectrum, which really makes the procedure stand out from others. Anion exchange could be done on these to obtain $CsPbCl_3$ and $CsPbI_3$ materials where both the size and shape of the starting particles are preserved. They compared the electronic and optical properties of the $CsPbBr_3$ NPLs with thickness smaller than the estimated Bohr radius (3.5 nm for $CsPbBr_3$) to those of non-quantum-confined samples. For the latter case, both polycrystalline thin films and cube-shaped NCs with large edge sizes were utilized. The major findings from this comparison was that, the magnitude of the band gap varied with the number of layers which means it is dependent on the extent of 2D confinement, but the excitonic dynamics in lead halide perovskites is virtually independent of it.

vi. Ligand Assisted Reprecipitation (LARP Method/Solution Phase Method)

LARP (ligand assisted reprecipitation) is one common synthetic method towards lead-halide perovskite NCs which depends on a large difference of the precursor solubility in their "good" (typical example is dimethyl formamide, DMF) and "bad" (typical example is toluene) solvents, wherein bad solvent induces supersaturation and thus triggers the nucleation and further growth of NCs (Figure 5). [42, 43] By utilizing this LARP method, an interesting observation regarding the size and shape control of perovskite NCs was made by Zhang et al. and they were able to get NPL. [44] If traces of a polar solvent (such as water) is introduced into the LARP reaction mixture, there would exist a solubility equilibrium between the dissolution and crystallization of perovskite NCs, which may allow to control their growth conditions and eventually lead to formation of different perovskite phases and/or result in formation of different NC sizes and shapes. Lead-halide perovskites are well known to decompose rapidly when exposed to polar solvents, such as water. Contrary to this familiar observation, they have found that through introducing a suitable minor amount of water into the reaction mixture, they can synthesize stable $CsPbBr_3$NCs with different shapes.

Formation of differently shaped perovskites, as for example, spherical NCs, rectangular NPLs, or nanowires are possible by adding suitable amounts of water in the reaction medium which change the crystallization environment. They also utilized these bright $CsPbBr_3$ NCs with the photoluminescence QY reaching 90% for fabrication of inverted hybrid inorganic/organic light-emitting devices, with the peak luminance of 4428 cdm^{-2} and external QY of 1.7%. An example in organometal halide perovskite can be provided by Zhu et al. who have demonstrated the systematic use of a variety of solvents and capping agents to produce $CH_3NH_3PbI_3$ and $CH_3NH_3PbBr_3$ NCs with different shapes (dot, rod, and plate or sheet morphologies) using LARP technique. [45] They were able to prepare $CH_3NH_3PbBr_3$ plates of dimension 150 nm×30 nm exhibiting band gap value of 2.30 eV and a QY of 0.43%.

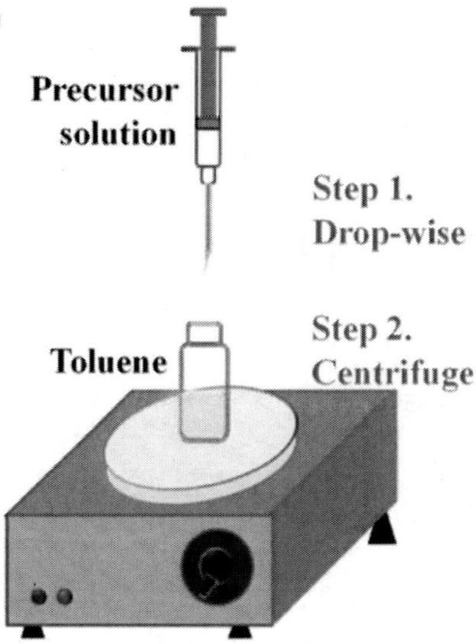

Figure 5. Schematic illustration of the reaction system and process for LARP technique (Reproduced with permission. [42] Copyright 2015, American Chemical Society).

vii. Solvothermal Method

The solvothermal method is a very easy and facile route to tune the morphology of NSs because of the high pressure/temperature inside the closed environment of the autoclave. In the 'colloidal synthetic route by hot injection' section we have discussed how to control the thickness of all-inorganic nanosheet ($CsPbBr_3$) precisely so that the relationship between dimensionality and quantum confinement could be understood. Several reports are there on tuning the morphology of the inorganic-based perovskites ($CsPbBr_3$) using the solvothermal method. Precise control of the thickness of large-area 2D organometal halide perovskite layers isrequired but it is extremely challenging due to the intrinsic instability of theorganic component. Towards this goal, a highly reproducible and facilesolvothermal route was reported by Parveen et al. to synthesize organic–inorganic NSs and tailor their thickness and optical band gap. [46] To achieve stable deep blue emission with near unity QY using at least partially lead-free perovskite, efforts were made and solvothermal method was useful in this purpose. Parveen et al. have developed a strategy to achieve stable and deep blue emission with absolute unity photoluminescence QY through Ce^{3+} and Tb^{3+} doping at high concentrations in 2D $CH_3NH_3PbBr_3$ NSs using a solvothermal method. [47] Their studies reveal that with Ce/Tb doping, the thickness of the NSs systematically goes down from 10 layers to bilayers (1.4 nm) at high doping levels and the bandgap of the 2D perovskite layer increases from 2.394 eV to 2.981 eV. The measured bandgap widening could be explained on the basis of the quantum confinement effect and lattice contraction. This study is notable asit reveals that byincorporating 70 mol% Ce^{3+} in the perovskite crystal, a deep blue emitting NPL with 100% QY could be achieved, thus making it one of the most efficientperovskite blue light emitters reported to date. The highly luminescent NPLswere also further demonstrated as a down converter white lightemittingdiode with CIE coordinates (0.334, 0.326), which closelymatches with those of pure white light. It can be concluded that the solvothermal approach is an efficient technique to develop rare-earth doped deep blue-emitting 2D perovskites with exceptionallyhigh PL QY and hence new insights into the structural stability and electronic structure of rare-earthdoped 2D perovskites could be realized.

viii. Single Crystal

MHP single crystals are potential candidates for photovoltaics and related optoelectronic devices. They show optical and charge transport properties superior to those of their polycrystalline counterparts. In this chapter, previous to this section, few examples are there mentioning single crystalline nature of the perovskite nanoplates and sheets. A detailed report on single crystal MHP could be found elsewhere. [48] Here, few important findings are mentioned.

Moving beyond methods based on colloidal chemistry, as shown in Figure 6, Dou et al. demonstrated a modified solid-state crystallization method for producing ultrathin 2D $(C_4H_9NH_3)_2PbBr_4$NPLs directly on flat substrates by carefully controlling the solvent composition, giving rise to a new way to study perovskites at an atomically thin level (Figure 7). [49] The direct growth of atomically thin sheets overcomes the limits of the conventional exfoliation and chemical vapor deposition methods, which usually produce comparatively thick perovskite plates. In 2016, Gao and co-workers produced $MAPbBr_3$ nanoplates with a narrow spectral band and a high PLQY of 85%; the light emitter based on these perovskite NPLs is one of the brightest perovskite LEDs reported to date. [12] $MAPbBr_3$ nanoplates were synthesized by a modified ligand-assisted precipitation method in which octylammonium-bromide was used as the long-chain capping ligand, which was found to be critical to the formation of the nanoplatelet morphology. These surface-passivated perovskite-nanoplatelet based emitter layers can be readily transferred onto a substrate by spin-coating, inkjet printing, and microcontact printing and other cost-effective deposition methods, thus could be potential for scalable, high-throughput production technology.

Till now, we have discussed about the various synthetic procedures for the preparation of nanosheets. It is clear that the stability of PNCs against external stimuli (e.g., water, polar solvents, light) is crucial to their efficacy for practical applications. Variety of approaches are available to offer improved stability of PNCs. Encapsulation of PNCs in bulk matrix (e.g., alumina, silica, etc.) [50, 51] is one such way, and they are often useful, especially when it is not required to trap individual NC with a shell. Disadvantage is, this method yields complex macroscopic hybrid materials instead of retaining well defined nanostructures and colloidal dispersity and processability of PNCs. PNCs with large structural anisotropy such as NSs are also required to have protective coating. In this case, mainly two factors pose significant synthesis challenge towards formation of inorganic shell at a single-nanocrystal level. One is uniform surface coating on such PNCs requires anisotropic growth of shell material, which is difficult and the other is ionic crystal nature of PNCs make them liable to degradation upon exposure to water and various polar molecules. He et al. reported a robust amphiphilic di block-copolymer-enabled strategy for crafting highly-stable anisotropic $CsPbBr_3$ NSs by in situ formation of a uniform inorganic shell (1st shielding) that is intimately ligated with hydrophobic polymers (2nd shielding). [52] The dual protected NSs display an array of remarkable stabilities (i.e., thermal, photostability, moisture, polar solvent, aliphatic amine, etc.) and find application in white-light-emitting diodes. This study implies that by attaching multidentate amphiphilic polymer ligands on the surface of PNCs, followed by templated-growth of shell materials on it, a wide range of multifunctional PNCs with evidently enhanced stabilities can be formed for use in real world applications.

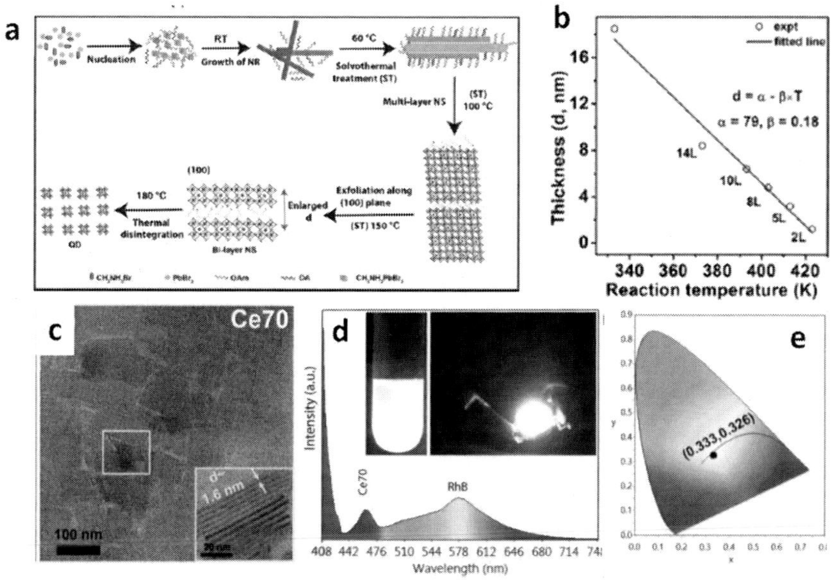

Figure 6. a) Schematic illustration of the growth mechanism and shape evolution of the perovskite NSs during different stages of solvothermal reaction at different temperatures. b) Change in the thickness of the 2D NS with solvothermal reaction temperature (Images a-b Reproduced with permission. [46] Copyright 2020, American Chemical Society). c) TEM image of Ce^{3+} doped 2D perovskite NSs with $CeBr_3$ concentrations at 70 mol%. d) EL spectra of the white light-emitting Ce doped 2D perovskite NS device fabricated with a UV LED chip, 70 mol% Ce doped 2D perovskite NSs, and Rhodamine B. The inset shows white light emission from the colloidal solution under UV excitation and digital photograph of the white light converter. (e) CIE chromaticity coordinates of the Ce doped 2D perovskite white light converter (Images c-e Reproduced with permission. [47] Copyright 2021, Royal Society of Chemistry).

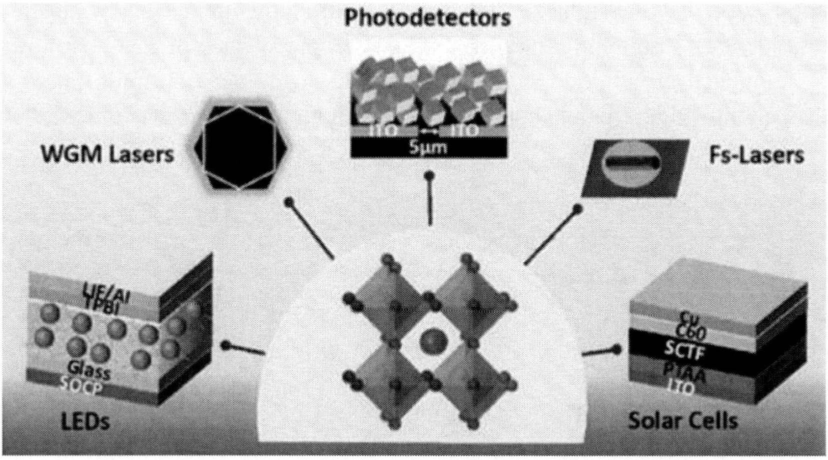

Figure 7. Schematic diagram showing different application of single crystals of perovskites (Reproduced with permission. [48] Copyright 2020, American Chemical Society).

7. Applications of Nanosheet/Nanoplate Structures

In the earlier sections, for example, in the introduction section and while discussing different synthetic methods, application of the NSs have also been mentioned with connection to that particular study. In this section, few examples of applications of NSs and NPLs are provided and their importance have been mentioned. All-inorganic perovskites NPLs are a fascinating system for the study of quantum confinement effects and are potential candidates for future devices, combining the excellent optoelectronic properties of perovskites with 2D geometry. [39]

The first example in the application is the lasing action of perovskite NSs. Ultrathin all-inorganic perovskite $CsPbX_3$ NSs have shown their higher exciton binding energy and carrier recombination efficiency than their bulk counterparts. In addition to this, owing to their naturally formed square shape, which may serve as a nearly perfect whispering gallery mode cavity, it is expected offer better light confinement and result in a high-quality factor. [9] Work is done by Jian-Yao Zheng et al., where LHP NSs grown from single crystalline lead halide possess advantages in device applications as they provide the possibility for control over morphology, composition, and crystallinity. [10] They first synthesized free-standing lead bromide ($PbBr_2$) single-crystalline NSs with sizes up to one centimetre from solution followed by their conversion to LHP while maintaining the NS morphology. Then these perovskite NSs can be processed directly for fabrication of photodetector (Figure 8) and laser arrays on a large scale (Figure 9). This strategy is advantageous as it will allow high yield synthesis of large-size perovskite NSs for functional devices in an integrated photonics platform. Perovskites can provide sufficient gain for lasing at room temperature. [53] As illustrated in figures, uniform arrays of $PbBr_2$ microcavities were fabricated on silicon substrates using focused ion beam (FIB) lithography techniques yielding more than one hundred microcavities in an area of 80 μm × 80 μm. These $PbBr_2$ microcavities can be converted to $MAPbBr_3$ microcavities by the gas-solid reaction. Room-temperature laser oscillation is achieved by optical pumping with a femtosecond (fs) laser. All the lasers characterized have similar emission wavelengths around 540 nm (Figure 9). Nonetheless, there is a huge opportunity in the future to improve the laser thresholds and photodetector efficiencies. All-inorganic perovskite nanosheets have also been reported to show lasing properties. Zheng et al. have synthesized high-quality 2D $CsPbI_3$ perovskite nanosheets with controllable morphology through space-confined vapor-phase epitaxial growth. [11] The $CsPbI_3$ perovskite nanosheets are typical square-like and they can be as thin as 6.0 nm with a lateral size of 10.0 μm. Here also, photoluminescence properties changed with the thickness of the NSs. These $CsPbI_3$ perovskite nanosheets demonstrate one-photon and two-photon excitation-pumped lasing with low-threshold and high-quality factor. High crystalline quality of the $CsPbI_3$ perovskite NSs are responsible for the brilliant multiphoton-pumped lasing properties. The next example is none other than solar cells and opto-electronic applications. $CH_3NH_3PbI_3$ perovskite has been recognized as suitable candidate for solar cell applications and one of the properties that accounts significantly is the long diffusion length of its charge carriers, which can be characterized by time-resolved photoluminescence spectroscopy. [31] The lifetime of the charge carriers in the perovskite is exceptionally long so that they can reach the electrodes of the cells before recombination and therefore reduce the loss in power conversion. When coming to the NS/NPL structure, the benefit is that $CH_3NH_3PbI_3$ and $CH_3NH_3PbI_xCl_{3-x}$ NPLs have a broad absorption covering the entire visible range (400–750 nm). No wonder why triiodide and iodide–chloride perovskites are more appropriate for solar cell applications. Again, the charge generation and transportation in the perovskite layer are well-correlated with the order and quality of its crystal network.

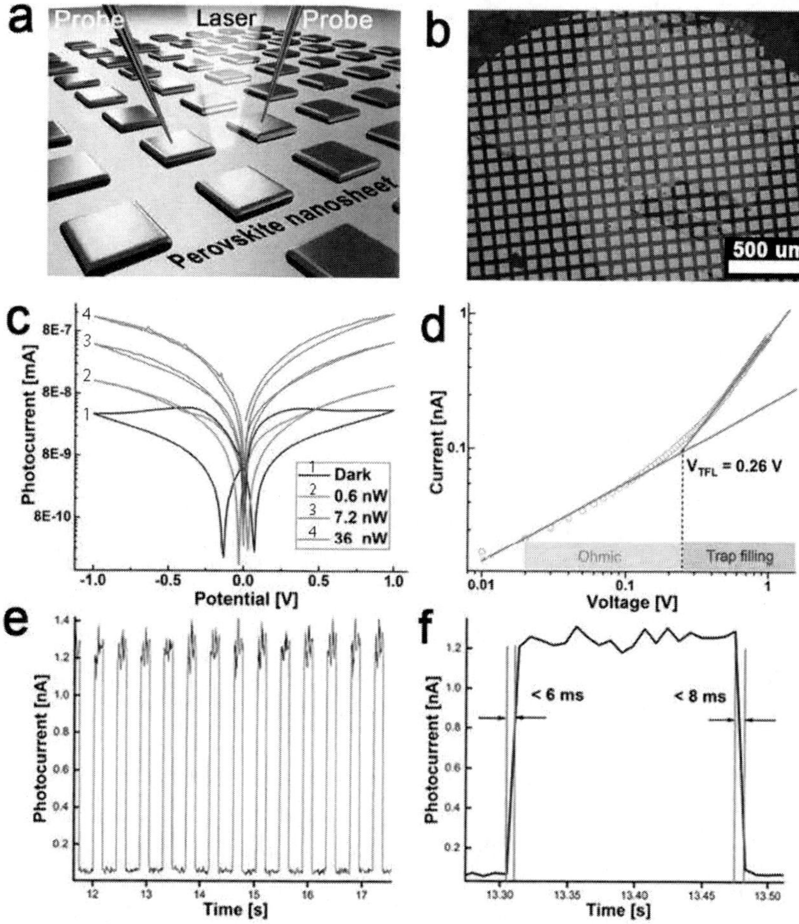

Figure 8. Photodetector measurements. (a) Schematic representation of the experimental setup for photodetector measurement, the laser spot was focused between the two electrodes. (b) Optical images of the device arrays on glass substrate. (c) *I-V* curves of the NSs perovskite-based device under laser irradiation with different powers, curves (1-4) corresponds to dark, 0.6 nW, 7.2 nW and 36 nW respectively. (d) Current-voltage trace of perovskite NSs. It shows a linear ohmic region followed by the trap-filled region (TFL) at V_{TFL}= 0.26 V. The trap density determined by $V_{TFL}$was calculated to be n_{trap}= 1.17×10^{10} cm^{-3}. (e) Time-dependent photocurrent measurement on the perovskite photodetector under the different power of a 405 nm laser with a voltage bias of 1 V. (f) Transient photocurrent of the photodetector (bias, 1 V; $\lambda = 405$ nm). The rise time (<6 ms) is defined as the time for the photocurrent to increase to 70% of the ON-state current. The fall time (<8 ms) is defined as the time for the photocurrent to decrease by 70% of the ON-state current (Image reproduced from Ref. 10. Open access article distributed under the terms of the Creative Commons CC BY license[10] Copyright 2019, Springer Nature).

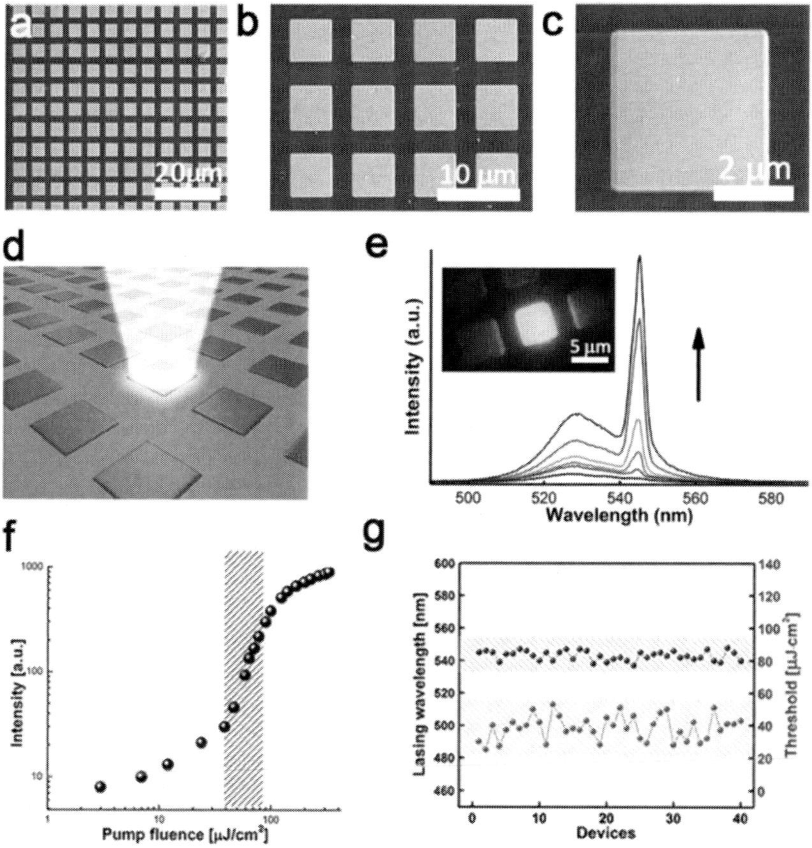

Figure 9. Room-temperature lasing characteristics. (a–c) SEM micrograph of the microcavity arrays. Low magnification SEM images of focused ion-beam milled section. (d) Schematic representation of the excitation method. (e) Spectra at six different optical pump fluences, showing the transition from spontaneous emission to amplified spontaneous emission and to lasing. Inset: Optical images showing emission from the microdisk. (f) Nonlinear response of laser output power with increasing pump fluence, showing threshold region as a 'kink' between the two linear regimes of spontaneous emission and lasing. The shadow area is the region of amplified spontaneous emission. (g) Variation in wavelength and threshold for 40 measured devices (Image reproduced from Ref. 10. Open access article distributed under the terms of the Creative Commons CC BY license [10] Copyright 2019, Springer Nature).

In a study done by Son Tung Ha et al. electron-diffusion length was estimated to be 210 ± 50 nm, which is longer than the minimal estimated values of at least 100 nm reported by another group. [31] This longer diffusion length could be attributed to the high crystal quality of the perovskite platelet prepared by this group. The synthesized perovskite platelets can be readily applied to ample applications, such as, single-crystal perovskite solar cells, LEDs, as well as other opto-electronic devices. Song et al. have reported the fabrication of atomically thin, 2D $CsPbBr_3$NSs having thickness of about 3.3 nm and edge length of about 1 μm and their application in solution-processed flexible photodetectors with high performance. [26] The NSs can be well dispersed in various solvents and thereby assembled into large-area, crack-free, high-quality films for flexible and ultrathin optoelectronic devices, such as ink-printing or roll-to-roll method. The flexible, 2D $CsPbBr_3$ visible-light photodetectors exhibit excellent electronic transport properties, a photosensitivity with an on/off ratio higher than 10^3, under 442 nm light at a power of 0.35 mW cm^{-2}, as well as a high stability and outstanding flexibility (>10000 cycles). These results demonstrate that all-inorganic perovskite $CsPbBr_3$ NSs, as a new class of 2D candidates, have huge potential for ultrathin and flexible solution-based optoelectronics, such as thin-film transistors, LEDs, and solar-cells.

8. Non-Pb Based Perovskite Nanosheets

Unfortunately, lead-based perovskites show poor stability in air, and lead is considered toxic, which can hinder commercialization. Germanium- and tin-based perovskites are nontoxic, but also suffer from stability issues. In contrast, bismuth halide perovskites are nontoxic, have excellent stability, and can be deposited using solution processing at relatively low temperatures. $Cs_3Bi_2I_9$ has been used in photovoltaic devices, as a supercapacitor, and as an electrode material in battery. Tiwari et al. have presented the first lead-free, bismuth-based, all-inorganic halide perovskite as an anode for a photo rechargeable battery that functions for both energy storage and photovoltaics. [54] Crystal structure of $Cs_3Bi_2I_9$ has been shown in Figure 10a. The devices can achieve a calculated photoconversion efficiency of ~0.43%, displaying competitive metrics with other photo batteries. This photobattery can be discharged and then photo-charged without any external current and then can power an external circuit for up to 2 h before charging again. Results obtained in this study establishes that $Cs_3Bi_2I_9$ perovskites represent a noteworthy step forward for photobatteries and advocate a promising future for lead free perovskites. Another report by Bai et al. presents a dissolution-recrystallization method to fabricate high-quality lead-free $Cs_3Bi_2I_9$ perovskite nanosheets film and explored its potential application in solar cells. [55] The solar cells based on ultrathin $Cs_3Bi_2I_9$ NS show a high power conversion efficiency of 3.20%, which is the highest efficiency value of Bi-based perovskite solar cells that has been reported so far. Not to mention, it also showed a long-term stability in ambient air. Definitely, these works validate the possibility of further development of the performance of lead-free perovskite solarcells.

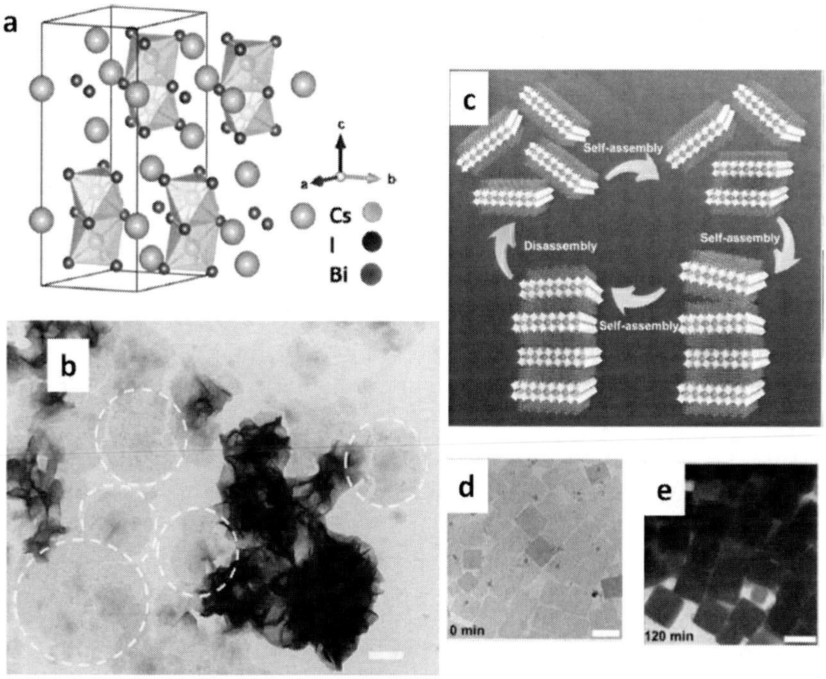

Figure 10. a) Crystal structure of $Cs_3Bi_2I_9$ shows hexagonal *P6₃/mmc* space group where BiI_6 octahedra share faces to form $[Bi_2I_9]^{3-}$ anions (Reproduced with permission. [54] Copyright 2021, American Chemical Society). b) TEM images of $(PEG6\text{-}NH_3^+)_nCs_{3-n}Bi_2Br_9$ nanosheets with scale bars of 200 nm. Yellow dashed circles indicate the presence of single-layer nanosheets (Reproduced with permission. [58] Copyright 2021, American Chemical Society). c) Schematic illustration of the layer-by-layer, self-assembly of $C_8H_{17}NH_3$-capped $CsPb_2Br_7$ nanosheets into layered $(C_8H_{17}NH_3)_2CsPb_2Br_7$ superlattice nanocrystals. The superlattice nanocrystals can be reversibly exfoliated back into monolayer building blocks by sonication in toluene. d) initial $C_8H_{17}NH_3$-capped $CsPb_2Br_7$ nanosheets. e) final $(C_8H_{17}NH_3)_2CsPb_2Br_7$ superlattice nanocrystals (Images c-e Reproduced with permission. [59] Copyright 2019, American Chemical Society).

We will now shift our attention to lead free double perovskites. Undoubtedly, the difficulty of manipulating dimensionality and probable phase segregation of multi-component compound makes the anisotropic growth of lead-free halide double perovskite NPLs/NSs challenging. But to create more possibilities for fundamental study and further real-world applications, an anisotropic growth method is important to investigate. [56, 57] Huang et al. demonstrated the synthesis of lead-free 2D halide double perovskite nanosheets. Air-free condition using standard Schlenk techniques which are being extensively utilized in the synthesis of inorganic LHP nanostructure, was also used for the synthesis of $Cs_2AgBiBr_6$ NSs. [56] To avoid the formation of unwanted phases such as $CsAgBr_2$ and $Cs_3Bi_2Br_9$ due to different precursor reactivities in solution resulting the phase segregation, simultaneous incorporation of both Ag and Bi cations into the structure is crucial during the synthesis of $Cs_2AgBiBr_6$ NSs. The synthesized $Cs_2AgBiBr_6$NSs exhibited thicknesses in the range of 3–5 nm and lateral dimensions of ~ 200 nm. The NSs showed a strong absorption peak centered at ~ 430 nm and the photoluminescence emission observed at ~ 630 nm. This study gives a guideline of controllable synthesis of multi-component compounds with tunable morphology, dimensionality and phase.

An excellent example of lead free, ultrathin 2D nanosheets has been demonstrated by Lee et al. using Bi^{3+} as the B-site cation forming the center of the perovskite octahedron and functional poly(ethylene glycols) (PEGs) as passivating ligands. [58] They reported a new Pb-free, perovskite [(PEG6-$NH_3^+)_nCs_{3-n}Bi_2X_9$ (X = Cl, Br, and I)] nanosheet synthetic route. "n" is an integer to represent the number of PEG6-NH_3^+ ligands inserting into the crystal lattice. These nanosheets form through the control of the inorganic lattice crystal structure, careful manipulation of the metal precursor concentration, and interplanar van der Waals interactions.

Appropriate selection of surface passivating ligands is crucial for the formation of these NSs. It is noteworthy that these nanosheets form a single layer (n = 1) which, in addition to the electronic benefits of PEG6 ligands, will provide ideal materials for charge separation (Figure 10b). Individual nanosheets can be self-assembled via mesoscale-directed growth resulting in the formation of quasi-Ruddlesden−Popper phase perovskite superlattices. For the 2D perovskite with Ruddlesden−Popper structure, the organic layer is composed of two univalent aromatic amines or alkylamine cations, such as phenylethylammonium, butylammonium, etc., and the inorganic layers are separated by the organic layer with a certain in-plane displacement. [29] These superlattices can be reversibly exfoliated to individual nanosheets via mechanical forces. Further studies need to be carried out on these Bi-based perovskite NSs to explore the 2D nanomaterial-based photocatalytic and optoelectronic applications with reduced toxicity.

Here is another example featuring that NS structure can assembled to form a distinct phase. Liu et al. synthesized 2D perovskite NSs of precise thickness. These NSs further assemble through a layer-by-layer process into 2D layered Ruddlesden−Popper phase superlattice NCs having composition of $(C_8H_{17}NH_3)_2Cs_{n-1}Pb_nBr_{3n+1}$ (where n represents the number of $[PbBr_6]^{4-}$octahedral layers). [59] Figure 10c schematically illustrates the layer-by-layer self-assembly process, from individual perovskite NSs with an alkylammonium ligand surface to many superlattice NCs matching the RP phase and Figures 10d-e show the corresponding TEM images. The nanosheet building blocks were synthesized by a modified, ligand-assisted, air-free, hot-injection method, [9] similar to the synthesis reported by Shamsi et al. [37] Like the last example, here also the weak van der Waals forces between the building blocks allows for separation of the assembly structure using mechanical force. Sonication in a weak polar solvent breaks the assembled superlattice and disassembles back into its building blocks. In a nutshell, layer-by-layer self-assembly process offers a simple yet robust approach to produce large-scale 2D layered halide perovskite superlattices with atomic scale precision. Thus, this system may offer a general pathway for studying the structure-property relationship in 2D layered superlattice nanomaterials which is beneficial for novel electronic and photonic applications.

Conclusion and Perspectives

Since the inception of the LHP NCs, optoelectronics and solar-cell domain undoubtedly has gotten a breakthrough. Continuous research is being carried out to overcome the difficulties associated with LHPs and to make them suitable for large-scale production and commercialization. Towards that goal, the NS structure of perovskite is definitely effective to a large extent. It has also been extended to other practical fields such as lasers, detectors, transistors and so on. Recently, MHP/2D material heterostructures, combining MHPs with diverse 2D materials, have demonstrated extraordinary enhancement in both performance and stability due to the characteristic features at hetero-interface and thus offer many possibilities to discover new functionalities within the 2D family. [60, 61] However, current research on 2D perovskite NS structure is still in the initial stage, there is a long way to go to comprehend high-performance devices. In the near future, it is expected that the following scientific questions will be addressed: 1) exploring new 2D lead-free perovskite materials by theoretical calculations, artificial intelligence and practical experiments; 2) understanding the guidelines of orientated growth for nanosheet structure and achieve certain degree of control over the synthetic procedure; 3) upgradation from lab-scale to commercial scale production by figuring out the loopholes in the synthetic schemes, both for lead and lead-free perovskites; 4) clear understanding of the exciton behaviors and structure-property correlation; 5) Combining with heterostructures. It is expected that solving these problems will create possibilities for new designs and functionalities and greatly increase the potential of 2D lead-free halide perovskite materials in practical applications.

References

[1] Wehrenfennig, C., Eperon, G. E., Johnston, M. B., Snaith, H. J., Herz, L. M. 'High Charge Carrier Mobilities and Lifetimes in Organolead Trihalide Perovskites'. *Adv. Mater.* 2013, 26, 1584-1589.

[2] Lee, M. M., Teuscher, J. T., Miyasaka, T., Murakami, N., Snaith, H. J. 'Efficient Hybrid Solar Cells Based on Meso-Superstructured Organometal Halide Perovskites', *Science* 2012, 338, 643-647.

[3] Liu, M., Johnston, M. B., Snaith, H. J. 'Efficient planar heterojunction perovskite solar cells by vapour deposition', *Nature* 2013, 501, 395-398.

[4] Manser, J. S., Christians, J. A., Kamat, P. V. 'Intriguing Optoelectronic Properties of Metal Halide Perovskites', *Chem. Rev.* 2016, 116, 12956-13008.

[5] Protesescu, L., Yakunin, S., Bodnarchuk, M. I., Krieg, Caputo, F. R., Hendon, C. H., Yang, R. X., Walsh, A., Kovalenko, M. V. 'Nanocrystals of Cesium Lead Halide Perovskites (CsPbX3, X = Cl, Br, and I): Novel Optoelectronic Materials Showing Bright Emission with Wide Color Gamut' *Nano Lett.* 2015, 15, 3692. https://pubs.acs.org/doi/10.1021/nl5048779

[6] Web link: https://scholar.google.co.in/citations?user=3nIi0RkAAAAJ&hl=en

[7] Lv, L., Xu, Y., Fang, H. Luo, W., Xu, Fangjie., Liu, L., Wang, B., Zhang, X., Yang, D., Hu, W., Dong, A. 'Generalized colloidal synthesis of high-quality, two-dimensional cesium lead halide perovskite nanosheets and their applications in photodetectors', *Nanoscale*, 2016, 8, 13589-13596.

[8] Zhou, Y., Chen, J., Bakr, O. M., Mohammed, O. F. 'Metal Halide Perovskites for X-ray Imaging Scintillators and Detectors', *ACS Energy Lett.* 2021, 6, 739–768.

[9] Zhang, Q., Su, R., Liu, X., Xing, J., Sum, T. C., Xiong, Q. 'High-Quality Whispering-Gallery-Mode Lasing from Cesium Lead Halide Perovskite Nanoplatelets', *Adv. Func. Mat.* 2016, 26, 6238-6245.

[10] Zheng, J., Manning, H. G., Zhang, Y., Wang, J. J., Purcell-Milton, F., Pokle, A., Porter, S., Zhong, C., Li, J., Meehan, R. Enright, R., Gun'ko, Y. K., Nicolosi, V., Boland, J. J., Sanvito, S., Donegan, J. F. 'Synthesis of centimeter-size free-standing perovskite nanosheets from single-crystal lead bromide for optoelectronic devices', *Scientific Reports* 2019, 9:11738.

[11] Zheng, Z., Wang, X., Shen, Y., Luo, Z., Li, L., Gan, L., Ma, Y., Li, H., Pan, A., Zhai, T. 'Space-Confined Synthesis of 2D All-Inorganic $CsPbI_3$ Perovskite Nanosheets for Multiphoton-Pumped Lasing', *Adv. Optical Mater.* 2018, 1800879.

[12] Ling, Y., Yuan, Z., Tian, Y., Wang, X., Wang, J. C., Xin, Y., Hanson, K., Ma, B., Gao, H. 'Bright Light-Emitting Diodes Based on Organometal Halide Perovskite Nanoplatelets', *Adv. Mater.* 2016, 28, 305–311.

[13] Bisquert, J., Juarez-Perez, E. J., 'The Causes of Degradation of Perovskite Solar Cells', *J. Phys. Chem. Lett.* 2019, 10, 19, 5889–5891.

[14] Tchounwou, P. B., Yedjou, CG., Patlolla, A. K., Sutton, D. J. 'Heavy metal toxicity and the environment', https://scholar.google.co.in/citations?user=3nIi0RkAAAAJ&hl=en.

[15] Jellicoe, T. C., Richter, J. M., Glass, H. F. J., Tabachnyk, M., Brady, R., Dutton, S. E., Rao, A., Friend, R. H., Credgington, D., Greenham, N. C., Böhm, M. L. 'Synthesis and Optical Properties of Lead-Free Cesium Tin Halide Perovskite Nanocrystals', *J. Am. Chem. Soc.* 2016, 138, 2941−2944.

[16] Luo, J., Wang, X., Li, S., Liu, J., Guo, Y., Niu, G., Yao, L., Fu, Y., Gao, L., Dong, Q., Zhao, C., Leng, M., Ma, F., Liang, W., Wang, L., Jin, S., Han, J., Zhang, L., Etheridge, J., Wang, J., Yan, Y., Sargent, E. H., Tang, J. 'Efficient and stable emission of warm-white light from lead-free halide double perovskites', *Nature* 2018, 563, 541−545.

[17] Creutz, S. E., Crites, E. N., De Siena, M. C., Gamelin, D. R. 'Colloidal Nanocrystals of Lead-Free Double-Perovskite (Elpasolite)Semiconductors: Synthesis and Anion Exchange To Access NewMaterials', *Nano Lett.* 2018, 18, 1118−1123.

[18] Wang, C.-Y., Liang, P., Xie, R.-J., Yao, Y., Liu, P., Yang, Y., Hu, J., Shao, L., Sun, X. W., Kang, F., Wei, G. 'Highly Efficient Lead-Free (Bi,Ce)-Codoped $Cs_2Ag_{0.4}Na_{0.6}InCl_6$ Double Perovskites for White Light-Emitting Diodes', *Chem. Mater.* 2020, 32, 7814−7821.

[19] Noculak, A., Morad, V., McCall, K. M., Yakunin, S., Shynkarenko, Y., Wörle, M., Kovalenko, M. V. 'Bright Blue and Green Luminescence of Sb(III) in Double Perovskite Cs_2MInCl_6 (M = Na, K) Matrices', *Chem. Mater.* 2020, 32, 5118−5124.

[20] Infante, I., Manna, L. 'Are There Good Alternatives to Lead Halide Perovskite Nanocrystals?', *Nano Lett.* 2021, 21, 1, 6–9.

[21] Goldschmidt, V. M. 'Die Gesetze der Krystallochemie [The Laws of Crystallochemistry]', *Naturwissenschaften* [Natural Sciences], 1926, 14, 477–485.

[22] C. Li, X. Lu, W. Ding, L. Feng, Y. Gao and Z. Guo, 'Formability of ABX_3 (X = F, Cl, Br, I) halide perovskites' *Acta Crystallogr., Sect. B: Struct. Sci.*, 2008, 64, 702–707.

[23] Travis, W., Glover, E. N. K., Bronstein, H., Scanlon, D. O., Palgrave, R. G. On the application of the tolerance factor to inorganic and hybrid halide perovskites: a revised system, *Chem. Sci.*, 2016, 7, 4548.

[24] Lu, C., Biesold-McGee, G. V., Liu, Y., Kang, Z., Lin, Z. 'Doping and ion substitution in colloidal metal halide perovskite nanocrystals' *Chem. Soc. Rev.*, 2020, 49, 4953-5007.

[25] Pham, P. V., Bodepudi, S. C., Shehzad, K., Liu, Y., Xu, Y., Yu, B., Duan, X. '2D Heterostructures for Ubiquitous Electronics and Optoelectronics: Principles, Opportunities, and Challenges', *Chem. Rev.* 2022, 122, 6, 6514–6613.

[26] Song, J., Xu, L., Li, J., Xue, J., Dong, Y., Li, X., Zeng, H. 'Monolayer and Few-Layer All-Inorganic Perovskites as a New Family of Two-Dimensional Semiconductors for Printable Optoelectronic Devices', *Adv. Mater.*, 2016, 28, 4861-4869.

[27] Stoumpos, C. C., Cao, D. H., Clark, D. J., Young, J., Rondinelli, J. M., Jang, J. I., Hupp, J. T., Kanatzidis, M. G. 'Ruddlesden–Popper Hybrid Lead Iodide Perovskite 2D Homologous Semiconductors', *Chem. Mater.* 2016, 28, 8, 2852–2867.

[28] Leung, T. L., Ahmad, I., Syed, A. A., Ng, A. M. C., Popović, J. Djurišić, A. B. 'Stability of 2D and quasi-2D perovskite materials and devices', *Commun. Mater.* 2022, 3, 63.

[29] Jiang, Y., Yuan, J., Ni, Y., Yang, J., Wang, Y., Jiu, T., Yuan, M., Chen, J. 'Reduced-Dimensional α-$CsPbX_3$ Perovskites for Efficient and Stable Photovoltaics', *Joule* 2018, 2, 1356–1368.

[30] Xing, G., Mathews, N., Sun, S., Lim, S. S., Lam, Y. M., Gratzel, M., Mhaisalkar, S., Sum, T. C. 'Long-Range Balanced Electron- and Hole-Transport Lengths', *Science*, 2013, 342, 344-347

[31] Ha, S. T., Liu, X., Zhang, Q., Giovanni, D., Sum, T. C., Xiong, Q. 'Synthesis of Organic–Inorganic Lead Halide Perovskite Nanoplatelets: Towards High-Performance Perovskite Solar Cells and Optoelectronic Devices', *Adv. Optical Mater.* 2014, 2, 833-844.

[32] He, Yeming; Ma, Yang; Li, X., Zhang, Y. 'All-Inorganic Perovskite Nanosheet Fabrication under Synergistic Effect for Integrated Optoelectronics with Strong Light−Matter Interactions', *ACS Appl. Nano Mater.* 2021, 4, 2634−2641.

[33] Wang, G., Li, D., Cheng, H., Li, Y., Chen, C., Yin, A., Zhao, Z., Lin, Z., Wu, H., He, Q. 'Wafer-scale growth of large arrays of perovskite microplate crystals for functional electronics and optoelectronics', *Sci. Adv.* 2015;1:e1500613.

[34] Burschka, J., Pellet, N., Moon, S. J., Baker, R. H., Gao, P., Nazeeruddin, M. K., Grätzel, M. 'Sequential deposition as a route to high-performance perovskite-sensitized solar cells' *Nature* 2013, 499, 316-319.

[35] Liang, K., Mitzi, D. B., Prikas, M. T. 'Synthesis and Characterization of Organic−Inorganic Perovskite Thin Films Prepared Using a Versatile Two-Step Dipping Technique', *Chem. Mater.* 1998, 10, 403-411.

[36] Mitzi, D. B. 'Thin-Film Deposition of Organic−Inorganic Hybrid Materials', *Chem. Mater.* 2001, 13, 3283-3298.

[37] Shamsi, J., Dang, Z., Bianchini, P., Canale, C., Di Stasio, F., Brescia, R., Prato, M., Manna, L. 'Colloidal Synthesis of Quantum Confined Single Crystal $CsPbBr_3$ Nanosheets with Lateral Size Control up to the Micrometer Range' *J. Am. Chem. Soc.* 2016, 138, 7240−7243.https://pubs.acs.org/doi/10. 1021/jacs.6b03166

[38] Li, Z., Hofman, E., Davis, A. H., Maye, M. M., Zheng, W. 'General Strategy for the Growth of $CsPbX_3$ (X = Cl, Br, I) Perovskite Nanosheets from the Assembly of Nanorods', *Chem. Mater.* 2018, 30, 3854−3860.

[39] Bekenstein, Y., Koscher, B. A., Eaton, S. W., Yang, P., Alivisatos, A. P. 'Highly Luminescent Colloidal Nanoplates of Perovskite Cesium Lead Halide and Their Oriented Assemblies' *J. Am. Chem. Soc.* 2015, 137, 16008−16011.

[40] Tyagi, P., Arveson, S. M., Tisdale, W. A. 'Colloidal Organohalide Perovskite Nanoplatelets Exhibiting Quantum Confinement', *J. Phys. Chem. Lett.* 2015, 6, 1911−1916.

[41] Akkerman, Q. A., Motti, S. G., Kandada, A. R. S., Mosconi, E., D'Innocenzo, V., Bertoni, G., Marras, S., Kamino, B. A., Miranda, L., De Angelis, F., Petrozza, A., Prato, M., Manna, L. 'Solution Synthesis Approach to Colloidal Cesium Lead Halide Perovskite Nanoplatelets with Monolayer-Level Thickness *Control' J. Am. Chem. Soc.* 2015, 137, 51, 16008–16011.

[42] Zhang, F., Zhong, H., Chen, C., Wu, X., Hu, X., Huang, H., Han, J., Zou, B., Dong, Y. 'Brightly Luminescent and Color Tunable Colloidal $CH_3NH_3PbX_3$ (X = Br, I, Cl) Quantum Dots: Potential Alternatives for Display Technology', *ACS Nano*, 2015, 9, 4533-4542.

[43] Zhang, Y. Siegler, T. D., Thomas, C. J., Abney, M. K., Shah, T., Gorostiza, A. D., Greene, R. M., Korgel, B. A. 'A "Tips and Tricks" Practical Guide to the Synthesis of Metal Halide Perovskite Nanocrystals', *Chem. Mater.*2020, 32, 13, 5410–5423.

[44] Zhang, X., Bai, X., Wu, H., Zhang, X., Sun, C., Zhang, Y., Zhang, W., Zheng, W., Yu, W. W., Rogach, A. L. 'Water-Assisted Size and Shape Control of $CsPbBr_3$ Perovskite Nanocrystals', *Angew. Chem. Int. Ed.* 2018, 57, 3337 –3342.

[45] Zhu, F., Men, L., Guo, Y., Zhu, Q., Bhattacharjee, U., Goodwin, P. M., Petrich, J. W., Smith, E. A., Vela, J. 'Shape Evolution and Single Particle Luminescence of Organometal Halide Perovskite Nanocrystals', *ACS Nano* 2015, 9, 3, 2948–2959.

[46] Parveen, S., Paul, K. K., Giri, P. K. 'Precise Tuning of the Thickness and Optical Properties of Highly Stable 2D Organometal Halide Perovskite Nanosheets through a Solvothermal Process and Their Applications as a White LED and a Fast Photodetector', *ACS Appl. Mater. Interfaces* 2020, 12, 6283−6297.

[47] Parveen, S., Prasanna, P. K., Chakraborty, S., Giri, P. K. 'Stable deep blue emission with unity quantum yield in organic–inorganic halide perovskite 2D nanosheets doped with cerium and terbium at high concentrations', *J. Mater. Chem. C*, 2021,9, 2437-2454.

[48] Feng, A., Jiang, X., Zhang, X., Zheng, X., Zheng, W., Mohammed, O. F., Chen, Z., Bakr, O. M. 'Shape Control of Metal Halide Perovskite Single Crystals: From Bulk to Nanoscale', *Chem. Mater.* 2020, 32, 7602−7617.

[49] Dou, L., Wong, A. B., Yu, Y., Lai, M., Kornienko, N., Eaton, S. W., Fu, A., Bischak, C. G., Ma, J., Ding, T., Ginsberg, N. S., WangWang, L., Alivisatos, A. P., Yang, P. 'Atomically thin two-dimensional organic-inorganic hybrid perovskites', *Science* 2015, 349, 1518-1521.

[50] Antony K. J., I., Jana, D. 'Stable Mn-Doped $CsPbCl_3$ Nanocrystals inside Mesoporous Alumina Films for Display and Catalytic Applications' *ACS Appl. Nano Mat.* 2020, 3, 2941-2951.

[51] Takhellambam, D., Meena, T. R., Jana, D. 'Room temperature synthesis of blue and green emitting $CsPbBr_3$ perovskite nanocrystals confined in mesoporous alumina film', *Chem. Commun.,* 2019, 55, 4785-4788.

[52] He, Y., Liang, Y., Liang, S., Harn, Y., Li, Z., Zhang, M., Shen, D., Li, Z., Yan, Y., Pang, X., Lin, Z. 'Dual-Protected Metal Halide Perovskite Nanosheets with an Enhanced Set of Stabilities', *Angew. Chem. Int. Ed.* 2021, 60, 7259–7266.

[53] Roh, K., Zhao, L., Gunnarsson, W. B., Xiao, Z., Jia, Y., Giebink, N. C., Rand, B. P. 'Widely Tunable, 'Room Temperature, Single-Mode Lasing Operation from Mixed-Halide Perovskite Thin Films', *ACS Photonics* 2019, 6, 12, 3331–3337.

[54] Tewari, N., Shivarudraiah, S. B., Halpert, J. E. 'Photo rechargeable Lead-Free Perovskite Lithium-Ion Batteries Using Hexagonal $Cs_3Bi_2I_9$ Nanosheets', *Nano Lett.* 2021, 21, 5578−5585.

[55] Baia, F., Hub, Y., Hua, Y., Qiua, T., Miaoa, X., Zhanga, S. 'Lead-free, air-stable ultrathin $Cs_3Bi_2I_9$ perovskite nanosheets for solar cells Solar Energy Materials and Solar Cells', *Sol. Energy Mater Sol. Cells* 2018, 184, 15–21.

[56] Huang, J., Zou, S., Lin, J., Liu, Z., Qi, M. 'Ultrathin lead-free double perovskite cesium silver bismuth bromide Nanosheets', *Nano Res*., 2021, 14, 4079–4086.

[57] Ashitha P. P, Joshi, M., Verma, D., Jadhav, S., Roy Choudhury, A., Jana, D. 'Layered $Cs_4CuSb_2Cl_{12}$ Nanocrystals for Sunlight-Driven Photocatalytic Degradation of Pollutants', *ACS Appl. Nano Mater*, 2021, 4, 1305-1313.

[58] Lee, J. T., Seifert, S., Sardar, R. 'Colloidal Synthesis of Single-Layer Quasi-Ruddlesden−Popper Phase Bismuth-Based Two-Dimensional Perovskite Nanosheets with Controllable Optoelectronic Properties', *Chem. Mater.* 2021, 33, 5917−5925.

[59] Liu, Y., Siron, M., Lu, D., Yang, J., Reis, R., Cui, F., Gao, M., Lai, M., Lin, J., Kong, Q., Lei, T., Kang, J., Jin, J., Ciston, J., Yang, P. 'Self-Assembly of Two-Dimensional Perovskite Nanosheet Building Blocks into Ordered Ruddlesden−Popper Perovskite Phase', *J. Am. Chem. Soc.* 2019, 141, 13028−13032.

[60] Zhang, Z., Wang, S., Liu, X., Chen, Y., Su, C., Tang, Z., Li, Y., Xing, G. 'Metal Halide Perovskite/2D Material Heterostructures: Syntheses and Applications', *Small Methods* 2021, 5, 2000937.

[61] Wang, J., Dong, J., Lu, F., Sun, C., Zhang, Q., Wang, N. 'Two-Dimensional Lead-Free Halide Perovskite Materials and Devices', *J. Mater. Chem. A*, 2019, 7, 23563-23576.

Chapter 5

Two-Dimensional Peptide-Based Supramolecular Nanosheets

Subramaniyam Sivagnanam and Priyadip Das*

Department of Chemistry, SRM Institute of Science and Technology, SRM Nagar, Potheri, Kattankulathur, Tamil Nadu, India

Abstract

Amino acids are naturally occurring biomolecules used to construct peptides and proteins. Recent advances in synthetic chemistry enabled researchers to mimic the natural process by synthesizing the peptides and proteins with different conformations and structural diversity as well as tunable properties. Peptides, especially short peptides are the most popular building blocks with programmable structural features. One of the most unique properties of these short peptides is their self-assembling ability into well-ordered supramolecular architectures. The non-covalent intermolecular forces like π-π interactions, hydrophilic interactions, hydrophobic interactions, electrostatic interactions, Van der Waals interactions, and intermolecular H-bonding direct the self-assembly process of the peptides to generate the supramolecular nanostructures with morphological individualities. These peptides self-assemble into nanostructures with various morphologies such as vesicles, tubes, fibrils, spherical, sheet, and rod-like structures. In this chapter, we have described the aspects and potential applications of peptide-based self-assembled "Nanosheets." These 2D nanosheets formed by peptide-based self-assembly find numerous applications including nanomedicine, tissue

* Corresponding Author's Email: priyadipcsmcri@gmail.com, priyadip@srmist.edu.in.

In: Properties and Applications of Nanosheets
Editors: Nanda Gopal Sahoo and Anirban Dandapat
ISBN: 979-8-89113-381-5

engineering, theranostics, biosensors, energy and environmental applications.

Keywords: Peptides, Self-assembly, Non-covalent interactions, 2D nanosheets

1. Introduction

Wide ranges of materials with nanometric dimensions in zero-dimensional (0D), one-dimensional (1D), two-dimensional (2D) and three-dimensional (3D) directions, gained considerable attention owing to their associated properties and applications. A material having a size within nanometers greatly differs from its conventional or bulk material. To date, many nanomaterials have been developed with unique optical, electrical, physical, chemical and mechanical properties [1, 2, 3]. In this regard, nanomaterials, especially 2D nanomaterials experienced several advances over the past few decades and utilized significantly. The unique features of 2D materials makes it more appropriate than 0D, 1D and 3D materials. The unique characteristics of 2D materials are the electronic confinement in the two-dimensional network, optical transparency, mechanical flexibility, and more importantly their large surface area [4]. These characteristic features make the employment of 2D materials in numerous applications including energy [5], bioimaging [6], sensors [7] and catalysis [8].

In this chapter, we briefly discussed the formation, properties and applications of peptide-based self-assembled nanosheets. Biomolecules like peptides have the potential ability to self-assemble into nanoscale-ordered supramolecular architectures and play an important role as a template for several applications. Recent advances in the synthetic chemistry and characterization technique enabled the researchers to study the molecular structure, structural stability and folding of the peptides. Compared to proteins and oligopeptides, the synthesis of short peptides avoids the associated synthetic complexities. Another important property of the short peptides is their proteolytic stability, which makes them stable over the enzymatic degradation and makes them ideal for biomedical applications. As a template to fabricate functional nano architectures, peptide-based self-assembly has been widely investigated, especially by supramolecular chemists [9]. Indeed, the self-assembly phenomena is not only limited to conventional supramolecular chemistry but also utilized to generate unique structures by involving other biomolecules [10] and inorganic compounds [11]. In one of their review article, He and Xu briefly discussed the instructed self-assembly method, where the well-ordered architectures are formed by at least one trigger event like Chemical reaction, enzymatic catalysis and ligand-receptor interaction [12].

2. Peptide-based self-assembly into 2D Nanosheets

I. W. Hamley et al., described the formation of 3nm thick nanosheets by the self-assembly of surfactant-like peptide **A_6R**. This Surfactant-like peptide, **A_6R** consists of a charged head group (Arginine) and neutral tail groups. For the first time, with the help of a scanning electron microscope and transmission electron microscope (SEM and TEM), they described the self-assembly of amphiphilic peptide **A_6R** into ultra-thin nanosheets under aqueous conditions. At higher concentration the sheets wrap into helical ribbons which further close into nanotube architectures. From the Figure 1, it is clear that the **A_6R** peptide molecules are organized in an inter-digitated monolayer structure where the **A_6R** molecules form dimers of opposed pairs [13]. Formation of Janus nanosheets by solvation-directed self-assembly (SDSA) using asymmetric peptide amphiphile **F6C11** was demonstrated by M. M. Stevens et al., This amphiphilic peptide **F6C11** comprised of hexaphenylalanine (Phe6), a hydrocarbon tail with eleven carbons and glutamic acids (Glu) (Figure 2a). This **F6C11** self-assembled into Janus single-layer nanosheets with ~5 nm thickness (Figure 2c). The precise combination of H-bonding, π-π stacking and the hydrophobic interactions maintains the thermodynamic energy balance, which directed in the formation of Janus nanosheets through the self-assembly process. The mutated form of the peptide **F6C11** self-assembled into nanobelt-like morphology by replacing the phenylalanine units with 3-cyclohexyl-L-alanine units. This structural alternation was energetically favored by reducing the phenylalanine units and weakening the π-π stacking interactions in the y-axial direction. Further, they hypothesized that the structural transformation from nanosheet to nanofibril can also achieved by introducing a cosolvent system during peptide self-assembly, without changing the phenylalanine units (Figure 2b). In order to examine this hypothesis, methanol was added to the **F6C11** self-assembly medium and then the morphological transformations were studied. As shown in the Figure 2c, the size of the nanosheets increased when the percentage of methanol was increased from 0% to 5%. Further, an increase in the methanol proportion to

10%, caused the elongation of the nanosheets and generated nanobelts-like morphology. When the percentage of methanol was increased up to 15%, amyloid-like nanofibrils were formed. Furthermore, the self-assembly of this **F6C11** peptide in different alcohols (e.g., ethanol, n-butanol, n-propanol and ethylene glycol) and polar aprotic solvents (e.g., formamide, N-methyl formamide, N, N-dimethylformamide, N, N-dimethylacetamide) were also studied to elucidate the morphological transformation from nanosheet to fibril like morphology. These finding clearly shows the solvation-directed self-assembly (SDSA) of **F6C11** amphiphilic peptide to generate various nanostructures with morphological individualities [14]. Another report about the fabrication of Janus nanosheets was achieved by the DNA-peptide diblock (DP) and DNA-peptide-DNA (DPD) triblock conjugates [15]. Diphenylalanine-rich peptide was chosen for its ability to form well-defined self-assembled nanostructures mainly through π-π stacking. Glutamic acid units were added at both ends of the peptide chain to balance the hydrophobic and hydrophilic interactions.

A Transmission electron microscope (TEM) was used to reveal the formation of Janus nanosheets with a lateral dimension of hundred nanometers to a few micrometers (Figure 3). This Janus nanosheet formation was further confirmed with the help of Circular Dichroism (CD) analysis by assuring the self-assembly mechanism through the formation of β-sheet confirmation of the self-assembled peptide. This Janus nanosheets are thermally stable up to 70oC. Computational investigations revealed that the formation of Janus nanosheets was only achieved by the parallel β-sheet confirmation when the DNA and peptide faces are opposite. On the other hand, anti-parallel β-sheet confirmation was observed when the peptide was aggregated in between two DNA faces and thus generated non-Janus nanosheets. Similarly, DNA-peptide-DNA (DPD) triblock conjugates also lead to the formation of Janus nanosheets, where the peptide was sandwiched between DNA1 and DNA2 faces. This DPD conjugate was used for the space-resolved enzyme cascade reaction. Glucose oxidase (GOx) and horseradish peroxidase (HRP) were functionalized with the DNA1 and DNA2 of the DPD conjugate. GOx oxidizes the glucose to gluconic acid, which is further converted into oxygen and hydrogen peroxide. This hydrogen peroxide was transformed into water and oxygen by the HRP, which was associated with the oxidation of 3,3',5,5'-tetramethylbenzidine (TMB) to its oxidized form of TMB2+ [15].

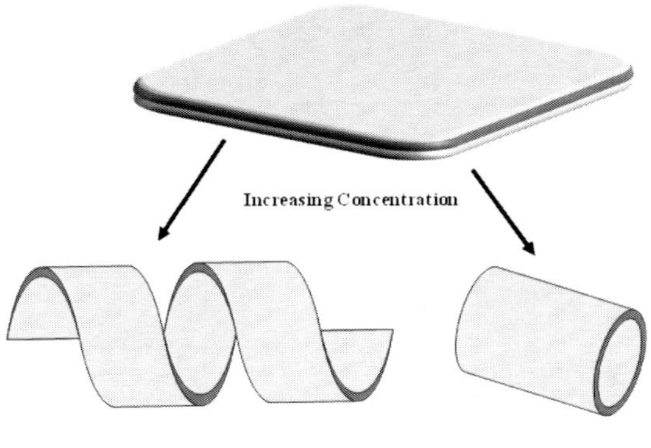

Figure 1. Gaussian curvature of self-assembled nano sheets (top) formed by **A₆R**, leads to the formation of helical ribbon and nanotube structures.

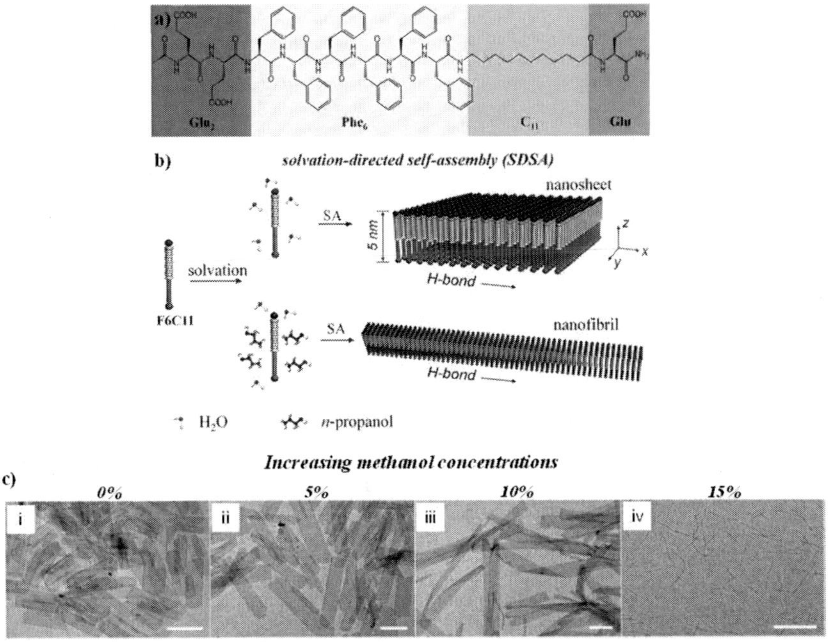

Figure 2. a) Chemical structure of the peptide **F6C11**. b) Schematic representation of solvation-directed self-assembly of **F6C11**. 2D nanosheets in aqueous solution is formed by the hydrogen bonding in the x-axial direction, π-π stacking in the y-axial direction, and hydrophobic effect in both directions. c) Transmission electron microscopy (TEM) images of nanosheets with increasing concentrations of methanol: (i-iv). Scale bar: (i–iii) 1 μm and (iv) 500 nm. This figure has been adapted from reference [14].

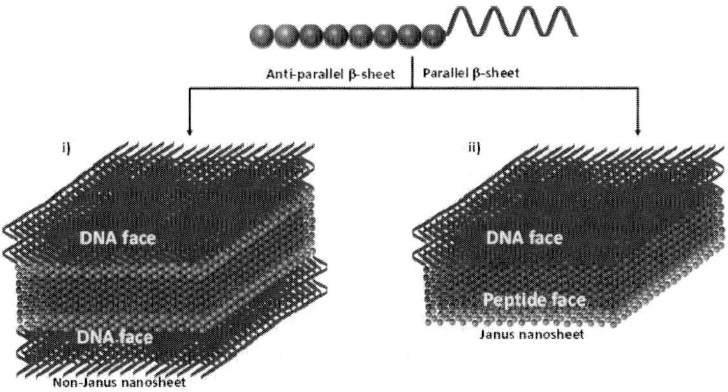

Figure 3. Two possible self-assembly modes of DP i) non-Janus nanosheets formed due to the Antiparallel or mixed β-sheet ii) Janus nanosheets formed due to the Parallel β-sheet assembly.

Amyloid-forming peptides can also be fabricated into nanosheets. B. Dai and coworkers discovered that the amyloid-forming unit of Alzheimer's (Aβ) self-assembled into nanosheets and they termed it as "amyloid-like nanosheets." The amyloid-forming unit of Italian mutant **KLVFFAK** (Aβ16-22), generates nanosheets in the aqueous medium (Figure 4a &b), which is distinguishable from the fibrils formed by most of the other amyloid-forming peptides of proteins. Interestingly, these nanosheets formed through the cross-β diffraction pattern found in amyloid fibrils, which confirms the presence of cross-β fibril spine, which is a crucial factor for amyloid formation. This nanosheet can be potentially used for retroviral transduction, as they notably boosted the lentiviral infection of the human embryonic kidney cell line HEK293T [16].

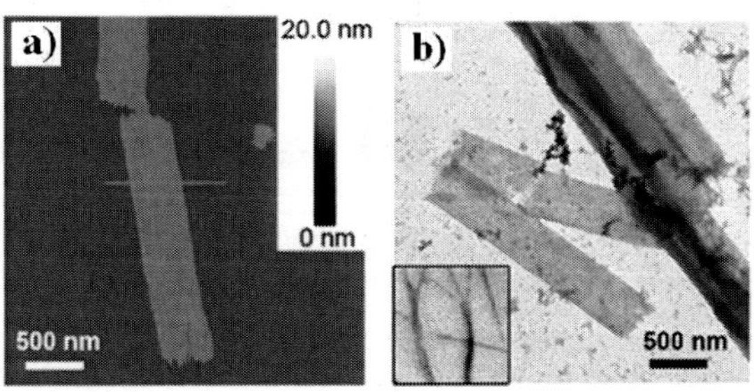

Figure 4. a) AFM micrograph of the **KLVFFAK** based nanosheet. b) TEM images of the **KLVFFAK** nanosheet and the amyloid fibril formed by peptide **VQIVYK** (Inset). This figure has been adapted from reference [16].

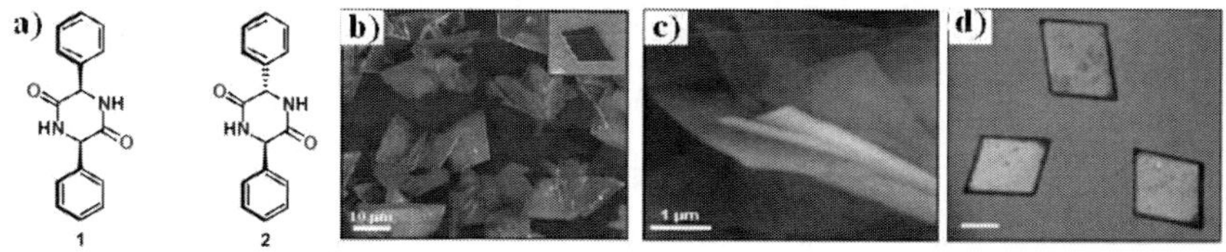

Figure 5. a) Chemical structures of cyclic dipeptides **1** and **2**. b) FE-SEM micrograph of 2D mesosheets, c) nanosheets formed from the solution of cyclic peptide **1** in $CHCl_3$–TFA and d) Single-crystalline rhomboid 2D sheets obtained from the cyclic peptide **2** (D-Phg- L-Phg) in 2-methoxyethanol. This figure has been adapted from reference [17].

Two different aromatic cyclic dipeptides **1** and **2** (L-Phg-L-Phg) and (D-Phg-L-Phg) (Figure 5a) self-assembled into two-dimensional nano as well as mesosheets with large lateral surface area (Figure 5b & c). This 2D sheet formation is associated with the strong intermolecular hydrogen bonding between N-H-O chains supported by π-π stacking between the aromatic units. On the other hand, cyclic peptide D-Phg-L-Phg forms 2D rhomboid sheets with non-crystalline nature (Figure 5d) [17].

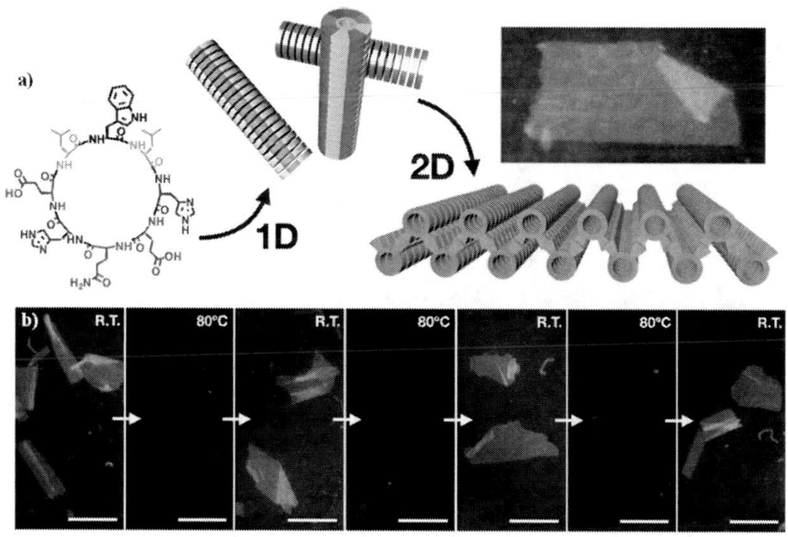

Figure 6. a) Pictorial representation of sequential self-assembly of cyclic peptide **CPx** from 1D-to-2D. b) Epifluorescence micrographs of a **CPx** solution during successive heating (80°C/1.5 h) and cooling (RT/1 h) steps. Scale bars = 30 μm. This figure has been adapted with permission from reference [18]. Copyright 2020, American Chemical Society.

Another cyclic peptide **CPx** consisting of alternating D and L amino acids subsequently self-assembled into a 1D nanotube which was further arranged into bilayers of tubular structures to form 2D sheets in the mesoscale. Thioflavin-T was used to visualize the nanosheets, as it intercalates with the β-sheets conformation during the self-assembly process (Figure 6a). These nanosheets formed by **CPx** are stable even outside of their solvated state. Thus, it is transparent and flexible in nature with the lateral dimension of 260 μm and 50 μm with 15000 μm^2 surface area. The self-assembly process is confirmed by studying the aggregation ability under high temperatures. The fluorescence of Thioflavin-T is quenched at 80oC, which was further restored upon cooling to room temperature. This was further supported with epifluorescence microscopy because at 80°C, Thioflavin-T stained nanosheets disappeared due to the removal of existing non-covalent interactions and reassembled upon cooling to room temperature (Figure 6b) [18].

a) (Pro-Yaa-Gly)$_4$-(Pro-Hyp-Gly)$_n$-(Xaa-Hyp-Gly)$_4$
CPII:(Pro-Arg-Gly)$_4$-(Pro-Hyp-Gly)$_4$-(Glu-Hyp-Gly)$_4$
NSI:(Pro-Amp-Gly)$_4$-(Pro-Hyp-Gly)$_4$-(Glu-Hyp-Gly)$_4$
NSII:(Pro-Amp-Gly)$_4$-(Pro-Hyp-Gly)$_7$-(Glu-Hyp-Gly)$_4$
NSIII:(Amp-Hyp-Gly)$_4$-(Pro-Hyp-Gly)$_4$-(Pro-Glu-Gly)$_4$

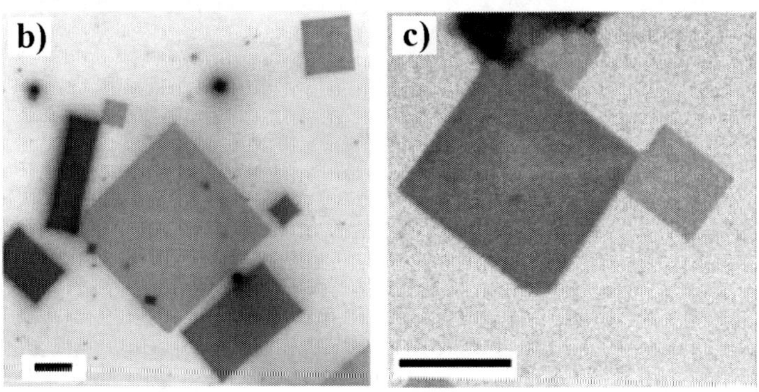

Figure 7. a) Amino acid sequences of peptides **CPII, NSI, NSII, NSIII**. b) TEM image of **NSI** nanosheets (scale bar = 1 μm). c) TEM image of **NSII** nanosheets (scale bar = 500 nm). This figure has been adapted with permission from reference [20]. Copyright 2014, American Chemical Society.

V. P. Conticello and coworkers designed two collagen-mimetic peptides **NSI** and **NSII** (Figure 7a), which self-assemble into sheets with nanoscale dimensions (Figure 7b & c). The design of **NSI** and **NSII** peptides were based on their earlier report about the collagen mimetic peptide **CPII,** which self-assembled into an oriented fibrillar network. In contrast to this, irregular sheet-like morphology was also formed by the **CPII** peptide in the buffered aqueous medium [19]. It is well-known that collagen proteins form triple helices and the positively charged arginine unit in the peptide sequence of **CPII** was replaced by noncanonical imino acid (2S,4R)-4-amino proline (Amp) to trigger the coulombic interactions between the triple helices. Circular dichroism (CD) spectra of **NSI** and **NSII** exhibited a positive peak at 224 nm and a negative peak at 198 nm revealing the formation of collagen triple helices conformation [20, 21].

The same group further developed the **NSIII** peptide by the chemical modification in the **NSI** sequence. In **NSIII**, the noncanonical imino acid (2S,4R)-4-amino proline (Amp) was replaced by (2S,4S)-4-amino proline (amp). It is important to note that, these two imino acids have preferences for the opposite ring puckers, Cγ-exo and Cγ-endo respectively. Crystallographic analysis of these peptides revealed that the common conformation of the Cγ-endo ring pucker is at Xaa position and Cγ-exo ring pucker is at Yaa position. Glutamic acid was moved to the Yaa position to compensate for the charge. The collagen triple helix conformation of the **NSIII** peptide was also confirmed by the Circular dichroism spectroscopy analysis. Compared to the peptide **NSI**, **NSIII** self-assembled into a more homogeneous population of nanosheets. The diagonal length distribution of the nanosheets formed by the **NSIII** was found to be 679 ± 57nm and for the **NSI,** it was 1-15μm. This size disparity clearly signifies the homogeneity of the nanosheets formed by **NSIII** [20, 21]. Few years later, V. P. Conticello and co-workers developed a new class of collagen mimetic peptides (CMP) with charge complementary heteromeric pairs (Figure 8a, b & c), which self-assembled into 1D ultra porous tubes (Figure 8f) or 2D bilayered nanosheets (Figure 8e) based on the self-assembly conditions. It is necessary to incorporate or encode the peptide sequence with the required structural information to achieve the desired well-ordered assembly through self-assembly. Two different CMP **R4P6** and **P6E4** designed and employed to reveal the complexity as well as associated difficulties in understanding the heteromeric peptide sequences involved in self-assembly (Figure 8a). The change in the pH of the self-assembly medium significantly altered the surface charge of these CMP triple helices and resulted in the morphological transition from tube to sheets. CD spectrum displayed the characteristic peaks (positive maximum at ca. 224 nm and a negative minimum at ca. 198 nm) corresponding to the triple helices. The co-assembly of these two **R4P6** and **P6E4** peptides was initiated by, mixing the one-week aged solution of **R4P6** in MOPS buffer with freshly prepared **P6E4** solution. Both solutions were prepared by thermal annealing method. This co-assembly of **R4P6** and **P6E4** produced tube-like structures rather than sheets and this tubular structure consists of a few layers of stacked lamellae. On the other hand, **R6P6/P6E4** co-assembled into sheet-like structure in MOPS buffer at neutral pH 7.0. This nanosheet exhibits a parallelogram-like morphology with a thickness of 17nm and suggests the presence of a bilayer structure [22].

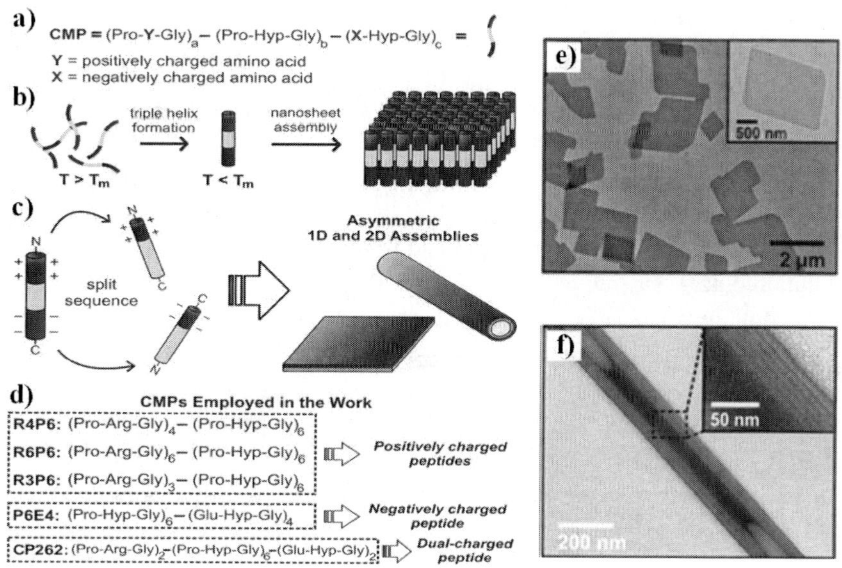

Figure 8. a) Chemical formula of the self-complementary CMP consisting of triblock with charge complementary triad segments. b) Self-assembly of CMPs into a 2D lattice. CMP triple helices are packed in an antiparallel orientation. c) Two homomeric collagen triple helices divided into oppositely charged units for the construction of 1D and/or 2D self-assemblies. d) Chemical formula of the CMP peptide sequences utilized for the present study. e) Negative stained TEM images of **R6P6/P6E4** self-assembled nanosheets (1 mg/mL) at pH 7.0. f) HR-TEM image revealing that the tube walls are constructed with lamellar arrangement of the layers. This figure has been adapted with permission from reference [22]. Copyright 2020, American Chemical Society.

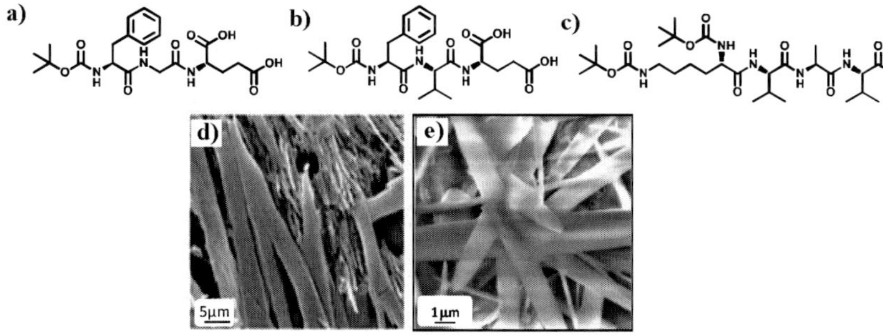

Figure 9. Chemical formula of a) **L1**, b) **L2**, and c) **L3**. d) HR-SEM micrographs of the self-assembled nano belt/sheets formed by **L1M** in 100% aqueous medium. e) FE-SEM micrographs of the self-assembled nanosheets formed by **L3** in 100% aqueous medium. These figures have been adapted with permission from reference [25], Copyright 2018, Royal Society of Chemistry, and from reference [26] Copyright 2021, American Chemical Society.

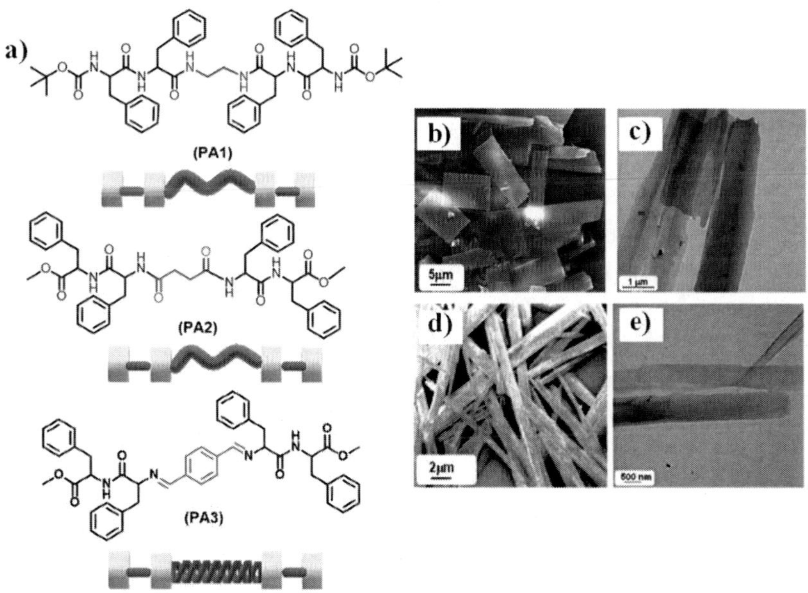

Figure 10. Chemical structures of **PA1, PA2** and **PA3** peptides with different chemical linkers. b) FE-SEM and c) TEM images of self-assembled nanosheets formed by **PA2** with succinic acid (SA) linker**.** d)FE-SEM and e) TEM images of self-assembled nanobelt like structures formed by **PA2.** The figure has been adapted with permission from reference [26]. Copyright 2021, Royal Society of Chemistry.

P. Das et al., designed and synthesized two tripeptides **L1** (Boc-NH-Phe-Gly-Glu-OH) and **L2** (Boc-NH-Phe-Val-Glu-OH) (Figure 9a & b) and their corresponding copper (Cu(II)) metal complexes **L1M** and **L2M** [23]. **L1** and **L2** self-assembled into flower-like structures and spherical assemblies respectively. Whereas, the Cu(II) metal complexes (**L1M** and **L2M**) are self-assembled into nanobelt/sheet-like structures (Figure 9d) and nano-flake like structures respectively. This difference between the self-assembled structures formed by both peptides was due to the presence of a central amino acid with a different hydrophobic nature and the additional metal-ligand binding interactions. **L1** consists of the less hydrophobic amino acid Glycine and **L2** consists of the more hydrophobic amino acid Valine. Further, they utilized the intrinsic fluorescence quenching property of the Cu(II) metal ion to develop a fluorescence "turn-on" displacement strategy assay to investigate the drug delivery capability of this **L1M** and **L2M**. For this purpose, **L1M-DOX** and **L2M-DOX** were prepared to study the delivery of Dox into mouse Neuroblastoma N2aM Cells. It was achieved by the concept of fluorescence dye displacement assay (FDDA), which was proposed by Anslyn and their co-workers [24]. P. Das et al. described the formation of sheet-like morphology from the self-assembly of tetrapeptide Lys-Val-Ala-Val (**L3**) (Figure 9c). The self-assembly property of this tetrapeptide **L3** was studied under different solvent conditions, such as 50% aqueous ethanol and 100% aqueous medium. Under 50% aqueous ethanol medium, **L3** self-assembled into spherical-like structures, and in 100% aqueous medium self-assembled into sheet-like structures with a high aspect ratio (Figure 9e). FT-IR analysis revealed that this sheet-like morphology is associated with the antiparallel β-sheet configuration. The precise balance between hydrophilic (Lys) and hydrophobic (Val-Ala-Val) interactions in different solvent conditions with varying polarity drives the self-assembly process and tunes the morphology of final self-assembled structures [25].

In another work by P. Das et at. described the fabrication of supramolecular nanostructures from diphenylalanine peptides with different rigid and flexible chemical linkers. Three different linkers namely ethylenediamine (EA), succinic acid (SA) and terephthalaldehyde (TA) were used to construct three different diphenylalanine peptide analogues **PA1** (BOC-FF-EA-FFBOC), **PA2** (OMe-FF-SA-FF-OMe) and **PA3** (OMe-FF-TA-FF-OMe) (Figure 10a) [26]. FE-SEM and TEM analysis were carried out to find the morphology of the self-assembled superstructures formed by these peptides under different solvent conditions. These three peptides with three different chemical linkers self-assembled into different supramolecular structures with distinct morphologies. Among them, peptide **PA2** with SA chemical linker self-assembled into nanosheets under 50% aqueous-ethanol medium (Figure 10b & c), whereas, in 100% aqueous medium self-assembled into nanobelt like morphology (Figure 10d & e). The delicate balance between the molecular flexibility of these linkers and geometrically restricted orientation of diphenylalanine units drives the self-assembly process of this peptides and generated different supramolecular nanostructures. EA and SA linkers are more flexible in nature compared to TA, which restrict the structural flexibility of **PA3** during self-assembly process. They further carried out the *in-silico* investigation to understand the mechanism behind the self-assembly process. From the results, it was found that, peptides **PA1** and **PA2** with more flexible EA and SA linkers respectively, exhibited several interactions between the diphenylalanine moieties on the both sides of connecting linker. In contrast, peptide **PA3** with rigid TA linkers, the diphenylalanine groups are far away from each other and preferred the intermolecular π-π stacking followed by lamellar molecular arrangement.

Redox-active Tyrosine consisting H-Tyr-Tyr-Ala-Cys-Ala-Tyr-Tyr-OH (**YYACAYY**) peptide self-assembled into vertically stacked nanosheets. Self-assembly of this peptide caused the faceting of the water droplets. Initially, facet formation was observed after 10 min with **YYCYY** peptide. In the case of a comparatively longer chain peptide **YYACAYY** (Figure 11a), the facet formation was observed within 1 min. The facet formation with respect to the incubation time is shown in Figure 11b. In this study, several peptides were synthesized to explore self-assembly with facet formation. Among these synthesized peptides, **YYACAYY** peptide displayed the facet forming capability with high quality (Figure 11c & d). TEM micrograph was used to analyze the morphology of the self-assembled nanosheets formed at the air and water interface (Figure 11e). AFM analysis provided the thickness of this nanosheets, which was 1.4 nm (Figure 11f). The minimum concentration required to form the nanosheets was found to be 0.22 mM. The mechanism behind the formation of nanosheets is curvature-induced self-assembly, as the facet formation is depending on the size and curvature of the water droplet. Elastic modulus of the flattened film was 8.4 GPa. The fabricated peptide nanosheets were strong enough to overcome the surface pressure of the water droplet [27].

To further understand the self-assembly mechanism, the same group designed another peptide sequence by introducing phenylalanine units. Five different sequences i.e., **YFCFY**, **FYCYF**, **YFCYF**, **FYCFY**, **FFCFF** were designed and utilized for the studies. The design is based on, introducing tyrosine as 2D structure forming unit, phenylalanine motif for strong aromatic π-π interactions and cysteine as a dimer forming unit. Among these peptides, **YFCFY** self-assembled into giant sheets, immediately after drop casting on a hydrophobic glass cover plate. This sheet formation is similar to the previous **YYACAYY** peptide. In this case, the self-assembly is associated with the dimerization between the cysteine units in aqueous conditions. Spectroscopic analysis supported the formation of dimer by the oxidation of the monomeric **YFCFY** units [28].

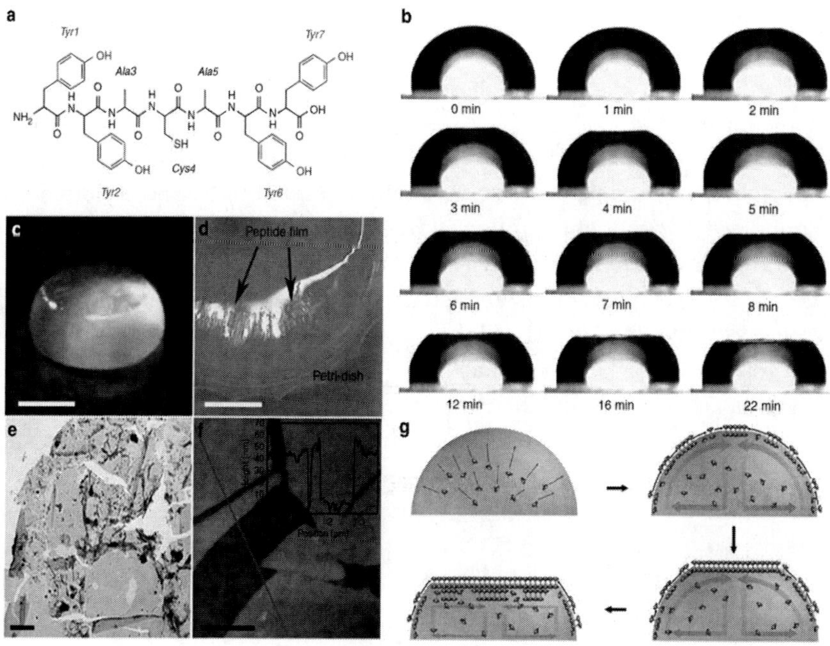

Figure 11. a) Chemical structure of the reported peptide **YYACAYY**. b) Optical microscopy photographs of a water droplet with respect to different incubation times. c) Single large facet formation of the water droplet. d) Peptide nanosheets formed at air/water interface in a Petri dish. e) TEM and f) AFM image of large 2D films formed by the **YYACAYY** peptide. g) Schematic representation of the facet-forming mechanism by the **YYACAYY** peptide solution. Scale bar, 0.5 cm (c), 1 cm (d), 5 μm (e), 1 μm (f). The figure has been adapted from reference [27].

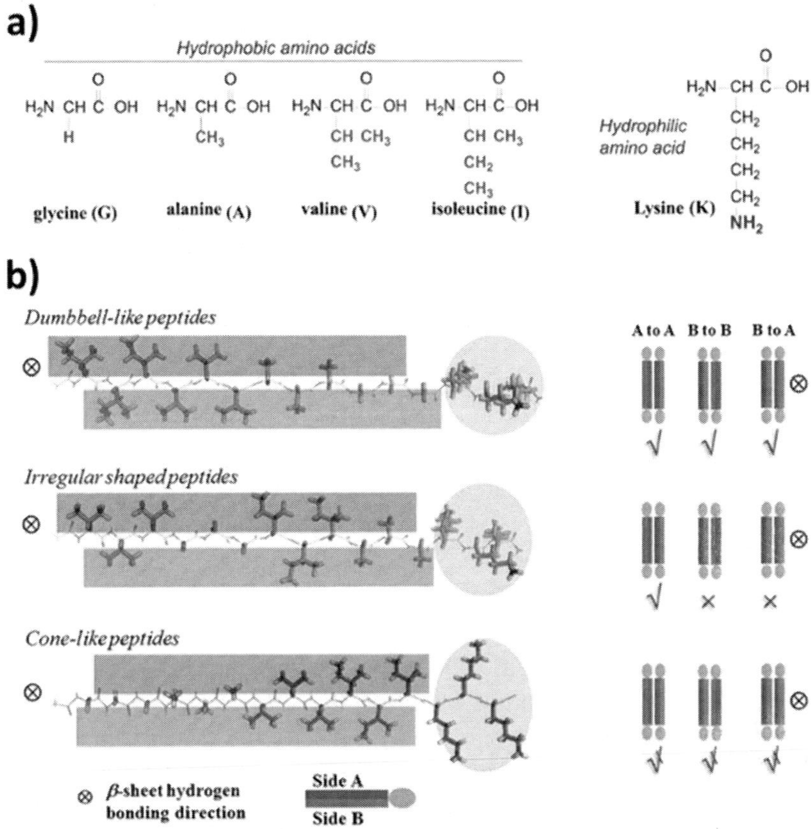

Figure 12. a) The amino acids used to construct the surfactant-like peptide. b) The molecular arrangements of the hydrophobic residues on the two faces of a β-strand configuration and the possible interaction between the β-sheet strands. The figure has been adapted with permission from reference [29]. Copyright 2017, American Chemical Society, copyright.

Surfactant-like peptides (SLPs) are a class of peptides that resemble with conventional surfactants, consists of hydrophobic amino acids in the tail, and corresponding hydrophilic amino acids as head group. These SLPs are generally having a short length with amphiphilicity, which provides better water solubility to self-assemble into several supramolecular architectures and makes them suitable for biomedical applications. J. R. Lu and co-workers synthesized six SLPs (**$G_3A_3V_3I_3K_3$, $K_3I_3V_3A_3G_3$, $I_3V_3A_3G_3K_3$, $K_3G_3A_3V_3I_3$, $V_3G_3I_3A_3K_3$**, and **$K_3A_3I_3G_3V_3$**) (Figure 12a & b) with the similar amino acid composition but with different sequential arrangement. Interestingly, except

$I_3V_3A_3G_3K_3$ and **$K_3G_3A_3V_3I_3$**, other peptides form antiparallel β-sheet conformation during the self-assembly process. These two peptides **$I_3V_3A_3G_3K_3$** and **$K_3G_3A_3V_3I_3$** only self-assembled into nanosheets with 4.0 ± 0.5 nm height and adopt dumbbell-like shape.

This result clearly states that the high order of the β-sheet configuration is required to achieve the final morphology. These dumbbell-like-peptides forms high order β-sheet configuration along the zippering axis with three possible pairing confirmations ("A to A," "B to B" and "A to B"), which drives the formation of nanosheets by self-assembly. This nanosheets formed by **I3V3A3G3K3** and **K3G3A3V3I3** showed the ability to encapsulate the dye Nile red with considerable loading capacity up to 17.1% and 22.6% in mole ratio respectively [29].

The self-assembly of peptides can be controlled by certain parameters. It is possible to fabricate different supramolecular nanostructures by single peptide backbone. L. Liu et al., described the morphological transformation of nanostructure from 1D fibril to 2D nanosheet from a single amphiphilic peptide, **KFAK** (H_2N-KKKFAFAFAFAKKK-COOH). At first, the self-assembly ability of this peptide under alkaline aqueous condition was tested. A 7 days aged solution of **KFAK** self-assembled into 1D nanofibrils like morphology with the height of 1.8 nm and Young's modulus of 2.1˘ 0.3GPa, which shows the hard/rigid nature of the formed fibrils. When a small organic molecule 4,4-bipyridyl (4Bpy) was introduced into the **KFAK** peptide sequence, yielded 2D nanosheets with 4nm height. The self-assembly of **KFAK** peptide is mainly driven by the intermolecular H-bonding between the peptide backbones and the hydrophobic side chain interactions provided by the aromatic moieties. On the other hand, the nitrogen of 4Bpy group binds with the carboxylic unit of the peptide through the H-bonding between and facilitated the formation of 2D nanosheets with Youngs modulus about 210˘8MPa, reflects the comparative soft nature compared to the 1D fibrils [30].

Conclusion

This chapter described the self-assembly of peptides, particularly short peptides into supramolecular 2D nanosheets. Molecular structure and aspects of the peptides were discussed with the aim to construct peptide-based 2D nanomaterials. The role of secondary conformations, mainly β-sheet configuration of the synthesized peptide's self-assembly have been highlighted. These 2D nanomaterials derived from peptides find potential applications in a wide range of areas including drug delivery and bio-imaging. However, there are numerous methods to fabricate peptide-based 2D nanosheets still leaves room for researchers towards future applications.

References

[1] Li, Y., Ziwei Li, Cheng Chi, Hangyong Shan, Liheng Zheng, and Zheyu Fang. 2017. "Plasmonics of 2D Nanomaterials: Properties and Applications." *Advanced Science* 4 (8): 1600430. https://doi.org/10.1002/advs.201600430.

[2] Patil, S. P., and Vishal V. Burungale. 2020. "Physical and Chemical Properties of Nanomaterials." In *Nanomedicines for Breast Cancer Theranostics*, 17–31. Elsevier. https://doi.org/10.1016/B978-0-12-820016-2.00002-1.

[3] Murty, B. S., P Shankar, Baldev Raj, B. B. Rath, and James Murday. 2013. "Unique Properties of Nanomaterials." In *Textbook of Nanoscience and Nanotechnology*, 29–65. Berlin, Heidelberg: Springer Berlin Heidelberg. https://doi.org/10.1007/978-3-642-28030-6_2.

[4] Yang, G., Chengzhou Zhu, Dan Du, Junjie Zhu, and Yuehe Lin. 2015. "Graphene-like Two-Dimensional Layered Nanomaterials: Applications in Biosensors and Nanomedicine." *Nanoscale* 7 (34): 14217–31. https://doi.org/10.1039/c5nr03398e.

[5] Han, R., Feng Liu, Xuefei Wang, Minghong Huang, Wenxian Li, Yusuke Yamauchi, Xudong Sun, and Zhenguo Huang. 2020. "Functionalised Hexagonal Boron Nitride for Energy Conversion and Storage." *Journal of Materials Chemistry A* 8 (29): 14384–99. https://doi.org/10.1039/D0TA05008C.

[6] Kumar, V., Kumar Nikhil, Partha Roy, Debrupa Lahiri, and Indranil Lahiri. 2016. "Emergence of Fluorescence in Boron Nitride Nanoflakes and Its Application in Bioimaging." *RSC Advances* 6 (53): 48025–32. https://doi.org/10.1039/C6RA05288F.

[7] Khan, A. F., Dale A. C. Brownson, Edward P. Randviir, Graham C. Smith, and Craig E. Banks. 2016. "2D Hexagonal Boron Nitride (2D-HBN) Explored for the Electrochemical Sensing of Dopamine." *Analytical Chemistry* 88 (19): 9729–37. https://doi.org/10.1021/acs.analchem.6b02638.

[8] Zhang, Y., Xuefei Weng, Huan Li, Haobo Li, Mingming Wei, Jianping Xiao, Zhi Liu, Mingshu Chen, Qiang Fu, and Xinhe Bao. 2015. "Hexagonal Boron Nitride Cover on Pt(111): A New Route to Tune Molecule–Metal Interaction and Metal-Catalyzed Reactions." *Nano Letters* 15 (5): 3616–23. https://doi.org/10.1021/acs.nanolett.5b01205.

[9] Ariga, K., Jonathan P. Hill, Michael V. Lee, Ajayan Vinu, Richard Charvet, and Somobrata Acharya. 2008. "Challenges and Breakthroughs in Recent Research on Self-Assembly." *Science and Technology of Advanced Materials* 9 (1): 014109. https://doi.org/10.1088/1468-6996/9/1/014109.

[10] Yang, J., and Jinfeng Li. 2018. "Self-Assembled Cellulose Materials for Biomedicine: A Review." *Carbohydrate Polymers* 181 (February): 264–74. https://doi.org/10.1016/j.carbpol.2017.10.067.

[11] Liu, Y. T., Peng Zhang, Ning Sun, Babak Anasori, Qi-Zhen Zhu, Huan Liu, Yury Gogotsi, and Bin Xu. 2018. "Self-Assembly of Transition Metal Oxide Nanostructures on MXene Nanosheets for Fast and Stable Lithium Storage." *Advanced Materials* 30 (23): 1707334. https://doi.org/10.1002/adma.201707334.

[12] He, H., and Bing Xu. 2018. "Instructed-Assembly (IA): A Molecular Process for Controlling Cell Fate." *Bulletin of the Chemical Society of Japan* 91 (6): 900–906. https://doi.org/10.1246/bcsj.20180038.

[13] Hamley, I. W., Ashkan Dehsorkhi, and Valeria Castelletto. 2013. "Self-Assembled Arginine-Coated Peptide Nanosheets in Water." *Chemical Communications* 49 (18): 1850–52. https://doi.org/10.1039/c3cc39057h.

[14] Lin, Y., Matthew Penna, Michael R. Thomas, Jonathan P. Wojciechowski, Vincent Leonardo, Ye Wang, E. Thomas Pashuck, Irene Yarovsky, and Molly M. Stevens. 2019. "Residue-Specific Solvation-Directed Thermodynamic and Kinetic Control over Peptide Self-Assembly with 1D/2D Structure Selection." *ACS Nano* 13 (2): 1900–1909. https://doi.org/10.1021/acsnano.8b08117.

[15] Albert, S. K., Sunghee Lee, Prasannavenkatesh Durai, Xiaole Hu, Byeongmoon Jeong, Keunwan Park, and So Jung Park. 2021. "Janus Nanosheets with Face-Selective Molecular Recognition Properties from DNA-Peptide Conjugates." *Small* 17 (12): 1–10. https://doi.org/10.1002/smll.202006110.

[16] Dai, B., Dan Li, Wenhui Xi, Fang Luo, Xiang Zhang, Man Zou, Mi Cao, Jun Hu, Wenyuan Wang, Guanghong Wei, Yi Zhang and Cong Liu. 2015. "Tunable Assembly of Amyloid-Forming Peptides into Nanosheets as a Retrovirus Carrier." *Proceedings of the National Academy of Sciences of the United States of America* 112 (10): 2996–3001. https://doi.org/10.1073/pnas.1416690112.

[17] Govindaraju, T., M. Pandeeswar, K. Jayaramulu, Garima Jaipuria, and Hanudatta S. Atreya. 2011. "Spontaneous Self-Assembly of Designed Cyclic Dipeptide (Phg-Phg) into Two-Dimensional Nano- and Mesosheets." *Supramolecular Chemistry* 23 (7): 487–92. https://doi.org/10.1080/10610278.2010.550685.

[18] Insua, I., and Javier Montenegro. 2020. "1D to 2D Self Assembly of Cyclic Peptides." *Journal of the American Chemical Society* 142 (1): 300–307. https://doi.org/10.1021/jacs.9b10582.

[19] Rele, S., Yuhua Song, Robert P. Apkarian, Zheng Qu, Vincent P. Conticello, and Elliot L. Chaikof. 2007. "D-Periodic Collagen-Mimetic Microfibers." *Journal of the American Chemical Society* 129 (47): 14780–87. https://doi.org/10.1021/ja0758990.

[20] Jiang, T., Chunfu Xu, Yang Liu, Zheng Liu, Joseph S. Wall, Xiaobing Zuo, Tianquan Lian, Khalid Salaita, Chaoying Ni, Darrin Pochan and Vincent P. Conticello. 2014. "Structurally Defined Nanoscale Sheets from Self-Assembly of Collagen-Mimetic Peptides." *Journal of the American Chemical Society* 136 (11): 4300–4308. https://doi.org/10.1021/ja412867z.

[21] Jiang, T., Chunfu Xu, Xiaobing Zuo, and Vincent P. Conticello. 2014. "Structurally Homogeneous Nanosheets from Self-Assembly of a Collagen-Mimetic Peptide."

Angewandte Chemie - International Edition 53 (32): 8367–71. https://doi.org/10.1002/anie.201403780.

[22] Merg, A. D., Gavin Touponse, Eric Van Genderen, Thorsten B. Blum, Xiaobing Zuo, Alisina Bazrafshan, Hew Ming Helen Siaw, Arthur McCanna, R. Brian Dyer, Khalid Salaita, Jan Pieter Abrahams and Vincent P. Conticello. 2020. "Shape-Shifting Peptide Nanomaterials: Surface Asymmetry Enables PH-Dependent Formation and Interconversion of Collagen Tubes and Sheets." *Journal of the American Chemical Society* 142 (47): 19956–68. https://doi.org/10.1021/jacs.0c08174.

[23] Das, P., Ieshita Pan, Ehud Cohen and Meital Reches. 2018. "Self-assembly of a metallo-peptide into a drug delivery system using a "switch on" displacement strategy." *Jouranl of Materials Chemistry B* 6(48): 8228-8237. https://doi.org/10.1039/C8TB01483C

[24] Nguyen, B. T., and Eric V. Anslyn. 2006. "Indicator–Displacement Assays." *Coordination Chemistry Reviews* 250 (23–24): 3118–27. https://doi.org/10.1016/j.ccr.2006.04.009.

[25] Sivagnanam, S., Madhuri Basak, Abilesh Kumar, Kiran Das, Tarun Mahata, Priya Rana, Abhishek Singh Sengar, Soumyajit Ghosh, Mahesh Subramanian, Adele Stewart, Biswanath Maity, Priyadip Das. 2021. "Supramolecular Structures Generated via Self-Assembly of a Cell Penetrating Tetrapeptide Facilitate Intracellular Delivery of a Pro-Apoptotic Chemotherapeutic Drug." *ACS Applied Bio Materials* 4 (9): 6807–20. https://doi.org/10.1021/acsabm.1c00530.

[26] Arul, A., Priya Rana, Kiran Das, Ieshita Pan, Debasish Mandal, Adele Stewart, Biswanath Maity, Soumyajit Ghosh, and Priyadip Das. 2021. "Fabrication of Self-Assembled Nanostructures for Intracellular Drug Delivery from Diphenylalanine Analogues with Rigid or Flexible Chemical Linkers." *Nanoscale Advances* 3 (21): 6176–90. https://doi.org/10.1039/D1NA00510C.

[27] Jang, H. S., Jung Ho Lee, Yong Sun Park, Young O. Kim, Jimin Park, Tae Youl Yang, Kyoungsuk Jin, Jaehun Lee, Sunghak Park, Jae Myoung You, Ki-Woong Jeong, Areum Shin, In-Seon Oh, Min-Kyung Kwon, Yong-Il Kim, Hoon-Hwe Cho, Heung Nam Han, Yangmee Kim, Yoon Ho Chang, Seung R. Paik, Ki Tae Nam and Yoon Sik Lee. 2014. "Tyrosine-Mediated Two-Dimensional Peptide Assembly and Its Role as a Bio-Inspired Catalytic Scaffold." *Nature Communications* 5 (May 2013): 1–11. https://doi.org/10.1038/ncomms4665.

[28] Lee, J., Ik Rang Choe, Nak-Kyoon Kim, Won-Je Kim, Hyung-Seok Jang, Yoon-Sik Lee, and Ki Tae Nam. 2016. "Water-Floating Giant Nanosheets from Helical Peptide Pentamers." *ACS Nano* 10 (9): 8263–70. https://doi.org/10.1021/acsnano.6b00646.

[29] Cao, M., Sha Lu, Wenjing Zhao, Li Deng, Meng Wang, Jiqian Wang, Peng Zhou, Dong Wang, Hai Xu, and Jian R. Lu. 2017. "Peptide Self-Assembled Nanostructures with Distinct Morphologies and Properties Fabricated by Molecular Design." *ACS Applied Materials and Interfaces* 9 (45): 39174–84. https://doi.org/10.1021/acsami.7b11681.

[30] Jin, H., Liwei Zhang, Yongxiu Song, and Lei Liu. 2019. "Modulation of Structure and Mechanical Properties of Self-Assembled Peptide Nanofibrils and Nanosheets." *Materials Letters* 255: 126540. https://doi.org/10.1016/j.matlet.2019.126540.

Chapter 6

Peptide, Peptoid and Protein-Based Nanosheets

Subhasish Roy*
Department of Chemistry, Birla Institute of Technology and Science-Pilani, K K Birla Goa Campus, Zuarinagar, Sancoale, Goa, 403726, India

Abstract

Self-assembling systems play a pivotal role in the functioning of living organisms and in emerging areas of science and technology research including food technology, cosmetics, energy, medicine and artificial intelligence. Peptides (connected through amide bonds between two/more amino acids) are the most sublime and elegant building blocks of supramolecular chemistry and programmable smart functional emerging advanced materials. If the side chain of the peptide, instead putting at α-carbon atom, placed on the nitrogen atom by replacing the hydrogen atom of the amide bond, produces an interesting class of molecules which are popularly known as peptoids. Proteins are the polymeric forms of peptides. Peptides form various supramolecular nanostructures through the involvement of various non-covalent interactions including hydrogen bonding, pi–pi stacking, electrostatic, hydrophobic, and van der Waals interactions. The bio-friendliness and viable nature of these peptide-based nanostructures open up the possibility to use these nanostructures in the field of biomedical applications and indeed to understand one of the most important phenomena of systems chemistry, 'origin of life'. The ease of synthetic flexibility of peptides and peptoids allow the possibilities to engineer the structure-function relationships. The self-assembly of peptide is the beauty of the biomolecular world and the peptides mainly driven by thermodynamics to construct various nanosctructures including nanovesicles, nanotubes, nanofibers, nanorods, nanowires, nanotapes,

* Correspodnign Author's Email: subhasishr@goa.bits-pilani.ac.in

In: Properties and Applications of Nanosheets
Editors: Nanda Gopal Sahoo and Anirban Dandapat
ISBN: 979-8-89113-381-5

nanoribbons, nanoflowers, nanomembrane and helical fibers. Moreover, peptide, peptoid and protein can also assemble to form two-dimensional (2D) nanosheets. The kinetic factor is also crucial factor in structural modulation and function integration for such assembling peptide, peptoid and protein-based nanostructures. This chapter focuses on the peptide, peptoid and protein-based two-dimensional nanoscale assemblies (nanosheets) and their potential applications at various targated fields.

Keywords: peptide, peptoid, protein, self-assembly, nanostructures, nanosheets

1. Introduction

Molecular self-assembly to form various supramolecular ordered micro/nanostructures from the basic building blocks of peptides is a spontaneous process under both thermodynamic and kinetic conditions by using various specific, local and collective non-covalent weak interactions such as hydrogen bonding, pi–pi stacking, electrostatic, hydrophobic, and van der Waals interactions [1-2]. The self-assembled nanostructures devised from biomolecules including amino acids and their derivatives (glycine derivatives), peptides, proteins, nucleobases, nucleic acids, peptide nucleic acids and lipids are of current science and technology's research interest due to their bio-friendliness, biological and chemical mutation, molecular and supramolecular recognition capability, cost-effectiveness, reproducibility, easy of synthesis and their purifications and bottom-up fabrication [3-4]. Among these, peptides which are made of amino acids are the most graceful, significant and versatile building blocks of supramolecular chemistry and for programmable smart advanced functional materials fabrication. The biological origin of peptide describes the applicability in the field of biological applications as well as in regenerative medicine. Peptide can also make the function and properties of the protein through self-assembly. The self-assembling capability of peptide building blocks allows the construction of various nanostructures. A number of peptide-based building blocks have been designed and discovered for the development of functional supramolecular nanostructures and the exploration of their possible applications in biology and nanotechnology [5]. To mimic the structures and functions relationship of natural protein, the amino acids derived systems including peptides, mimic proteins and N-alkylated glycine derived system like peptoids with definite amino acids sequences/N-protected glycine derivatives have been used for the development of the robust and tunable biomimetic bioinspired supramolecular materials [6]. However, the excel between unnatural amino acids and the natural amino acids for the synthesis of peptides and proteins and various hydrophobic and charged N-protected glycines for the synthesis of peptoids allow the possibility to understand the synthesized biomimetic materials with hierarchical structures and unique functions. Moreover, in contrast to natural sequence-dependent natural peptides/proteins, these synthetic peptides/proteins/peptoids have higher structural and functional tunability due to their advantages in synthetic methods by installing various unnatural amino acids, orthogonal functionalizations and derivatives of N-protected glycines respectively. Therefore, sequence-defined synthetic peptides/ proteins and peptoids have received increasing attention as building blocks for the assembly and orginisation of hierarchically supramolecular advanced functional smart materials development. Self-assembly of sequence-defined synthetic biopolymers has become an emerging platform for developing novel supramolecular smart materials [7]. Among these sequence-defined synthetic foldamers, peptides/peptoids/proteins have received particular attention because they combine advantages from both biopolymers and bulk polymers. Peptoids have identical backbones as peptides, while the side chains are attached to the backbone amide nitrogen (Scheme 1) [8]. Due to this significant structural variation, inter- and intra-backbone hydrogen bonding interactions are eliminated in peptoids, resulting in a class of non-natural polymers where structures and properties exclusively controlled by the side-chain's chemistry [8-9]. The facile solid-phase synthetic approach, solution-phase methodology, orthogonal functionalizations, various protection-deprotection chemistry and native chemical ligation enables peptides/peptoids/proteins as potential building blocks of supramolecular chemistry with enormously accessible side-chain chemistries and sequences [10-16]. In addition, the readily available primary amines as the side-chain precursors bring tremendous flexibility to the design and synthesis of peptoid sequences. Self-assembly of peptoids has demonstrated exceptional versatility in forming various nanostructures, such as nanomembranes, nanosheets, nanotubes, helices, and nanoflowers [8]. These features make peptoids an ideal platform to study the relationship between sequential patterns and self-assembled structures. Peptoids have become one of the most extensively studied sequence-defined synthetic polymers in the hierarchicy of nanomaterials with large-scale periodicity, robustness and processability [8-9]. Over the past decades, Zuckermann and co-workers have been reviewed peptoids under various topics e.g., peptoid-based nanosheets, peptoid polymers, and applications of various peptoid-based materials [8]. In addition,

Luxenhofer and co-workers presented a complete picture of peptoid history (Scheme 2 describes the history of peptide and peptoid), synthesis, physicochemical characteristics, and applications in the field of supramolecular engineering [9]. This chapter includes perspective on the design and synthesis of these sequence defined peptides, peptoids and proteins, their characterizations, their applications in supramolecular chemistry to build high-information-content hierarchical smart functional two-dimensional materials with programmability, predictability and a summary of the established structure-property relationships.

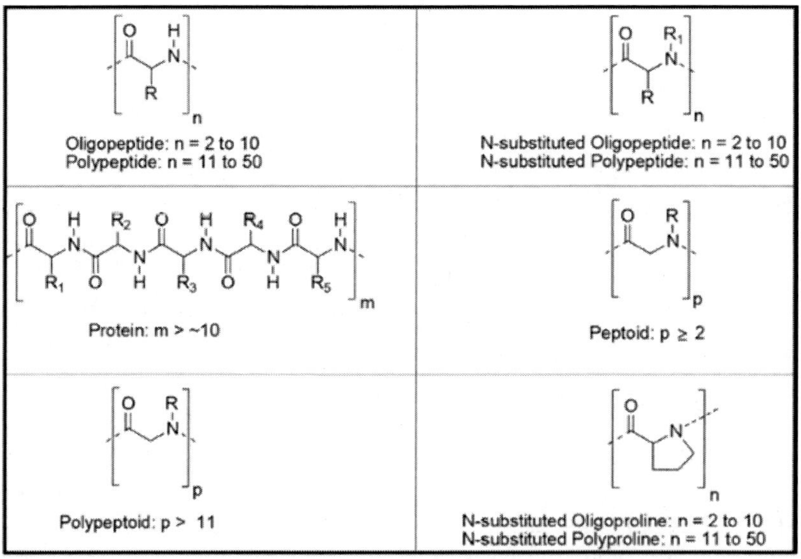

Scheme 1. General Chemical Structures of oligopeptide, polypeptide, peptoid, polypeptoid and protein.

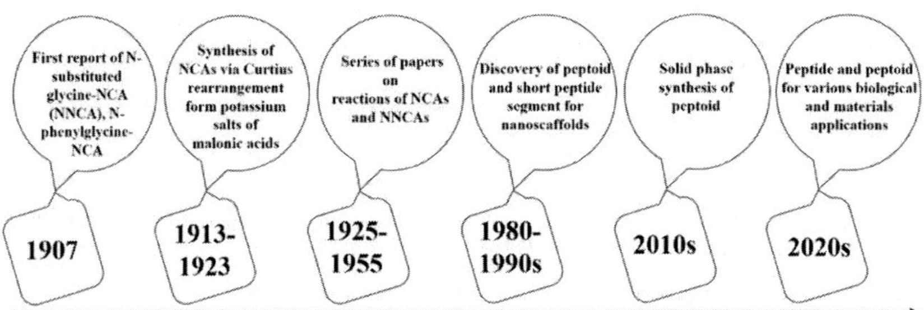

Scheme 2. History of peptide and peptoid (See for more details: Reference [3] *Chem. Soc. Rev.*, 2012, *41*, 4736-4754 and Reference [9] *Chem. Rev.* 2016, *116*, 1753−1802).

The two-dimensional (2D) nanoscale assemblies, the nanosheets are the missing link between low-dimensional assemblies (nanoparticles, nanofibers, nanotubes, and nanoribbons) and extended three-dimensional structures (crystalline solids) in terms of smart intellectual and physical properties. The designing and devising of controlled fabrication of two-dimensional assemblies from environmental friendly materials on the nanoscale describes a real challenge. The inspiration to focus a chapter on nanosheets derived from the basic building blocks of life (based on the 'origin of life' concept) came from the nature as the natural two-dimensional bilayer cell membranes composed of both hydrophilic heads and hydrophobic tails of phospholipids. Nevertheless, the development of effective strategies for the creation of two-dimensional assemblies is essential to the success of the emergent field of 2D nanostructures, in which chemical and physical principles are applied to engineer nanosheet's structures for integration into functional devices. I believe this book chapter can shed the light on the development of programmable, functional and robust two-dimensional nanostructured smart functional materials derived from peptides, peptoids and proteins. There are many reports on other two-dimensional nanosheets however, this chapter's focus is mainly dedicated on the naturally occurring chiral, achiral and unnatural amino acids derived supramolecular building blocks including peptides, peptoids and proteins for successful nanosheet's design, synthesis, characterizations and their applications in various targeted fields. The engineering of these two-dimensional nanostructure materials for the development of smart functional nanomaterials will also be discussed. The synthesis of peptides, peptoids and proteins can be achieved by solid phase peptide and peptoid synthetic protocol, native chemical ligation and also by following solution phase peptide synthetic protocol where applicable [10-16].

2. Peptide Nanosheets

The 'origin of life' is absolutely related to the world of chirality of biomolecules. The chirality of life is asymmetric as during the creation of life from the prebiotic molecular evolution, nature has selected all amino acids in L-forms to construct the proteins and all sugars in D-forms to construct ribonucleotides. Nevertheless, symmetries are indeed ubiquitous in nature including human body structures, butterfly wings and also in man-made architectures such as India Gate in Delhi, Gate way of India in Mumbai, Taj Mahal in Agra, Hawa Mahal in Jaipur and so on. A huge effort has been dedicated previously to understand the self-assembly behavior of the peptides (building blocks), the constructions of various nanostructures, their applications and emergence of the chirality with spontaneous self-assembly from chiral peptides with specific L- and D-amino acids sequences for hierarchical self-assembly in the field of nanotechnology and nano-biotechnology. These building blocks are self-replicate to form assembled structures following the "bottom-up" approach with several repetitions of same units in three-dimensions produces novel nanostructures for potential uses. Self-assembly of peptide to construct nanostructures depends meticulously on several factors including proper balances between the hydrophilicity and hydrophobicity of the peptide, understanding among the peptide's building blocks mainly among the interacting amino acids, amide, side chains and so on. The field of peptide chemistry is become very reasonable due to their ease of synthesis and purifications following solution and solid phase peptide synthetic protocols, cost-effectiveness, ease of amino acid mutations, chirality of the molecules, very good water solubility, tunable solubility based on protection-deprotection chemistry, ease of various moieties functionalization, self-assembling in nature to construct various secondary structures followed by formation of various nanostructures (Scheme 3) at various conditions.

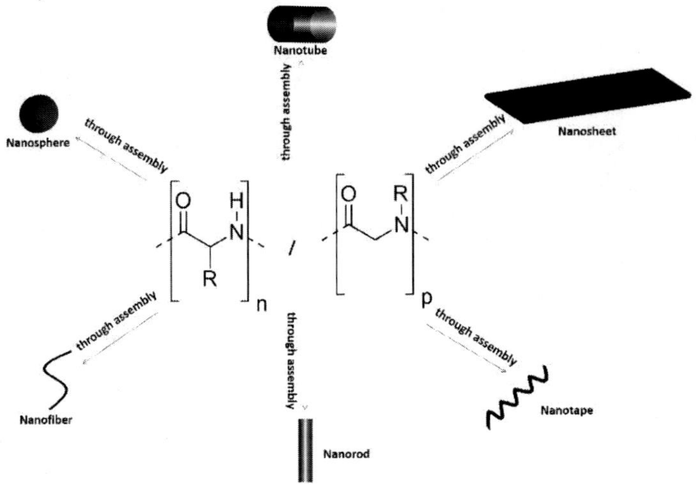

Scheme 3. Supramolecular assembly of peptides / (proteins) and peptoids to form various nanostructures.

By varying the amino acids sequences including hydrophobic and charged amino acids in a specific length scale of a peptide, changing the pH of the medium and triggering various inputs including sonication, light, mechanical force etc. various nanostructures can be constructed. The formation of peptide-based nanostructures (Scheme 3) has been first identified by Zhang and his research group by scanning electron microscopic imaging that actually motivates peptide chemists to invistigate peptide-based nanostructures formations, mechanism and their applications which becomes an emerging area of current research [3].

3. Designing and Synthesis of Peptide Nanosheets (PNSs) and their Applications

The construction of peptide nanosheets through self-assembly and co-assembly of peptides have been studied extensively during the last few decades and applied in various potential purposes [17-43]. The linear peptides and cyclic-peptides have been used for the construction of supramolecular two-dimensional nanostructures in the literature and that is why it will be attentive for the readers to discuss the nanosheets formation from linear and cyclic peptides separately and their applications [17-43].

3.1. Linear Peptide-Based Nanosheets

In the year 2013, Hamely and his research group has reported for the first time that surfactant-like peptide's $(Ala)_6(Arg)$ [where Ala is L-Alanine and Arg is L-Arginine, a basic amino acid, Scheme 4A] self-assembly to form 3 nm thick sheets in aqueous solution [17]. Layers structures with anti-parallel dimers have been confirmed by using scanning electron microscopic imaging study at lower concentrations. The formation of helical ribbon like structures and nanotubes at higher concentrations of the peptide has also been established in this study [17]. They have also reported the spontaneous assembly of a bolaamphiphilic peptide, RFL_4FR (R = L-Arginine; F = L-Phenylalanine and L = L-Leucine, Scheme 4B) in water into nanosheets [18]. This is the first report of bolaamphiphilic peptide as a coating substrate to tune the wettability of solid surfaces and for cell culture applications [18]. In this report, it has been found that the hydrophobic leucine residues are located at the core of the nanosheets and the hydrophilic charged arginine moieties were exposed to the water at the surface of the nanosheets. They also have discovered a surfactant-like peptide $(Ala)_6$-(Asp) [where Ala is L-Alanine and Asp is L-Aspartic acid which is an acidic amino acid, Scheme 4C] which self-assemble into ultrathin nanosheets in aqueous solution above a critical aggregation concentration [19]. Liu and co-workers have discovered amyloid-like-nanosheets from Italian familial amyloid protein Aβ 16–22 (KLVFFAK, Scheme 4D) in aqueous medium [20]. These nanosheets have been exploited in boosting retroviral gene transduction [20]. Du and research group have demonstrated synthesis and characterization of antibacterial peptide-grafted hyperbranched polymer-based nanosheet [21]. The hyperbranched polymer was synthesized by following sequential Michael addition-based thio-ene reaction followed by free radical thiol-ene reaction and ring opening polymerization of N-carboxyanhydride [21]. These nanosheets endowed with very good antimicrobial activity [21]. Stevens and her co-workers have devised peptide amphiphile containing hexa-phenylalanine, a hydrophobic tail, and glutamic acids as charged amino acids for janus 2D structures with single-layer thickness, which enable a predetermined surface heterofunctionalization (Scheme 4E) [22]. These nanosheets exhibit enhanced enzymatic activity on interacting with guest component including proteins and nanoparticles [22]. Ma and co-workers have discovered an amyloid nanosheet which has formed by an Aβ (16–22) variant, KLVXFAK, (where X is *p*-Cyanophenylalanine, Scheme 4F) by using an infrared probe technique [23]. They have established by theoretical calculation that the nanosheets are the stack of class 7-type steric-zipper-like amyloid structures and the beta-sheets are stacked along the zipper axis with each individual beta-sheet sheared corresponding to its adjacent beta-sheets by two residues through the cyano–lysine inter-sheet hydrogen bonding interactions (Scheme 4F) [23]. Oliveira and research group have formulated polypeptides with hexaalanines and one arginine (A_6R) in three different nanosheet structures (Scheme 4A) [24]. The polyalanine and arginine with maximum, intermediate and minimal spatial overlaps lead to different porosity in water. The molecular dynamics simulation study shows that with maximal special overlap, the porosity became least and vice versa [24]. Two-dimensional peptide nanosheets have been derived previously by self-assembling motif specific peptide molecule (LLVFGAKMLPHHGA, Scheme 4G). These nanosheets exhibit biofunctionality including the biomimetic mineralization of hydroxyapatite (HA) minerals [25]. The covalent conjugation of peptide nanosheets with three-dimensional graphene foam has been utilized for the growth of hydroxyapatite (HA) minerals. These hybrid nanosheets are biocompatible in nature and applied for bone tissue engineering and biomedical engineering.

Conticello and co-workers have made significant contribution in the field of peptide-based nanosheets design, synthesis and their structure-function relationship [26-31]. Two sequence variants [(Pro-Amp-Gly)$_4$-(Pro-Hyp-Gly)$_n$-(Glu-Hyp-Gly)$_4$, where n = 4 and 7; Pro = L-Proline; Gly = L-Glycine; Glu = L-Glutamic acid; Amp = (2S,4R)-4-Aminoproline and Hyp = (2S,4R)-4-Hydroxyproline] of the collagen proteins with a symmetric sequence of three blocks of four triads, in which the triad sequence within each block is responsible for different electrostatic properties have been assembled to form two-dimensional layered assemblies [26]. The nanosheet thickness has been regulated by changing the peptide length and terminal functionality. The surface chemistry has also been engineered by installing functional capping groups [26]. (Pro-Amp-Gly)$_4$-(Pro-Hyp-Gly)$_4$-(Glu-Hyp-Gly)$_4$ and (amp-Hyp-Gly)$_4$-(Pro-Hyp-Gly)$_4$-(Pro-Glu-Gly)$_4$ where amp = (2S,4S)-4-Aminoproline with three segments of having positive, neutral and negative charges have undergone nanosheets formation in aqueous medium [27]. The stereoelectronic effects of the different aminoprolines on the local triple helical conformation of the peptides have been found to be responsible for different structural nanosheets assemblies [27]. The compositional and structural control of multilayer collagen nanosheets have also been reported by them [28]. These complementary oppositely charged collagen-mimetic-proteins with their triple helical cohesive ends at the layer interfaces propagated through the z-direction of the nanosheet with a compositional gradient. Their approach is applicable to devise two-dimensional nanostructures with compositional control [28]. Sequence-specific α-helical peptides-based self-assembly to form structurally defined nanoscale objects including nanofibers, nanotubes, and nanosheets have also been developed by them recently [29]. These nanosheets have been found to be thermally stable at lower concentrations in solution. This specific α-helical sequence self-assembly to form hexagonal honeycomb like structure resulting the formation of porous materials has been explored previously and such kind of nanomaterials can be applied as vechile and encapsulations of appropriate size metal ions [29]. Co-assembly of collagen-mimetic peptide and DNA origami-based nanosheets have also been studied extensively to form nanowires with repeating periodicity of 10 nm [30]. The face-to-face stacking of DNA nanosheets followed by perpendicular attachment of the peptide on the nanosheet's surface have also been evolved recently [30]. The tuning of the size of the DNA has been done by edge-to-edge and face-to-face collagen-mimetic peptides and DNA association [30]. Two-dimensional core−shell nanostructures by heteroepitaxial seeded growth of collagen-mimetic peptide triple helices have also been emerged recently for functional nanoscale device fabrication application [31]. This report describs a general working methodology for the synthesis of two-dimensional nanosheets with collagen mimetic peptide sector that can be co-assembled through temperature-controlled heterogeneous nucleation [31]. A general protocol of manipulating the structural hierarchy of two-dimensional nanomaterials that employ rigid, cylindrical structural units of helical peptide building blocks through homogeneous nucleation has also been discovered [32]. The pH-dependent interconversion by collagen-mimetics triple helices from nanotubes to nanosheets has been well understood recently [33]. In this work, they have proposed a mechanism of forming two-dimensional nanosheets through the surface stresses and the morphological transition which have been identified through the development of differential surface charge between two structurally distinct interfaces [33]. Two and three component collagen-mimetic peptide systems have been utilized and the morphological transition has been observed for all of these peptides. This report is an example of dynamic responsive material through the construction of higher order structural transition through the incorporation of the molecular mechanism [33].

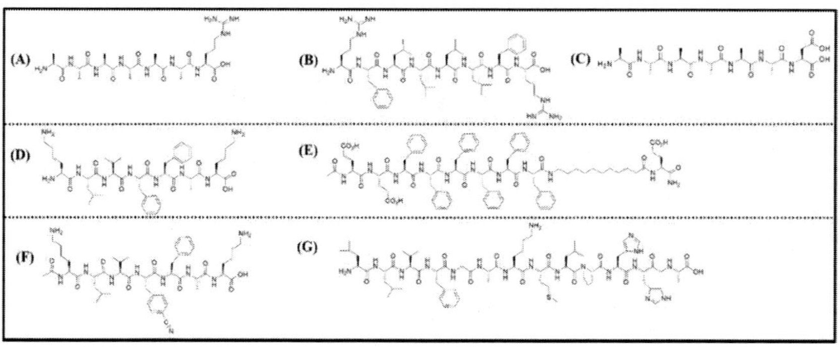

Scheme 4. Nanosheets forming peptides reported in the literature.

3.2. Cyclic Peptide-Based Nanosheets

The self-assembly to form one-dimensional nanotube through the amide hydrogen bonding interactions in antiparallel fashion from D/L-alternating cyclic octapeptide has been first reported by Ghadiri and his co-workers in the year 1993 [34].

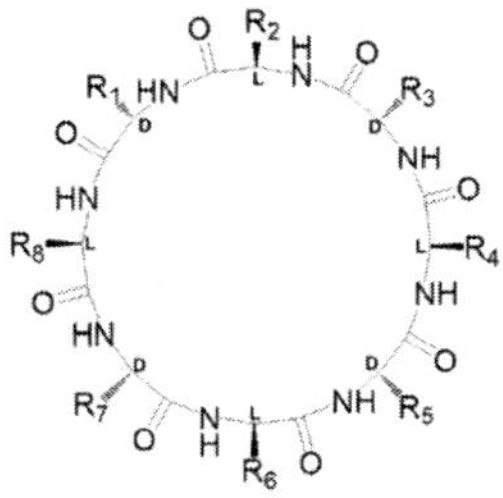

R_1	R_2	R_3	R_4	R_5	R_6	R_7	R_8
Glu	His	Leu	Trp	Leu	Glu	His	Gln
Glu	His	Leu	**Ala**	Leu	Glu	His	Gln
Glu	His	Leu	**Leu**	Leu	Glu	His	Gln
Glu	His	Leu	**Phe**	Leu	Glu	His	Gln
Glu	His	**Phe**	Trp	**Phe**	Glu	His	Gln
Glu	His	**Phe**	**Phe**	**Phe**	Glu	His	Gln
Glu	His	Leu	Trp	Leu	**His**	**Glu**	Gln

Scheme 5. Nanosheets forming general amino acid sequences of alternating L/D-cyclic octapeptides.

There are various reports on D/L-alternating cyclic octapeptide-based nanotube formation in the literature [35-37]. The one-dimensional cyclic peptide nanotubes to two-dimensional supramolecular nanosheets formation from various cyclic peptide sequences in water have also been well studied recently [38]. The structure-assembly relationship of the two-dimensional nanosheets has been established by specific amino acid mutations on D/L-alternating cyclic octapeptides [38]. Among these studied nine octapeptides modifications, six cyclic octapeptides undergone nanosheets formation where the central hydrophobic sector is occupied by tryptophan followed by mutated with alanine, leucine and phenylalanine which were responsible for the two-dimensional nanostructures formation [38]. It has also been identified that the presence of dyads of negatively charged glutamic acid side chain and positively charged histidine side chain on the cyclic peptides lead to the formation of nanosheets. This suggests the proper balance between the hydrophobic and alternating positive and negative charge balance is responsible for the nanosheets formation in water. A general protocol for the designing and synthesis of D/L-alternating cyclic octapeptide-based nanosheets and their uses (Scheme 5) has been thoroughly discussed in this study [38]. The largest fully organic two-dimensional assembled nanostructure from the amino acids present in the D/L-alternating cyclic octapeptide with sufficient coded chemical information has been discovered recently which forms two-dimensional nanosheets in aqueous medium [39]. The importance of negatively charged glutamic acid and the next amino acid as positively charged histidine on a D/L-alternating cyclic octapeptide sequence for the formation of nanosheets in water has again been well supported by Montenegro and co-workers discovery [39]. Govindaraju and his co-workers have made significant contribution in the field of self-organization into nano/mesosheets from cyclic-dipeptides [40-43]. Self-assembly of designed simplest aromatic cyclic dipeptides of (L-Phg-L-Phg) and (D-Phg-L-Phg) to form highly stable two-dimensional (2D) nano- and mesosheets with large lateral surface area have been reported previously [40]. Nuclear magnetic resonance study in solution and solid phase revealed the presence of strong (N-H-O) hydrogen-bonded molecular chains supported by aromatic pi–pi interactions which are responsible for the nanosheet structure. Cyclic dipeptide-based nanosheets formation and the use of such two-dimensional nanosheets in drug delivery applications has also reported recently [41]. These cyclic dipeptide amphiphiles have been synthesized from glycine and aspartic acid attached varying lengths of alkyl chains (C8–C18) [41]. These synthesized cyclic dipeptides form well-defined and diverse nanoarchitectures including robust nanotubes, nanospheres, nano/microsheets, and flowers. It has been identified that the non-covalent interactions between the cyclic dipeptide amphiphiles is the driving force for such asemblies. The co-assembly behavior of these cyclic dipeptides with nucleosides has also been studied for potential drug delivery vehicles. These cyclic dipeptides have been found to be biocompatible and cytocompatible in nature. An anticancer drug 5-bromo-2′-deoxyuridine (BrdU) has also been used for the co-assembly study with these cyclic dipeptide amphiphiles and the cytotoxicity has also been successfully investigated in HeLa cells [42]. This report allows these nanostructures for potential use as drug delivery vechicle [42]. The assembly of naphthalene bisimide appended amino acids derivatives into nanosheets has also been well established [42-43].

It has been well understood that the amphiphilic and bolaamphilic peptides, amyloid protein-based peptide sequences, collagen protein mimetic peptides, D/L-alternating cyclic octapeptides and cyclic dipeptides are impressive building blocks for supramolecular assembly to form two-dimensional nano/mesosheets.

4. Peptoid Nanosheets and Their Applications

The side chain of the peptide is instead putting at α-carbon atom, placing on the nitrogen atom by replacing the hydrogen at nitrogen atom of the amide bond produces an interesting class of molecules which are popularly known as peptoids. The ease of synthetic flexibility of these peptoids opens up the possibilities to engineer the structure-function relationships. Peptoids have been discovered by a start-up company called Protos Corp. in Emeryville, CA, USA in the late-1980s [10]. Zuckermann and his co-workers have pioneering contribution in the field of peptoids syntheses and their potential uses [44-61]. Zuckermann and his research group actually have laid the foundation stone of the peptoid chemistry, their functions and then on words it became an emerging field of research [10]. They have established the solid phase detail synthetic protocols for the synthesis of polypeptoid and their self-assembly behavior into highly ordered two-dimensional nanosheets [46]. Nanosheets formation by self-assembling polypeptoids have been meticulously understood by using fluorescence microscopic imaging (Figure 1) and scanning electron microscopic imaging (Figure 2) [46]. The periodic amphiphilic, electrostatic and aromatic interactions governed self-assembly of two peptoids into extremely thin two-dimensional crystalline sheets in aqueous medium have been reported by Zuckermann and co-workers. These two peptoids on mixing in equimolar ratio in aqueous solution produces sheets with 2.7 nm of thickness which has been characterized by using aberration-corrected transmission electron microscopy [47]. The binding of Streptavidin peptide sequence by using a hydrophilic linker conjugated to the peptoid sheet forming sequence has also been carried out in this work [47]. The free-floating nanosheets formation in aqueous medium from two oppositely charged peptoid strands [$(Nae\text{-}Npe)_{18}$ and $(Nce\text{-}Npe)_{18}$ where, Nae = *N*-(2-aminoethyl)glycine, Npe = *N*-(2-phenylethyl)glycine and Nce = *N*-(2-carboxyethyl)glycine] has also been explored (Scheme 6A and 6B) [48-49]. The peptoid nanosheet formation with the vial rotation technique has been identified with a conclusion that the nanosheets are the stable structure and did not form spontaneously, suggesting large activation energy during the materialization [48]. They have also developed the protocols for nanosheets formation from antibody-like architecture and information rich linear peptoids [49]. Their protocols of developing nanosheets has followed the adsorption-compression of the peptoids at the air-water interface. They have also established in this work that the incorporation of multifunctionality within the peptoid structures for binding with inorganic and organic recognition sites to gain extra stability of the nanosheets. These loop-displaying peptoid nanosheets are also promising backbone for sensing and catalysis purposes [49]. They have also developed nacre mimetic materials by using peptoid-based nanosheets (Scheme 6C) [50]. These free floating zwitterionic nanosheets have been mineralized with amorphous calcium carbonate for the formation of planner nacre synthons with 10 nm layers of amorphous calcium carbonate by using CO_2 diffusion method [50]. The structure-function relationship by synthesizing varying hydrophobic domain of the amphiphilic peptoids have also been established previously (Scheme 6D, E, F, G, H, I, J, K and L) [51]. Fluorescence spectroscopic study and optical microscopic study of these peptoid nanosheets clearly suggest that the mutation of the hydrophobic core with meta and para variations can be tolerated however, other sector's mutations can be resulted instability in the two-dimensional nanostructures [51]. The position of the methyl substituent in the aromatic ring plays a critical role in the aggregation process (Scheme 6D, E, F, G, H, I, J, K and L) [51]. The structure-rheology relationship has also been established by using peptoid forming monolayers [52]. The chemical and mechanical stability of the peptoid nanosheets have been achieved by cross linking photo labile group(s) on the hydrophobic core of the peptoid backbone [53]. This is a general protocol to enhance the mechano / chemical stability of the assembled peptoid nanosheets from specific peptoid oligomers which have been synthesized following solid phase synthetic protocol with photolabile handles. These nanosheets are stable on treatment with UV light, sonication, lyophilization, pelleting and in organic solvents [53]. The glycosylated peptoid nanosheets for protein recognition have been discovered recently [54]. The glycosylated peptoids have been synthesized by copper catalyzed azide−alkyne cycloaddition reaction by conjugating the saccharide moiety at the middle of the hydrophilic peptoid backbone [54]. The existence of *cis*-amide bonds in peptoid nanosheets has been discovered recently [55]. Experimental evidences, molecular dynamics and quantum mechanical calculations of peptoid nanosheets suggest that the conformations depend on the inter and intramolecular interactions [56]. Saccharide functionalized peptoid nanosheets have been utilized previously to bind with specific multivalent proteins and with bacterial cells [57]. The protein binding ability under the physiological conditions has been achieved by using this lipid functionalization of the peptoid nanosheets [57]. This approach also enabling multivalent systems recognition sites by the fastening ligand with high selectivity and specificity [57]. Due to the micron scale length of these nanosheets open up the possibility of binding with the bacterial cells and inhibiting the cell growth and thus resulting antimicrobial properties [57]. The biocompatible and antimicrobial properties of such two-dimensional nanosheets indicate the feasibility to use as biomedicines in future. The atomic scale resolution imagining of polypeptoid nanosheets has been successfully investigated recently [58]. The design, synthesis and screening of applying the combinatorial libraries of sequence-defined peptoid polymers engineered to fold into ordered supramolecular nanosheets with a loop on the surface of the nanosheets. These polyvalent, loop-functionalized proteolytically stable nanosheets have been utilized by using Förster resonance energy transfer (FRET) study in order to understand the binding nature to various proteins. It has been well understood that these peptoids are specific for binding to the heptameric protein and the binding is dependent on the loop functionality [59]. The selective assembly between the peptoid chains have been done by using shape complementarity of sequence-defined peptoids [60].

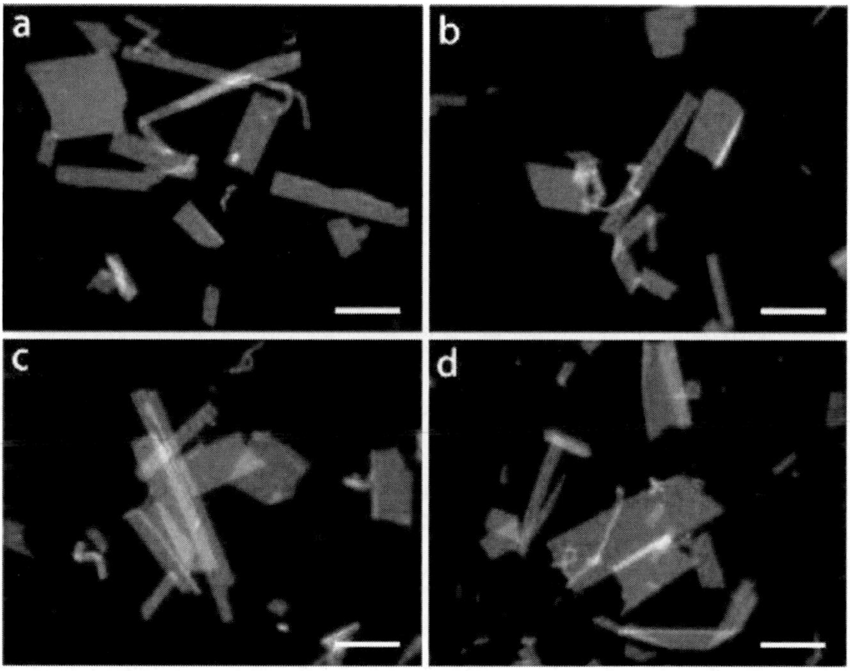

Figure 1. Fluorescence microscopic images of peptoid nanosheets (Reprinted from J. Vis. Exp., 2011, 57, e3373, doi:10.3791/3373, reference number in this chapter as 45. H. Tran, S. L. Gael, M. D. Connolly, R. N. Zuckermann, Solid-phase Submonomer Synthesis of Peptoid Polymers and their Self-Assembly into Highly-Ordered Nanosheets, *J. Vis. Exp.*, **2011**, *57*, e3373. JoVE copyright/reprint permissions: Article 3373, Figure 2 (Scale bar 100 μm) from JoVE, 2023).

Trapezoid, comb and zig-zag shaped self-complimentary diblock copolypeptoids have been synthesized and characterized thoroughly by differential scanning calorimetry and X-ray diffraction [60]. These peptoids have been self-assembled in aqueous medium into crystalline monolayer nanosheets with a known rectangular crystal lattice motif. The complementary molecular shaped peptoids self-assembly and crystallizations can be utilized to construct more advanced hierarchically-ordered peptoid nanomaterials which is found in biomacromolecular nanostructure [60]. The discovery of self-assembling behavior of a dipeptide, diphenylalanine has created tremendous contribution to the emerging area of various aspects of structure-function relationship of Phe-Phe dipeptide [61]. An aromatic peptoid (Nbn [N-benzylglycine]-based dipeptoid) analogue of the diphenylalanine has been synthesized and the self-assembly has been studied [62]. The crystallization of this compound has also been successfully achieved which is the first crystal structure for such kind of peptoid system stabilized through hydrogen bonding interactions. This dipeptoid has undergone two-dimensional nanosheets formation in solution by using cooperative intermolecular pi-pi stacking interactions [62].

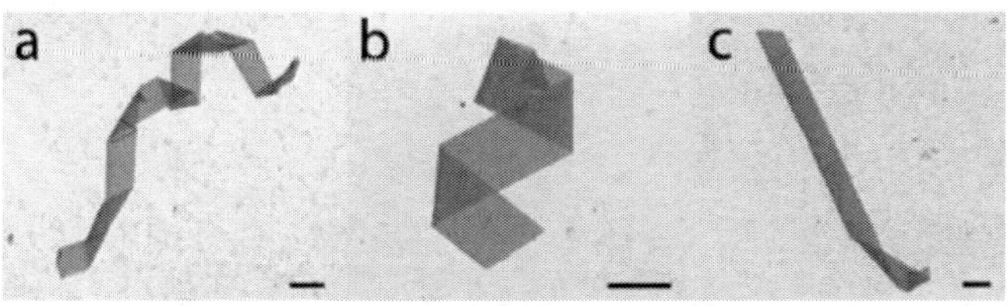

Figure 2. Scanning electron microscopic images of peptoid nanosheets (Reprinted from J. Vis. Exp., 2011, 57, e3373, doi:10.3791/3373, reference number in this chapter as 45. H. Tran, S. L. Gael, M. D. Connolly, R. N. Zuckermann, Solid-phase Submonomer Synthesis of Peptoid Polymers and their Self-Assembly into Highly-Ordered Nanosheets, *J. Vis. Exp.*, 2011, *57*, e3373. JoVE copyright/reprint permissions: Article 3373, Figure 3 (Scale bar 5 μm) from JoVE, 2023).

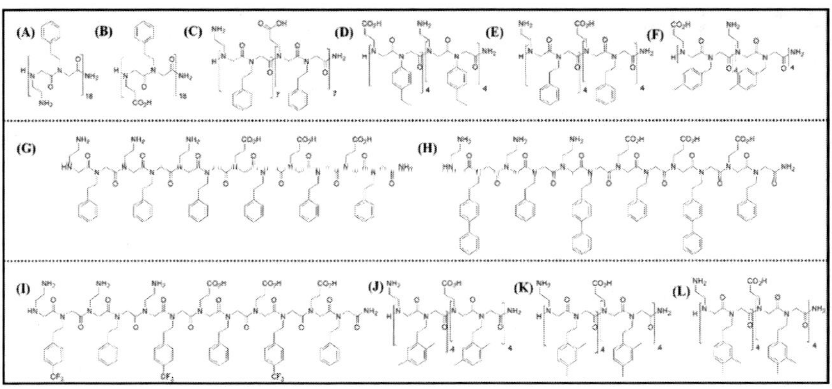

Scheme 6. Nanosheets forming peptoid sequences reported in the literature.

Li and co-workers have developed two-dimensional nanodisk and nanosheet-like structures from a triblock copolymer containing both polypeptide and polypeptoid [63]. The hierarchical supramolecular nanosheet formation from the self-assembly of polypeptoid-based diblock copolymers driven by the process of crystallization has also been invented by them recently [64]. They have also discovered the crystallization driven nanofilaments to nanosheet transformation by poly(ethyleneglycol)-b-poly(N-(2-phenylethyl)glycine) diblock copolymer appended hydrophobic crystallizable polypeptoid [65]. Hu and co-workers have investigated label-free serum-based Alzheimer's disease diagnosis by using antibody mimetic peptoid nanosheets [66]. Yang and co-workers have developed a sensor based on peptoid nanosheets for the diagnosis of the surveillance of the Alzheimer's disease and Amnestic mild cognitive impairment [67]. They have nicely fabricated highly sensitive peptoid nanosheet-based surface plasmon resonance imaging (SPRi)- detection tool which is highly efficient to distinguish between the plasma and serum from normal individuals. The patients suffering from Alzheimer's disease and Amnestic mild cognitive impairment can be detected early by this SPRi technique with high sensitivity and specificity. The functionalized SPRi chip made out of these peptoid nanosheets can improve the diagnostic performance of the plasma amyloid $A\beta_{42}$-recognizing molecule and the antibody specific to amyloid $A\beta_{42}$ [68]. Fluorescence microscopic and atomic force microscopic imaging show nanosheets formation up on the self-assembly of the peptoid with a loop [68]. Ferguson and co-workers have developed amphiphilic peptoids appended with NH_2-capped block of six hydrophobic N-((4-bromophenyl)ethyl)glycine residues conjugated to a polar $NH_3(CH_2)_5CO$ tail to disorder aggregates to one-dimensional helical rods to crystalline two-dimensional nanosheets [69] formation. They have also studied the assembly behavior in organic-water mixture solvents with slow evaporation of organic solvents. These two-dimensional (*2D*) nanosheets have also been utilized for catalysis and for molecular separation [69] purposes. The experimental observations *(1D rods by using:* small-angle X-ray scattering, circular dichroism and atomic force microscopy; *2D crystalline sheets by using:* X-ray diffraction and atomic force microscopy) have been computationally verified and both are in good agreement with the structural evolution of the *1D* rods into *2D* sheets [69] transformation. Buuren and co-workers have well studied the assembly of three different peptoids which exhibit distinctly different crystallization kinetics by performing X-ray scattering and molecular dynamics simulations studies. The *in-situ* measurements reveal that the peptoids aggregate first form a nascent phase that is less crystalline than the assembled peptoid membrane [70]. It has been understood that anisotropic aromatic interactions are determined to be the dominant driving force in the early stages of membrane formation. These results answer how the peptoid assembly can be manipulated during the early stages of assembly, nucleation and growth [70]. Baneyx and co-workers have synthesized peptoid nanosheets (3.3 nm thick) from equimolar mixtures of unmodified and maleimide functionalized versions of peptoid oligomers [71]. They have figured out the conditions to engineer the location of the cysteine residues in superfolder green fluorescent protein (silica binding derivative) for maleimide conjugation in order to form protein−peptoid hybrids with homogenous protein coverage on both sides of the nanosheets. This approach can be applied for bio-imaging and photo-catalysis applications [71]. The above discussion covers every aspect of the peptoid nanosheets designing, synthesis, structure determination, proterties, uses and their structure-function relationship chemistry which is an emerging area of current scientific research.

5. Protein Nanosheets and Their Possible Applications

The self-assembly of protein is also known in the literature to form various architectonics [72]. Takeoka and co-workers have developed free standing albumin-nanosheets with hetero-surfaces in the year 2007 [73]. Champion and co-workers have developed nanosheets from two leucine functionalized proteins through leucine zipper formation [74]. These two proteins held together by high-affinity interactions of two leucine zippers to form bolaamphiphiles. Nanosheets formation have been observed when the proteins were mixed at 4 °C in aqueous solutions followed by heating to 25°C in a rotating container at the air−water interface. These synthesized protein nanosheets are robust than the globular protein [74]. The small fluorescent proteins and large enzymes incorporation into the protein nanosheets has also been established [74]. Liu and co-workers have discovered a template-free, highly ordered two-dimensional light-harvesting nanosheets by covalent-induced co-assembly of EBFP2 (energy donor) and EGFP (energy acceptor) (EBFP2 share almost the same structure and amino acid sequence with EGFP) [75]. They have also introduced approximate square planar binding sites on the side surface of protein in order to sluggish the assembly pattern. The self-assembly has been extended in orthogonal directions to achieve monolayered and protein nanoarrays. The light harvesting nature was explored for co-assembling the fluorescent proteins by excellent anti self-quenching property of fluorescent proteins [75]. In future various emerging areas of protein-based nanosheets must be developed and FRET-based energy transfer systems to devise novel light harvesting bioantennas.

Conclusion

In this chapter, the historical development of peptide and peptoid's supramolecular assemblies to form various nanostructures have been described thoroughly. The backbone of the chemical structures and various non-covalent interactions were identified as most important factors for specific nanostructure formation. The design, synthesis and potential applications of peptide, protein and peptoid-based nanosheets have been discussed extensively with various examples. The emergence of sequence-dependent peptide, peptoid and protein-based nanosheets formation and their mechanism have also been reviewed. Overall, this chapter would be helpful for the reader to understand the historical development of peptide, peptoid and protein-based two-dimensional nanosheets, applications and their structure-function relationships.

Acknowledgments

I would like to acknowledge DST-Inspire Faculty Award Research Grant, DST, New Delhi, India (Project no. DST/INSPIRE/04/2017/001084, IFA17-CH269).

References

[1] J. M. Lehn, Toward complex matter: Supramolecular chemistry and self-organization, *Proc. Natl. Acad. Sci. USA.*, 2022, *99*, 4763-4768.

[2] G. M. Whitesides and B. Grzybowski, Self-Assembly at All Scales, *Science*, 2002, *295*, 2418-2421.

[3] Z. Luo and S. Zhang, Designer nanomaterials using chiral self-assembling peptide systems and their emerging benefit for society, *Chem. Soc. Rev.*, 2012, *41*, 4736-4754.

[4] J. Wang, K. Liu, R. Xing and X. Yan, Peptide self-assembly: thermodynamics and kinetics, *Chem. Soc. Rev.*, 2016, *45*, 5589-5604.

[5] X. Yan, P. Zhua and J. Li, Self-assembly and application of diphenylalanine-based nanostructures, *Chem. Soc. Rev.*, 2010, *39*, 1877–1890.

[6] X. J. Wang, J. Cheng, L. Y. Zhang and J. G. Zhang, Self-assembling peptides-based nano-cargos for targeted chemotherapy and immunotherapy of tumors: recent developments, challenges, and future perspectives, *Drug Delivery*, 2022, *29*, 1184–1200.

[7] E. Gatto, C. Toniolo and M. Venanzi, Peptide Self-Assembled Nanostructures: From Models to Therapeutic Peptides, *Nanomaterials*, 2022, *12*, 466.

[8] E. J. Robertson, A. Battigelli, C. Proulx, R. V. Mannige, T. K. Haxton, L. Yun, S. Whitelam, and R. N. Zuckermann, Design, Synthesis, Assembly, and Engineering of Peptoid Nanosheets, *Acc. Chem. Res.*, 2016, *49*, 379−389.

[9] N. Gangloff, J. Ulbricht, T. Lorson, H. Schlaad and R. Luxenhofer, Peptoids and Polypeptoids at the Frontier of Supra- and Macromolecular Engineering, *Chem. Rev.*, 2016, *116*, 1753−1802.

[10] R. N. Zuckermann, Commentary Peptoid Origins, *Biopolymers (Pep. Sci.)*, 2010, *96*, 545-555.

[11] L. Ferrazzano, D. Corbisiero, G. Martelli, A. Tolomelli, A. Viola, A. Ricci and W. Cabri, Green Solvent Mixtures for Solid-Phase Peptide Synthesis: A Dimethylformamide-Free Highly Efficient Synthesis of Pharmaceutical-Grade Peptides, *ACS Sustainable Chem. Eng.*, 2019, *7*, 12867−12877.

[12] J. M. Palomo, Solid-phase peptide synthesis: an overview focused on the preparation of biologically relevant peptides, *RSC Adv.*, 2014, *4*, 32658–32672.

[13] M. Amblard, J. A. Fehrentz, J. Martinez and G. Subra, Methods and Protocols of Modern Solid Phase Peptide Synthesis, *Mol. Biotech.*, 2006, *33*, 239-254.

[14] B. Merrifield, Solid Phase Synthesis, *Science*, 1986, *232*, 341-347.

[15] Coin, M. Beyermann and M. Bienert, Solid-phase peptide synthesis: from standard procedures to the synthesis of difficult sequences, *Nat. Protocol*, 2007, *2*, 3247-3256.

[16] L. Raibaut, O. E. Mahdi and O. Melnyk, Solid Phase Protein Chemical Synthesis, Protein Ligation and Total Synthesis II, 2015, *363*, 103–154.

[17] W. Hamley, A. Dehsorkhi and V. Castelletto, Self-assembled arginine-coated peptide nanosheets in water, *Chem. Commun.*, 2013, *49*, 1850-1852.

[18] E. R. d. Silva, M. N. M. Walter, M. Reza, V. Castelletto, J. Ruokolainen, C. J. Connon, W. A. Alves and I. W. Hamley, Self-Assembled Arginine-Capped Peptide Bolaamphiphile Nanosheets for Cell Culture and Controlled Wettability Surfaces, *Biomacromolecules*, 2015, *16*, 3180−3190.

[19] W. Hamley, J. Hutchinson, S. Kirkham, V. Castelletto and A. Kaur, Nanosheet Formation by an Anionic Surfactant-like Peptide and Modulation of Self-Assembly through Ionic Complexation, *Langmuir*, 2016, *32*, 10387−10393.

[20] B. Daia, D. Lia, W. Xi, F. Luo, X. Zhang, M. Zou, M. Cao, J. Hu, W. Wang, G. Wei, Y. Zhang and C. Liu, Tunable assembly of amyloid-forming peptides into nanosheets as a retrovirus carrier, *Proc. Natl. Acad. Sci. USA.*, 2015, *112*, 2996-3001.

[21] J. Gao, M. Wang, F. Wang and J. Du, Synthesis and Mechanism Insight of a Peptide-Grafted Hyperbranched Polymer Nanosheet with Weak Positive Charges_but Excellent Intrinsically Antibacterial Efficacy, *Biomacromolecules*, 2016, *17*, 2080–2086.

[22] Y. Lin, M. R. Thomas, A. Gelmi, V. Leonardo, E. T. Pashuck, S. A. Maynard, Y. Wang and M. M. Stevens, Self-Assembled 2D Free-Standing Janus Nanosheets with Single Layer Thickness, *J. Am. Chem. Soc.*, 2017, *139*, 13592–13595.

[23] B. Jia, Y. Sun, L. Yang, Y. Yu, H. Fan and G. Ma, A structural model of the hierarchical assembly of an amyloid nanosheet by an infrared probe technique, *Phys. Chem. Chem. Phys.*, 2018, *20*, 27261-27271.

[24] B. Proenca, L. B. A. Oliveira and G. Colherinhas, Stability and Structural Analysis of A6R Polypeptide Nanosheets: A Theoretical Study Using the Classical Molecular Dynamics Simulation, *J. Phys. Chem. C*, 2018, *122*, 24445–24453.

[25] K. Li, Z. Zhang, D. Li, W. Zhang, X. Yu, W. Liu, C. Gong, G. Wei and Z. Su, Biomimetic Ultralight, Highly Porous, Shape-Adjustable, and Biocompatible 3D Graphene Minerals via Incorporation of Self-Assembled Peptide Nanosheets, *Adv. Funct. Mater.*, 2018, 1801056.

[26] T. Jiang, C. Xu, Y. Liu, Z. Liu, J. S. Wall, X. Zuo, T. Lian, K. Salaita, C. Ni, D. Pochan and V. P. Conticello, Structurally Defined Nanoscale Sheets from Self-Assembly of Collagen-Mimetic Peptides, *J. Am. Chem. Soc.*, 2014, *136*, 4300–4308.

[27] T. Jiang, C. Xu, X. Zuo and V. P. Conticello, Structurally Homogeneous Nanosheets from Self-Assembly of a Collagen-Mimetic Peptide, *Angew. Chem. Int. Ed.*, 2014, *53*, 8367–8371.

[28] T. Jiang, O. A. Vail, Z. Jiang, X. Zuo and V. P. Conticello, Rational Design of Multilayer Collagen Nanosheets with Compositional and Structural Control, *J. Am. Chem. Soc.*, 2015, *137*, 7793–7802.

[29] E. L. Magnotti, S. A. Hughes, R. S. Dillard, S. Wang, L. Hough, A. Karumbamkandathil, T. Lian, J. S. Wall, X. Zuo, E. R. Wright and V. P. Conticello, Self-Assembly of an α-Helical Peptide into a Crystalline Two-Dimensional Nanoporous Framework, *J. Am. Chem. Soc.*, 2016, *138*, 16274–16282.

[30] T. Jiang, T. A. Meyer, C. Modlin, X. Zuo, V. P. Conticello and Y. Ke, Structurally Ordered Nanowire Formation from Co-Assembly of DNA Origami and Collagen-Mimetic Peptides, *J. Am. Chem. Soc.*, 2017, *139*, 14025–14028.

[31] D. Merg, E. v. Genderen, A. Bazrafshan, H. Su, X. Zuo, G. Touponse, T. B. Blum, K. Salaita, J. P. Abrahams and V. P. Conticello, Seeded Heteroepitaxial Growth of Crystallizable Collagen Triple Helices: Engineering Multifunctional Two-Dimensional Core–Shell Nanostructures, *J. Am. Chem. Soc.*, 2019, *141*, 20107–20117.

[32] D. Merg, G. Touponse, E. V. Genderen, X. Zuo, A. Bazrafshan, T. Blum, S. Hughes, K. Salaita, J. P. Abrahams and V. P. Conticello, 2D Crystal Engineering of Nanosheets Assembled from Helical Peptide Building Blocks, *Angew. Chem. Int. Ed.*, 2019, *58*, 13507–13512.

[33] D. Merg, G. Touponse, E. v. Genderen, T. B. Blum, X. Zuo, A. Bazrafshan, H. M. H. Siaw, A. McCanna, R. B. Dyer, K. Salaita, J. P. Abrahams and V. P. Conticello, Shape-Shifting Peptide Nanomaterials: Surface Asymmetry Enables pH-Dependent

Formation and Interconversion of Collagen Tubes and Sheets, *J. Am. Chem. Soc.*, 2020, *142*, 19956−19968.

[34] M. R. Ghadiri, J. R. Granja, R. A. Milligan, D. E. McRee and N. Khazanovich, Self-assembling organic nanotubes based on a cyclic peptide architecture, *Nature*, 1993, *366*, 324-327.

[35] N. Ashkenasy, W. S. Horne and M. R. Ghadiri, Design of Self-Assembling Peptide Nanotubes with Delocalized Electronic States, *small*, 2006, *2*, 99–102.

[36] O. Silberbush, M. Engel, I. Sivron, S. Roy and N. Ashkenasy, Self-Assembled Peptide Nanotube Films with High Proton Conductivity, *J. Phys. Chem. B*, 2019, *123*, 9882−9888.

[37] S. Roy, L. Zheng, O. Silberbush, M. Engel, Y. Atsmon-Raz, Y. Miller, A. Migliore, D. N. Beratan and N. Ashkenasy, Mechanism of Side Chain-Controlled Proton Conductivity in Bioinspired Peptidic Nanostructures, *J. Phys. Chem. B*, 2021, *125*, 12741−12752.

[38] S. Díaz, I. Insua, G. Bhak and J. Montenegro, Sequence Decoding of 1D to 2D Self-Assembling Cyclic Peptides, *Eur. J. Chem.*, 2020, *26*, 14765-14770.

[39] Insua and J. Montenegro, 1D to 2D Self Assembly of Cyclic Peptides, *J. Am. Chem. Soc.*, 2020, *142*, 300−307.

[40] T. Govindarajua, M. Pandeeswara, K. Jayaramulub, G. Jaipuriac and H. S. Atreya, Spontaneous self-assembly of designed cyclic dipeptide (Phg-Phg) into two-dimensional nano- and mesosheets, *Supramolecular Chemistry*, 2011, *23*, 487–492.

[41] K. Pandurangan, B. Roy, K. Rajasekhar, Y. V. Suseela, P. Nagendra, A. Chaturvedi, U. R. Satwik, N. A. Murugan, U. Ramamurty and T. Govindaraju, Molecular Architectonics of Cyclic Dipeptide Amphiphiles and Their Application in Drug Delivery, *ACS Appl. Bio Mater.*, 2020, *3*, 3413−3422.

[42] M. B. Avinash and T. Govindaraju, Engineering Molecular Organization of Naphthalenediimides: Large Nanosheets with Metallic Conductivity and Attoliter Containers, *Adv. Funct. Mater.*, 2011, *21*, 3875–3882.

[43] M. Pandeeswar and T. Govindaraju, Green-fluorescent naphthalene diimide: conducting layered hierarchical 2D nanosheets and reversible probe for detection of aromatic solvents, *RSC Adv.*, 2013, *3*, 11459–11462.

[44] R. J. Simon, R. S. Kania, R. N. Zuckermann, V. D. Huebner, D. A. Jewell, S. Banville, S. NG, L. Wang, S. Rosenberg, C. K. Marlowe, D. C. Spellmeyer, R. Tans, Alan D. Frankel, D. V. Santi, F. E. Cohen and P. A. Bartlett, Peptoids: a modular approach to drug discovery. *Proc. Natl. Acad. Sci. USA.*, 1992, *89*, 9367-9371.

[45] H. Tran, S. L. Gael, M. D. Connolly, R. N. Zuckermann, Solid-phase Submonomer Synthesis of Peptoid Polymers and their Self-Assembly into Highly-Ordered Nanosheets, *J. Vis. Exp.*, 2011, *57*, e3373.

[46] K. T. Nam, S. A. Shelby, P. H. Choi, A. B. Marciel, R. Chen, L. Tan, T. K. Chu, R. A. Mesch, B. C. Lee, M. D. Connolly, C. Kisielowski and R. N. Zuckermann, Free-floating ultrathin two-dimensional crystals from sequence-specific peptoid polymers, *Nat. Mater.*, 2010, *9*, 454-460.

[47] Sanii, R. Kudirka, A. Cho, N. Venkateswaran, G. K. Olivier, A. M. Olson, H. Tran, R. M. Harada, L. Tan and R. N. Zuckermann, Shaken, Not Stirred: Collapsing a

Peptoid Monolayer To Produce Free-Floating, Stable Nanosheets, *J. Am. Chem. Soc.*, 2011, *133*, 20808–20815.

[48] R. Kudirka, H. Tran, B. Sanii, K. T. Nam, P. H. Choi, N. Venkateswaran, R. Chen, S. Whitelam, R. N. Zuckermann, Folding of a single-chain, information-rich polypeptoid sequence into a highly ordered nanosheet, *Pep. Sci.*, 2011, *96*, 586-595.

[49] G. K. Olivier, A. Cho, B. Sanii, M. D. Connolly, H. Tran and R. N. Zuckermann, Antibody-Mimetic Peptoid Nanosheets for Molecular Recognition, *ACS Nano*, 2013, *7*, 9276-9286.

[50] J. M. V. Jun, M. V. P. Altoe, S. Aloni and R. N. Zuckermann, Peptoid nanosheets as soluble, two-dimensional templates for calcium carbonate mineralization, *Chem. Commun.*, 2015, *51*, 10218-10221.

[51] E. J. Robertson, C. Proulx, J. K. Su, R. L. Garcia, S. Yoo, E. M. Nehls, M. D. Connolly, L. Taravati and R. N. Zuckermann, Molecular Engineering of the Peptoid Nanosheet Hydrophobic Core, *Langmuir*, 2016, *32*, 11946−11957.

[52] E. J. Robertson, E. M. Nehls and R. N. Zuckermann, Structure−Rheology Relationship in Nanosheet-Forming Peptoid Monolayers, *Langmuir*, 2016, *32*, 12146−12158.

[53] D. Flood, C. Proulx, E. J. Robertson, A. Battigelli, S. Wang, A. M. Schwartzberg and R. N. Zuckermann, Improved chemical and mechanical stability of peptoid nanosheets by photo-crosslinking the hydrophobic core, *Chem. Commun.*, 2016, *52*, 4753-4756.

[54] A. Battigelli, J. H. Kim, D. C. Dehigaspitiya, C. Proulx, E. J. Robertson, D. J. Murray, B. Rad, K. Kirshenbaum and R. N. Zuckermann, Glycosylated Peptoid Nanosheets as a Multivalent Scaffold for Protein Recognition, *ACS Nano*, 2018, *12*, 2455−2465.

[55] C. Hudson, A. Battigelli, M. D. Connolly, J. Edison, R. K. Spencer, S. Whitelam, R. N. Zuckermann and A. K. Paravastu, Evidence for cis Amide Bonds in Peptoid Nanosheets, *J. Phys. Chem. Lett.*, 2018, *9*, 2574−2578.

[56] J. R. Edison, R. K. Spencer, G. L. Butterfoss, B. C. Hudson, A. I. Hochbaum, A. K. Paravastu, R. N. Zuckermann, and S. Whitelam, Conformations of peptoids in nanosheets result from the interplay of backbone energetics and intermolecular interactions, *Proc. Natl. Acad. Sci. USA.*, 2018, *115*, 5647–5651.

[57] J. H. Kim, E. M. Grzincic, L. Yun, R. K. Spencer, M. A. Klinea and R. N. Zuckermann, Lipid-anchor display on peptoid nanosheets via co-assembly for multivalent pathogen recognition, *Soft Matter*, 2020, *16*, 907-913.

[58] X. Jiang, S. Xuan, R. K. Spencer, R. N. Zuckermann, K. H. Downing and N. P. Balsara, Imaging of Polypeptoid Nanosheets with Atomic Scale Precision, *Microsc. Microanal.*, 2019, *25*, 1356-1357.

[59] J. H. Kim, S. C. Kim, M. A. Kline, E. M. Grzincic, B. W. Tresca, J. Cardiel, M. Karbaschi, D. C. Dehigaspitiya, Y. Chen, V. Udumula, T. Jian, D. J. Murray, L. Yun, M. D. Connolly, J. Liu, G. Ren, C. L. Chen, K. Kirshenbaum, A. R. Abate and R. N. Zuckermann, Discovery of Stable and Selective Antibody Mimetics_from Combinatorial Libraries of Polyvalent, Loop-Functionalized Peptoid Nanosheets, *ACS Nano*, 2020, *14*, 185−195.

[60] S. Xuan, X. Jiang, N. P. Balsara and R. N. Zuckermann, Crystallization and self-assembly of shapecomplementary sequence-defined peptoids, *Polym. Chem.*, 2021, *12*, 4770–4777.

[61] M. Reches and E. Gazit, Casting Metal Nanowires within Discrete Self-Assembled Peptide Nanotubes, *Science*, 2003, *300*, 625–627.

[62] V. Castelletto, A. M. Chippindale, I. W. Hamley, S. Barnett, A. Hasan and K. H. A. Lau, Crystallization and lamellar nanosheet formation of an aromatic dipeptoid, *Chem. Commun.*, 2019, *55*, 5867-5869.

[63] Y. Ni, J. Sun, Y. Wei, X. Fu, C. Zhu and Z. Li, Two-Dimensional Supramolecular Assemblies from pH-Responsive Poly(ethyl glycol)-b-poly(L-glutamic acid)-b-poly(N-octylglycine) Triblock Copolymer, *Biomacromolecules*, 2017, *18*, 3367−3374.

[64] Z. Shi, Y. Wei, C. Zhu, J. Sun and Z. Li, Crystallization-Driven Two-Dimensional Nanosheet from Hierarchical Self-Assembly of Polypeptoid-Based Diblock Copolymers, *Macromolecules*, 2018, *51*, 6344−6351.

[65] Y. Wei, J. Tian, Z. Zhang, C. Zhu, J. Sun and Z. Li, Supramolecular Nanosheets Assembled from Poly(ethylene glycol)-b-poly(N-(2-phenylethyl)glycine) Diblock Copolymer Containing Crystallizable Hydrophobic Polypeptoid: Crystallization Driven Assembly Transition from Filaments to Nanosheets, *Macromolecules*, 2019, 52, 1546−1556.

[66] L. Zhu, Z. Zhao, P. Cheng, Z. He, Z. Cheng, J. Peng, H. Wang, C. Wang, Y. Yang, Z. Hu, Antibody-Mimetic Peptoid Nanosheet for Label-Free Serum-Based Diagnosis of Alzheimer's Disease, *Adv. Mater.*, 2017, 1700057.

[67] H. Gao, M. Liu, Z. Zhao, C. Yang, L. Zhu, Y. Cai, Y. Yang, and Z. Hu, Diagnosis of Mild Cognitive Impairment and Alzheimer's Disease by the Plasma and Serum Amyloid-beta 42 Assay through Highly Sensitive Peptoid Nanosheet Sensor, *ACS Appl. Mater. Interfaces*, 2020, *12*, 9693−9700.

[68] H. Gao, J. Wang, J. Liu, S. Ye, X. Meng, S. Song, C. Wang, X. Yu, L. Zhu, H. Wang, S. Lei and Y. Yang, Peptoid Nanosheet-Based Sensing System for the Diagnosis and Surveillance of Amnestic Mild Cognitive Impairment and Alzheimer's Disease, *ACS Chem. Neurosci.*, 2021, *12*, 4257−4264.

[69] M. Zhao, K. J. Lachowski, S. Zhang, S. Alamdari, J. Sampath, P. Mu, C. J. Mundy, J. Pfaendtner, J. J. De Yoreo, C. L. Chen, L. D. Pozzo and A. L. Ferguson, Hierarchical Self-Assembly Pathways of Peptoid Helices and Sheets, *Biomacromolecules*, 2022, *23*, 992−1008.

[70] J. A. Hammons, M. D. Baer, T. Jian, J. R. I. Lee, T. M. Weiss, J. J. De Yoreo, A. Noy, C. L. Chen and A. V. Buuren, Early-Stage Aggregation and Crystalline Interactions of Peptoid Nanomembranes, *J. Phys. Chem. Lett.*, 2021, *12*, 6126−6133.

[71] J. Ma, B. Cai, S. Zhang, T. Jian, J. J. D. Yoreo, C. L. Chen and F. Baneyx, Nanoparticle-Mediated Assembly of Peptoid Nanosheets Functionalized with Solid-Binding Proteins: Designing Heterostructures for Hierarchy, *Nano Lett.*, 2021, *21*, 1636−1642.

[72] A. Solomonov and U. Shimanovich, Self-Assembly in Protein-Based Bionanomaterials, *Isr. J. Chem.*, 2019, *59*, 1152-1170.

[73] Y. Okamura, T. Goto, D. Niwa, Y. Fukui, M. Otsuka, N. Motohashi, T. Osaka, S. Takeoka, Fabrication of free-standing albumin-nanosheets having heterosurfaces, *J. Biomed. Mater. Res. A*, 2007, *89*, 233-241.

[74] R. Dautel and J. A. Champion, Self-Assembly of Functional Protein Nanosheets from Thermoresponsive Bolaamphiphiles, *Biomacromolecules*, 2022, *23*, 3612−3620.

[75] X. Li, S. Qiao, L. Zhao, S. Liu, F. Li, F. Yang, Q. Luo, C. Hou, J. Xu and J. Liu, Template-Free Construction of Highly Ordered Monolayered Fluorescent Protein Nanosheets: A Bioinspired Artificial Light Harvesting System, *ACS Nano*, 2019, *13*, 1861−1869.

Chapter 7

Graphene-Based Nanosheets for New Generation Flame Retardant Composites

Chetna Tewari[1]
Yebom Kim[1,2]
and Yong Chae Jung[1,*]
[1]Institute of Advanced Composite Materials,
Korea Institute of Science and Technology (KIST) 92 Chudong-ro,
Bongdong-eup, Wanju-gun, Jeonllabuk-do, 55324, Republic of Korea
[2]Department of Chemical Engineering, Jeonbuk National University,
Jeonju-si, Jeonllabuk-do, 54896, Republic of Korea

Abstract

Graphene has recently attracted significant educational and commercial interest due to its excellent performance in electrical, mechanical, and thermal properties. Incorporation of a small quantity of graphene can remarkably enhance the various properties, including the flame-retardant property, of the polymer composites. On the other hand, epoxy composites are a fascinating class of polymers that have various applications and uses in various places like industries, homes, and other places. The charming properties and wide range of epoxy composites increase their applicability, but at the same time, their easily flammable nature is a serious safety concern, as fire causes several losses every year. Therefore, this book chapter discusses the effect of graphene and its derivatives as flame retardant probes for epoxy composites, along with a detailed explanation of the charring, anti-dripping, and thermal stability of the composites after incorporation of graphene-based materials (GBMs). This book chapter also concludes the synergistic effects of

[*] Corresponding Author's Email: ycjung@kist.re.kr.

In: Properties and Applications of Nanosheets
Editors: Nanda Gopal Sahoo and Anirban Dandapat
ISBN: 979-8-89113-381-5

GBMs in epoxy matrices with their flame-retardant applications, as well as the challenges associated with the industrial use of graphene and its derivatives.

Keywords: composites; epoxy; flame retardant; graphene; synthesis

1. Introduction

Graphene is a wonderful material, having a carbon lattice in a hexagonal pattern with sp^2 hybridization [1]. Its sp^2 hybridized carbon network and nanoscale size are responsible for its flexibility, superior thermal, and electrical conductivity [2]. Presently, graphene and derivatives of graphene are applied in various fields of research due to the ease of modification and functionalization. On the other side, epoxy and its composites are widely sold and used as high-performance materials that are commonly applied in aerospace, automotive, construction, marine, adhesives, and other common products. Dr. Greenlee of the USA and Dr. Castan of Switzerland first discovered epoxies in 1936 as having superlative thermal stability, mechanical properties, solvent resistance, and ease of handling/modification [3]. The charming properties of epoxy lead to its presence everywhere, in markets, industries, and homes. This fact enhances the necessity to study the safety concerns of epoxy composites, as every year fire causes significant loss of resources and humankind [4, 5]. According to the research papers, graphene can act as a great filler because its small amount enhances various properties of epoxy composites, including their flame-retardant properties, but the proper dispersion of graphene in an epoxy matrix is a difficult task. A better dispersion might be shown by graphene oxide (GO) due to the presence of the oxygen containing functional groups, but it has a reduced thermal stability and weakens the intumescence ability of the composites. As a result, GO is frequently reduced following the addition of composites or textiles [6-8]. To provide ever-better fire-retardant property, functionalization of GO with additional flame-retardant functional groups is a possibility. Several common tests, including the limiting oxygen index test, TGA, the vertical flame test, and cone calorimetry, can be used to describe the performance [5].

Graphene and its derivatives improve the fire safety concern of the material by enhancing charring and heat release rates, anti-dripping properties, self-extinguishing ability, high limited oxygen demand and ignition time, and thermal stability [5]. In order to prevent or minimize the adverse effects caused by the fire, flame-retardant materials (paints, adhesives, sprays, and composites) gain the attraction of the scientific organization. This book chapter summarizes the development, properties, and applications of graphene-based epoxy composites, highlighting the important properties of graphene for flame retardant applications. Also, this book chapter concludes the challenges interconnected with the industrial use of GBMs.

2. Synthesis of Graphene-Based Epoxy Composites

2.1. Solvent Processing

The functional groups present on the surface of the graphene allows for the proper dispersion of graphene in aquas and several organic solvents. This is the quickest and most popular method for manufacturing epoxy/graphene nanocomposites. This enhances the chemical or physical interaction of functionalized graphene with polymeric matrices. Incorporating different functional groups like amine, silane, organic phosphate, and others modified the graphene surface to easily disperse in a suitable solvent using bath sonicator. The suitable solvent is one that has functionality that is suitable for dispersion of both epoxy and graphene. The results claim that graphene with solvent dispersed well in epoxy matrix compared to the powder because of the viscous nature of the epoxy [3].

Many solvents have been studied to enhance the distribution of functionalized graphene in epoxy composites. Rafiee et al., used tip sonication to disperse graphene platelets in acetone before incorporating them in epoxy metrics, and the remaining solvent was removed by heating at 70 °C [9]. Fang et al., modified graphene with amine and disseminated graphene in DMF using bath sonication, which created a mechanical adhesion at graphene-epoxy interface. The load transfer efficiency between the interface was improved in the nanocomposites, and they also had a better ability to dissipate strain energy during fracture [10]. Tang et al., used a ball milling to get the proper distribution of reduced graphene oxide (rGO) in epoxy resin using ethanol as a solvent. They also discovered that highly distributed rGO produced significantly more intricate, fine riverlike formations on the shattered surface [11]. A functionalized graphene-filled epoxy nanocomposite's mechanical and thermal reinforcements were studied by Chatterjee et al., in 1990. Expanded graphene nanoplatelets (GNP) that had been amine functionalized were distributed into epoxy resins using a high-pressure processor, followed by three roll milling in acetone. An advantageous interaction at the graphene/epoxy interface may be seen from the remarkable gains in mechanical and thermal conductivity that the resultant nanocomposite showed. For solvent processing, a variety of solvents like water, acetone, tetrahydrofuran, ethanol, dimethylformamide, dichloromethane, and others, have been employed to get proper graphene disperse. For homogeneous dispersion, dispersion techniques such as bath and tip sonication, mechanical and shear mixing, and three roll calendaring have been mainly used, and the majority of these approaches have produced positive results [12].

2.2. Resin Impregnation

This technique entails the impregnation of epoxy resin directly into the freshly made graphene filter sponge/cake. Im et al., [13] used this technique for the first timc to create a very high 60 wt% nanocomposite material in 2012. They used ultrasonication to suspend GO particles in distilled water, and they then poured the resulting mixture into a glass mold that was set on a silicon oxide (SiO_2) membrane. Vacuum filtration was used to modify the liquid before it was placed into the glass mold. Following filtration, the filter cake, which had been detached from the membrane, was heated during an annealing process to drive out any remaining water. Finally, the curing agent-containing epoxy was put over the filter cake and heated to cure it. Using this technique, epoxy resin is infused in graphene sheet using capillary-driven wetting force, making highly concentrated nanocomposites with enhanced mechanical qualities. Li et al., [14] created an 11.84 wt% graphene/epoxy nanocomposite using a similar method with ethanol and acetone as solvents. After proper dispersion, the solvent was removed using vacuum filtering. Self-assembly of the aligned graphene took place throughout the filtration process, and following that, they submerged this aligned graphene in epoxy monomer and curing agent. Han et al., [15] described the creation of the epoxy/graphene nanocomposite (0.1 wt%) by chemical vapour deposition of epoxy resin into a 3D graphene-nickel foam, followed by polymer curing and Ni template etching. Excellent fracture toughness was achieved by this nanocomposite and the glass transition temperature rose by 31 °C in comparison to pure epoxy. Furthermore, they claimed that the 3D interconnected network of graphene acts as quick charge-carrier transport routes and resulting extraordinary electrical conductivity of the composite.

2.3. Other Methods

Solvent processing is the method most frequently employed to create epoxy/graphene nanocomposites. But some derivative techniques have also been used. By mechanically mixing, Martin et al., [16] disseminated graphene in epoxy monomer. Following the addition of a photoinitiator, the mixture was exposed to UV light to cure. They noted that UV curing improved the nanocomposite's thermal and mechanical properties. Similar UV-cured epoxy/graphene nanocomposites were created by Sangermano et al., [17], who also demonstrated improved characteristics. In order to create the epoxy/graphene nanocomposite that showed multiple-fold increases in thermal conductivity, Yu et al., employed hot pressing during the curing process. Without a solvent, spreading graphene in an epoxy matrix is probably less effective because of the viscose nature of the epoxy [18]. Hsu et al., used three-roll milling to combined epoxy monomer, graphene, and curing agent at the room temperature. Because of the higher viscosity of epoxy resin, uniform graphene dispersion was hampered, making mixing without solvent a less successful dispersion method [19].

However, many routes were tried to develop epoxy/GBM composites for different applications, but the solution processing method is more prominent due to its advantages like simplicity and a lower time requirement.

3. Flame Retardant Applications/Properties of Graphene and Its Derivatives

The unique properties of graphene makes them a promising building block for creating the next generation smart epoxy composites for improving their flame retardants properties. It has an intrinsic tensile strength of 130 GPa, a high surface area of 2,629 m^2/g, and a Young's modulus of 1 TPa. It has a high electric resistivity of 10^{-6} Ω cm, and a high thermal conductivity (5000 $Wm^{-1}K^{-1}$). The main challenge is to control the morphology of the graphene, as it has low bending stiffness and high surface energy. The layered structure of GBMs can act as a physical barrier to prevent matrix disintegration and hindered heat transfer to the interior matrix. Qu et al., described development of functionalized GO (FGO) by covalent grafting of octa (propyl glycidyl ether) polyhedral oligomeric silsesquioxane on the surface of GO with γ-aminopropyl triethoxysilane as a chemical bridge [20]. They described the

FGO-added epoxy resin's flame-retardant mechanism as shown in figure 1. The tortuous path effect of uniformly distributed FGO can impede heat diffusion and flammable gas vaporization. Additionally, Si present on the γ-aminopropyl triethoxysilane strengthen the char layer during burning to stop the mixing of oxygen and pyrolysis gas, and heat transfer [20].

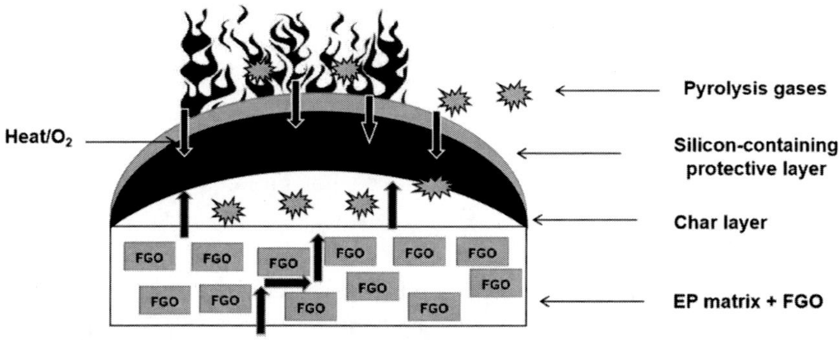

Figure 1. The mechanism of FGO/epoxy flame-retardancy.

3.1. Limiting Oxygen Index and Ignition Time

The ignition time is the amount of time between exposure to a heat source and combustion, whereas the limiting oxygen index (LOI) is minimum oxygen quantity needs for combustion. Both of them have to do with how quickly a material can catch fire. Neat GO or rGO only slightly improves the level of LOI, but when functionalized with nanohybrides of iron, copper, and molybdenum, they show better results. The nanohybrid of iron, i.e., iron phenylphosphinate-functionalized graphene oxide (FeHP@GO), demonstrates better flame retardant properties compared to the neat epoxy. The epoxy nanocomposite of FeHP@GO (2.0 wt%) shows a 42.5% improvement in the LOI and 46.2% and 23.5% reductions in the PHRR and the THR, respectively [21]. In a different study, GO was coated with uniformly sized copper organophosphate nanosheets through the self-assembly method. The composite (1 wt%) with these functionalized nanosheets showed 147% improved strength over the composite with GO. The study showed that the presence of copper- and organophosphate-containing components in GO promotes the physical obstacle of GO, enhances the char layer's aromatization, and increases the fire resistance property of the epoxy matrix [22].

Free-radical quenching in the gas phase

$$PO\cdot + H\cdot \rightarrow HPO\cdot$$

$$HPO\cdot + H\cdot \rightarrow H_2 + PO\cdot$$

$$PO\cdot + OH\cdot \rightarrow HPO\cdot + O\cdot$$

Figure 2. Flame-retardant mechanism of DOPO–VTES–GO/epoxy composite.

The hybrid of mybdenum-based metal organic framework (Mo-MOF) and GO was functionalized with [BMIM]BF4 i.e., boron-containing ionic liquid by Li et al., Epoxy composite with this hybrid (3wt %) showed decreases in PHRR, carbon monoxide production rate, and peak smoke release rate by 52.10%, 55.88%, and 40.12%, respectively. The physical barrier effect of GO, gas phase dilution (NH_3 and H_2O in gas generation), and catalytic carbonization were all combined to create the flame retardant mechanism [23].

3.2. Thermal Stability

Thermal stability primarily refers to the capacity of the material to resist heat and keep its mass and other properties. The thermal stability might be increased by adding graphene, either as a coating or filler. The impact of graphene, however, is still being researched and is quite complicated. It has been observed that cotton fabrics coated with GO or functionalized GO have lower breakdown temperatures, T5% or Tmax, than plain fabrics. This could be a result of the functional groups being pyrolyzed, which also causes a higher maximal decomposition rate and weight loss [24]. In contrast, high stability has been obtained in epoxy composites reinforced by functionalized graphene [24, 25]. Phosphorus and silicon modified GO was prepared by Zhi et al., By using solvothermal and hydrolysis-condensation reactions, respectively, they were able to successfully graft vinyltriethoxysilane (VTES) and 9,10-Dihydro-9-oxa-10-phosphaphenanthrene-10-oxide (DOPO) onto the surface of GO, and then incorporate these materials into epoxy resin using the solution blending technique. Figure 2 illustrates how free-radical quenching in the gas phase and catalytic carbonization in the condensed phase work together to dramatically increase the flame-retardant property of epoxy resin [26].

When epoxy molecules break down, the addition of GO can lessen the permeability for combustible gases. A phosphorus-rich residue is created during the breakdown of DOPO groups of DOPO-VTES-GO, which encourages the production of char residues. The transmission of heat and the movement of gases can be effectively blocked by a constant and packed char layer. The chain reaction that fuels combustion can be stopped in the gas phase by the production of PO free radicals during the decomposition of DOPO, which can then trap OH and H free radicals produced by the thermal disintegration of the epoxy chains. In addition, VTES groups frequently form the thermostable Si-O-Si bridging structure, which stimulates the formation of the graphitic structure and improves the thermostability. Consequently, these epoxy composites have improved flame retardancy [26]. This is most likely caused by the addition of graphene flakes, which restrict the mobility of polymer chains in a same manner like in graphene-reinforced composites. The char layer development and barrier qualities of the materials based on graphene are to blame for this [26].

3.3. Heat Release Rate and Combustion

Graphene inclusion has a significant impact on the combustion process. The combustion process can be characterized by a number of features. The after-glow time and after flame time in the VFT test refer to the length of time the flame remains lit after the heat source has been removed, respectively. The THR and associated PHRR indicate the maximum and cumulative values of the HRR, respectively. The HRR denotes the rate at which heat is generated during the process of combustion. Graphene significantly reduces HRR, PHRR, and THR. This is due to the obstacle qualities of graphene and the larger yield of char residue following the integration of graphene, which results in a decrement in the release of gases [27]. The physical barrier qualities of GBMs and the development of tortuous paths among the flakes, which increase in size with the number of coatings, are related to the reduction in THR and HRR [28, 29].

3.4. Charring and Self-Extinguishing

The time required for a fire to go out after the heat source has been turned off is known as the self-extinguishing time. At high temperatures, when volatile organic compounds and water get dehydrated, graphene-based materials have the capacity to generate char. Before combustion, the material is given additional stability by the char layer, which acts as a protective layer. This layer enhances the stability of the material, promotes its self-extinguishing capabilities, stops the spread of fire, and lessens damage to textiles by preventing the interchange of heat, oxygen, and fuel between the material and the environment. Qu et al., studied that the graphene nanosheets (GNS) enhanced the thermal stability by changing the decomposition pathway of epoxy resin at high temperatures. The presence of graphene nanosheets promoted the formation of char residue and enhanced the flame-retardant property of the composites [30].

3.5. Anti-Dripping Properties

Graphene has long been known to have an impact on the anti-dripping qualities of polymer composites, where it can greatly increase the viscosity of polymers so, that the flow of polymer at high temperatures can be prevented. The composite of graphene nanosheets/epoxy resin can significantly drop the melt flow and stop the flammable drips of epoxy resin during the burning process. Incorporation of GNS (3 wt%) shows increment in the LOI value of epoxy resin from 15.7 to 21.0 and also reduced the THRR from 33.37 to 28.20 kJ/m^2 [30]. The epoxy composite with graphene nanosheets shows anti-dripping properties. This is because of the unique lamellar structure of graphene and its derivatives, which can endow epoxy resin with good flame-retardant properties [30].

3.6. Synergistic Effect with Commercial Fire Retardant

GBMs can be utilized to make fabrics that are only fire resistant, but they can also be mixed with another commercial fire-retardants to create a synergistic effect that could result in even greater flame resistance and multifunctionality [31, 32]. Commercial flame-resistants in particular could raise the UL-94 and the LOI values requirements, which are normally challenging for graphene to meet. Graphene can be added either alongside or directly on the top of current fire retardants. This is especially helpful for already available fabrics in market that already have flame retardant coatings because they may be treated with graphene without having to remove the coatings first. Cu(0) and Cu(II) modified graphene hybrids (Cu(II)-GO, CuO-GNS, and Cu-rGO) were created in order to investigate the synergism and mechanism of Cu-decorated graphene on enhancing the flame retardant efficiency of ammonium polyphosphate based polyamide-cured epoxy resin. According to the findings, CuO-rGO has stronger synergistic effects for reducing smoke and harmful gases, while Cu(II)-GO has synergistic effects for boosting flame retardancy. LOI and UL-94 tests were used to assess the flame retardancy of EPs. EP has a low LOI value (20%), making it a particularly combustible substance, while the LOI value of the ammonium polyphosphate APP composite is dramatically raised to 35.5% by the addition of 28 wt% APP; however, it is still unable to achieve UL-94 V-0 certification. By including an additional 2 wt% of the synergists in the EP/28% APP system, the UL-94 V-0 rating and higher LOI values can be attained [32]. The Cu^{2+}-GO and other hybrids increase the carbonization rate while also halting the spread of fire and dripping. This is most likely because the crosslinking of viscous pyrolytic-liquid and the conversion of it into char are accelerated by these hybrids. As a result, all hybrids significantly increase the development of char when combined with APP, and different copper oxidation states and its compound structures produce varying synergistic flame-retardant efficacy when combined with APP [32].

4. Opportunities and Challenges

Epoxy-derived materials are used almost everywhere, and their fire safety is important. On the other hand, GBMs surely have excellent scope in the field of fire safety applications. That's why science and technology conduct lots of research that focuses on the use of GBMs as flame retardants for epoxy composites [33]. Although GBMs have many advantages, like a large surface area and superior surface modification capabilities but there are still many challenges associated with them that hinder their use at the industrial level, like their cost-effectiveness, quality control synthesis, up-scaling, and environmental impacts. The real world requires bulk production of these carbon-based materials in a cost-effective manner with quality control. Also, the quality of carbon-based materials affected by several factors, including the nature of the precursor. Nevertheless, large-scale processing and the use of fabrication techniques that consume less energy and less expensive feedstock can lower the cost of development. In this regard, various researchers have created carbon-based materials in bulk quantities from lignite, wood, coconut shells, agricultural waste, and plastic waste [34-36]. Because of the excellent properties of GBMs there is a hiking demand for these materials in terms of quality, quantity, dependability, and cost. The adoption of these materials may be slowed down for a variety of reasons. The limited availability of these GBMs can hinder their commercial adoption. For 2D nanomaterials, Ahmad et al., estimated that 1 kiloton of graphene nanoplatelets and a million square meters of graphene films per year would be sufficient for short-term product development [37]. This data shows the demand of GBM increases day by day and importance to find new route of synthesis of nanomaterials with high quality and low cost.

Conclusion

Graphene is the smartest material of the 21st century, having a range of applications in energy, environmental, biological, and other fields. Combination of graphene and GBMs in the polymeric matrix shows enhancement in the thermal properties, electric, and mechanical properties of the composites. Even the presence of GBMs can also improve the flame-retardant properties. The functionality of the GBMs plays an key role in the development of epoxy nanocomposites. Their laminar structure can provide tortuous path that creates a barrier between the polymer matrix and the fire. Also, the ability of graphene-based material to form char at high temperature, help in suspension of fire and act as a self-fire extinguisher.

Additionally, to achieve a synergistic flame-retardant action, the GBMs can function by themselves or in combination with other commercially available flame-retardant substances or nanoparticles. A protective char layer that serves as a physical barrier helps graphene layers function in a condensed phase mechanism. For epoxy-based polymer nanocomposites, graphene and its composites present particularly intriguing and secure flame-retardant materials. For the upcoming ten years, graphene sheets may also be a smart, sustainable flame-retardant material for polymeric/textile materials.

Declaration of Competing Interest

There are no known competing interests.

Acknowledgment

Authors acknowledge Korea Institute of Science and Technology, South Korea for financial support.

References

[1] Pandey, N., Tewari, C., Dhali, S., Bohra, B. S., Rana, S., Mehta, S. P. S., Singhal, S., Chaurasia, A. and Sahoo, N. G., 2021. Effect of graphene oxide on the mechanical and thermal properties of graphene oxide/hytrel nanocomposites. *Journal of Thermoplastic Composite Materials*, *34*(1), pp.55-67. https://doi.org/10.1177/0892705719838010.

[2] Tatrari, G., Karakoti, M., Tewari, C., Pandey, S., Bohra, B. S., Dandapat, A. and Sahoo, N. G., 2021. Solid waste-derived carbon nanomaterials for supercapacitor applications: a recent overview. *Materials Advances*, *2*(5), pp.1454-1484. https://doi.org/10.1039/D0MA00871K.

[3] Wei, J., Vo, T. and Inam, F., 2015. Epoxy/graphene nanocomposites–processing and properties: a review. *Rsc Advances*, *5*(90), pp.73510-73524. https://doi.org/10.1039/C5RA13897C.

[4] Kim, Y. N. and Jung, Y. C., 2022. *Recycling Studies of Epoxy Fiber-Reinforced Composites*. In Handbook of Epoxy/Fiber Composites (pp. 373-392). Singapore: Springer Nature Singapore. https://doi.org/10.1007/978-981-19-3603-6_46.

[5] Ge, Q., Chu, J., Cao, W., Yi, F., Ran, Z., Jin, Z., Mao, B., Li, Z. and Novoselov, K. S., 2022. Graphene-Based Textiles for Thermal Management and Flame Retardancy. *Advanced Functional Materials*, *32*(42), p.2205934. https://doi.org/10.1002/adfm.202205934.

[6] Jamali, N., Khosravi, H., Rezvani, A., Tohidlou, E. and Poulis, J. A., 2021. Viscoelastic and dry-sliding wear properties of basalt fiber-reinforced composites based on a surface-modified graphene oxide/epoxy matrix. *Journal of Industrial Textiles*, *50*(6), pp.939-953. https://doi.org/10.1177/1528083719850839.

[7] Oh, C. R., Lee, D. I., Park, J. H. and Lee, D. S., 2019. Thermally healable and recyclable graphene-nanoplate/epoxy composites via an in-situ Diels-Alder reaction on the graphene-nanoplate surface. *Polymers*, *11*(6), p.1057. https://doi.org/10.3390/polym11061057.

[8] Liu, Z., Chen, Y., Li, Y., Dai, W., Yan, Q., Alam, F. E., Du, S., Wang, Z., Nishimura, K., Jiang, N. and Lin, C. T., 2019. Graphene foam-embedded epoxy composites with significant thermal conductivity enhancement. *Nanoscale*, *11*(38), pp.17600-17606. https://doi.org/10.1039/C9NR03968F.

[9] Rafiee, M. A., Rafiee, J., Wang, Z., Song, H., Yu, Z. Z. and Koratkar, N., 2009. Enhanced mechanical properties of nanocomposites at low graphene content. *ACS nano*, *3*(12), pp.3884-3890. https://doi.org/10.1021/nn9010472.

[10] Fang, M., Zhang, Z., Li, J., Zhang, H., Lu, H. and Yang, Y., 2010. Constructing hierarchically structured interphases for strong and tough epoxy nanocomposites by amine-rich graphene surfaces. *Journal of Materials Chemistry*, *20*(43), pp.9635-9643. https://doi.org/10.1039/C0JM01620A.

[11] Tang, L. C., Wan, Y. J., Yan, D., Pei, Y. B., Zhao, L., Li, Y. B., Wu, L. B., Jiang, J. X. and Lai, G. Q., 2013. The effect of graphene dispersion on the mechanical properties of graphene/epoxy composites. *Carbon*, *60*, pp.16-27. https://doi.org/10.1016/j.carbon.2013.03.050.

[12] Chatterjee, S., Wang, J. W., Kuo, W. S., Tai, N. H., Salzmann, C., Li, W. L., Hollertz, R., Nüesch, F. A. and Chu, B. T. T., 2012. Mechanical reinforcement and thermal conductivity in expanded graphene nanoplatelets reinforced epoxy composites. *Chemical Physics Letters*, *531*, pp.6-10. https://doi.org/10.1016/j.cplett.2012.02.006.

[13] Im, H. and Kim, J., 2012. Thermal conductivity of a graphene oxide–carbon nanotube hybrid/epoxy composite. *Carbon*, *50*(15), pp.5429-5440. https://doi.org/10.1016/j.carbon.2012.07.029.

[14] Li, X., Feng, Y., Chen, C., Ye, Y., Zeng, H., Qu, H., Liu, J., Zhou, X., Long, S. and Xie, X., 2018. Highly thermally conductive flame retardant epoxy nanocomposites with multifunctional ionic liquid flame retardant-functionalized boron nitride nanosheets. *Journal of Materials Chemistry A*, *6*(41), pp.20500-20512. https://doi.org/10.1039/C8TA08008A.

[15] Han, N. M., Wang, Z., Shen, X., Wu, Y., Liu, X., Zheng, Q. and Kim, J. K., 2017, August. Effects of graphene sheet size on electrical conductivity and percolation of graphene aerogel/epoxy composites. In *21st International Conference on Composite Materials, ICCM* (Vol. 2017).

[16] Martin-Gallego, M., Verdejo, R., López-Manchado, M. A. and Sangermano, M., 2011. Epoxy-graphene UV-cured nanocomposites. *Polymer*, *52*(21), pp.4664-4669. https://doi.org/10.1016/j.polymer.2011.08.039.

[17] Sangermano, M., Calza, P. and Lopez-Manchado, M. A., 2013. Graphene oxide–epoxy hybrid material as innovative photocatalyst. *Journal of Materials Science*, *48*, pp.5204-5208. https://doi.org/10.1007/s10853-013-7308-7.

[18] Yu, L., Park, J. S., Lim, Y. S., Lee, C. S., Shin, K., Moon, H. J., Yang, C. M., Lee, Y. S. and Han, J. H., 2013. Carbon hybrid fillers composed of carbon nanotubes directly grown on graphene nanoplatelets for effective thermal conductivity in epoxy composites. *Nanotechnology*, *24*(15), p.155604. http://dx.doi.org/10.1088/0957-4484/24/15/155604.

[19] Hsu, C. H., Hsu, M. H., Chang, K. C., Lai, M. C., Liu, P. J., Chuang, T. L., Yeh, J. M. and Liu, W. R., 2014. Physical study of room-temperature-cured epoxy/thermally reduced graphene oxides with various contents of oxygen-containing groups. *Polymer international*, *63*(10), pp.1765-1770. https://doi.org/10.1002/pi.4763.

[20] Qu, L., Sui, Y., Zhang, C., Li, P., Dai, X., Xu, B. and Fang, D., 2020. POSS-functionalized graphene oxide hybrids with improved dispersive and smoke-suppressive properties for epoxy flame-retardant application. *European Polymer Journal*, *122*, p.109383. https://doi.org/10.1016/j.eurpolymj.2019.109383

[21] Chen, Q., Liu, L., Zhang, A., Wang, W., Wang, Z., Zhang, J., Feng, J., Huo, S., Zeng, X. and Song, P., 2023. An iron phenylphosphinate@ graphene oxide nanohybrid enabled flame-retardant, mechanically reinforced, and thermally conductive epoxy nanocomposites. *Chemical Engineering Journal*, *454*, p.140424. https://doi.org/10.1016/j.cej.2022.140424.

[22] Rao, W., Tao, J., Yang, F., Wu, T., Yu, C. and Zhao, H. B., 2023. Growth of copper organophosphate nanosheets on graphene oxide to improve fire safety and

mechanical strength of epoxy resins. *Chemosphere*, *311*, p.137047. https://doi.org/10.1016/j.chemosphere.2022.137047.

[23] Li, J., Wu, W., Duan, R., Bi, X., Meng, W., Xu, J. and Qu, H., 2023. Boron-containing ionic liquid functionalized Mo-MOF/graphene oxide hybrid for improving fire safety and maintaining mechanical properties for epoxy resin. *Applied Surface Science*, *611*, p.155736. https://doi.org/10.1016/j.apsusc.2022.155736.

[24] Wang, W., Wang, X., Pan, Y., Liew, K. M., Mohamed, O. A., Song, L. and Hu, Y., 2017. Synthesis of phosphorylated graphene oxide based multilayer coating: Self-assembly method and application for improving the fire safety of cotton fabrics. *Industrial & Engineering Chemistry Research*, *56*(23), pp.6664-6670. https://doi.org/10.1021/acs.iecr.7b01293.

[25] Bhattacharjee, S., Macintyre, C. R., Wen, X., Bahl, P., Kumar, U., Chughtai, A. A. and Joshi, R., 2020. Nanoparticles incorporated graphene-based durable cotton fabrics. *Carbon*, *166*, pp.148-163. https://doi.org/10.1016/j.carbon.2020.05.029.

[26] Zhi, M., Liu, Q., Chen, H., Chen, X., Feng, S. and He, Y., 2019. Thermal stability and flame retardancy properties of epoxy resin modified with functionalized graphene oxide containing phosphorus and silicon elements. *ACS omega*, 4(6), pp.10975-10984. https://doi.org/10.1021/acsomega.9b00852.

[27] Liu, S., Yan, H., Fang, Z. and Wang, H., 2014. Effect of graphene nanosheets on morphology, thermal stability and flame retardancy of epoxy resin. *Composites Science and Technology*, 90, pp.40-47. https://doi.org/10.1016/j.compscitech.2013.10.012.

[28] Yu, B., Shi, Y., Yuan, B., Qiu, S., Xing, W., Hu, W., Song, L., Lo, S. and Hu, Y., 2015. Enhanced thermal and flame retardant properties of flame-retardant-wrapped graphene/epoxy resin nanocomposites. *Journal of Materials Chemistry A*, *3*(15), pp.8034-8044. https://doi.org/10.1039/C4TA06613H.

[29] Qu, L., Sui, Y., Zhang, C., Li, P., Dai, X., Xu, B. and Fang, D., 2020. POSS-functionalized graphene oxide hybrids with improved dispersive and smoke-suppressive properties for epoxy flame-retardant application. *European Polymer Journal*, *122*, p.109383. https://doi.org/10.1016/j.eurpolymj.2019.109383.

[30] Qu, L., Sui, Y., Zhang, C., Li, P., Dai, X. and Xu, B., 2020. Compatible cyclophosphazene-functionalized graphene hybrids to improve flame retardancy for epoxy nanocomposites. *Reactive and Functional Polymers*, *155*, p.104697. https://doi.org/10.1016/j.reactfunctpolym.2020.104697.

[31] Q., Zhao, Y., Gao, S., Yang, X., Fan, R., Zhi, M. and Fu, M., 2021. Recent advances in the flame retardancy role of graphene and its derivatives in epoxy resin materials. *Composites Part A: Applied Science and Manufacturing*, 149, p.106539. https://doi.org/10.1016/j.compositesa.2021.106539.

[32] Ye, T. P., Liao, S. F., Zhang, Y., Chen, M. J., Xiao, Y., Liu, X. Y., Liu, Z. G. and Wang, D. Y., 2019. Cu (0) and Cu (II) decorated graphene hybrid on improving fireproof efficiency of intumescent flame-retardant epoxy resins. *Composites Part B: Engineering*, *175*, p.107189. https://doi.org/10.1016/j.compositesb.2019.107189.

[33] Wang, X., Kalali, E. N., Wan, J. T. and Wang, D. Y., 2017. Carbon-family materials for flame retardant polymeric materials. *Progress in Polymer Science*, *69*, pp.22-46. https://doi.org/10.1016/j.progpolymsci.2017.02.001.

[34] Tewari, C., Tatrari, G., Kumar, S., Pandey, S., Rana, A., Pal, M. and Sahoo, N. G., 2022. Green and cost-effective synthesis of 2D and 3D graphene-based nanomaterials from Drepanostachyum falcatum for bio-imaging and water purification applications. *Chemical Engineering Journal Advances*, *10*, p.100265. https://doi.org/10.1016/j.ceja.2022.100265.

[35] Pandey, S., Karakoti, M., Dhali, S., Karki, N., SanthiBhushan, B., Tewari, C., Rana, S., Srivastava, A., Melkani, A. B. and Sahoo, N. G., 2019. Bulk synthesis of graphene nanosheets from plastic waste: an invincible method of solid waste management for better tomorrow. *Waste management*, *88*, pp.48-55. https://doi.org/10.1016/j.wasman.2019.03.023.

[36] Tewari, C., SanthiBhushan, B., Srivastava, A. and Sahoo, N. G., 2021. Metal doped graphene oxide derived from Quercus ilex fruits for selective and visual detection of iron (III) in water: Experiment and theory. *Sustainable Chemistry and Pharmacy*, *21*, p.100436. https://doi.org/10.1016/j.scp.2021.100436.

[37] Ahmed, M., Giwa, A. and Hasan, S. W., 2019. Challenges and opportunities of graphene-based materials in current desalination and water purification technologies. *Nanoscale Materials in Water Purification*, pp.735-758. https://doi.org/10.1016/B978-0-12-813926-4.00033-1.

Chapter 8

Nanosheets for Supercapacitor Applications

Buddhodev Chowdhury[1,2]
Arnab Mukherjee[3]
Amrit Sahis[3]
Sudipta Pal[2]
and Debasis Dhak[3,*]
[1]Department of Physics, University of Kalyani, Kalyani, Nadia, WB, India
[2]Sponsored Teachers' Training College, Purulia, WB, India
[3]Department of Chemistry, Sidho-Kanho-Birsha University, Purulia, WB, India

Abstract

The present book chapter discusses various types of energy storage devices and comparisons among them. The versatile characteristics of the supercapacitor as an energy storage device differ from other energy storage devices. A supercapacitor's working principle is defined by its capacitance, energy, and power density. Electric double-layer capacitors (EDLC), pseudocapacitors, and hybrid capacitors are the energy storage mechanisms used by various types of supercapacitors, according to research. Nanosheets with various properties are used as electrodes in supercapacitors. Synthesis processes with various types of nanosheets and, the study of the surface morphology of different types of prepared nanosheets with the help of SEM images are discussed. A comparative study on the electrochemical properties of various types of nanosheets related to supercapacitor electrodes with the given results is reported in this chapter.

---

* Corresponding Author's Email: debasis.chem@skbu.ac.in.

In: Properties and Applications of Nanosheets
Editors: Nanda Gopal Sahoo and Anirban Dandapat
ISBN: 979-8-89113-381-5

Keywords: energy storage devices, supercapacitor, nanosheets, synthesis methods, surface morphology, electrochemical performances

1. Introduction

The primary sources of energy used today are fossil fuels, which are depleting gradually and will be depleted very soon [1–4]. Furthermore, fossil fuels emit harmful gases that pollute the environment, rendering it unfit for human habitation. On the other hand, expanding modern civilization necessitates a massive amount of energy. So, we need another source of energy that will be almost endless and environmentally friendly. Today, some usual energy sources, such as wind energy, solar energy, etc. [5–10], are engaged to reduce the deficiency of energy, but it is not sufficient because those types of energy can't always be used due to the absence of the source. The electrolytic capacitor and electrochemical battery are both used as energy storage devices, but they're not completely efficient. The capacity of an electrolytic capacitor is very low, for which it is not suitable as a large storage device, and though the conventional battery can store a huge amount of energy, nevertheless, it is not suitable where high power density is needed due to its low power density (1 kW/kg) [11–13]. In general, the energy density of electrochemical batteries is nearly 150 Wh/Kg, which is roughly 10 times greater than that of an electrolytic capacitor, whereas the power density is up to 200 W/Kg, which is roughly 20 times lower [14–16]. That's why some researchers are concentrating on finding one such energy storage device that can store a huge amount of energy with high power density and is also environmentally friendly. A newly invented energy storage device called a supercapacitor can solve this problem. It confines all the requirements that need to be in a modern civilization. In a supercapacitor, both the energy density and power density are comparatively high. Its energy density lies between that of an electrolytic capacitor and that of an electrochemical battery [17–20]. A supercapacitor has a higher specific capacitance than an electrolytic capacitor and a higher power than a conventional battery. As a result, it remains in the middle of the electrolytic capacitor and the conventional battery. It has many benefits, including a long cycle life (>100,000 cycles), fast charging and discharging, a wide range of working temperatures (-40 °C to 70 °C), low maintenance costs, and environmental friendliness [14, 20-30]. The use of supercapacitors is competent where mainly power bursts are required with independent energy storage capacity [4]. An analogy of different storage devices namely capacitor, supercapacitor, and battery is given in Table 1 along with an analogy of supercapacitor and batteries pertaining to different variable parameters such as storage mechanism, specific power, storage of energy, charging rate, life cycle, etc. are mentioned in Table 2.

Table 1. Analogy of said energy storage devices [8]

Features	Capacitors	Supercapacitors	Batteries
Specific energy (Wh/kg)	<0.10	1 to 10	10 to 100
Specific power (W kg $^{-1}$)	$\gg 10,000.00$	50 to 10,000	<1000.00
Discharging times	10^{-6} - 10^{-3}	From second to minute	0.3 to 3 h
Charging times	10^{-6}-10^{-3}	From second to minute	1 to 5 h
Efficiency (%)	Approx 100	85 to 98	70 to 85
Lifecycle	Close to infinite	$>500,000.00$	Approx 1000

Table 2. Analogy of batteries as well as supercapacitors [8]

Analogy variables	Batteries	Supercapacitors
Storage mechanism	Chemically	Physically
Specific power	Reaction kinetics, mass transport	Conductivity of electrolytes
Storage of energy	Exalted (bulk)	Confined surface area
Charging rate		Exalted, similar as discharging case
Lifecycle	Kinetically limited	Kinetically limited

2. Characteristics Measurement Parameter of a Supercapacitor

Two electrodes are separated by an ion-permeable membrane, such as tissue paper or filter paper, to form a supercapacitor. These electrodes are also associated with electrolytes. The electrode materials of the supercapacitor are carbon-like materials such as activated carbon, graphene, and carbon nanotubes for their large surface area and low cost, and transition metal oxide or conducting polymers for large storage. When an external voltage is applied to a supercapacitor, most of the charges accumulate on the exterior surface of the electrodes, while oppositely charged ions from the electrolyte accumulate near the electrodes, resulting in the formation of two capacitors near both electrodes. Figure 1 depicts the inner structure of a supercapacitor and two capacitors connected in series so that their equivalent capacitance can be measured using the equation below.

$$C_{total} = \frac{C_1 . C_2}{C_1 + C_2} \tag{1}$$

where C_1 and C_2 represent the capacitance of the electrodes separately. C_{total} denotes the total capacitance of the supercapacitor cell, and for symmetric supercapacitor electrodes, $C_1 = C_2$ then equivalent capacitance becomes $C = 0.5\ C_1$ which means the equivalent capacitance of the supercapacitor cell is equal to half of their individual capacitor [31].

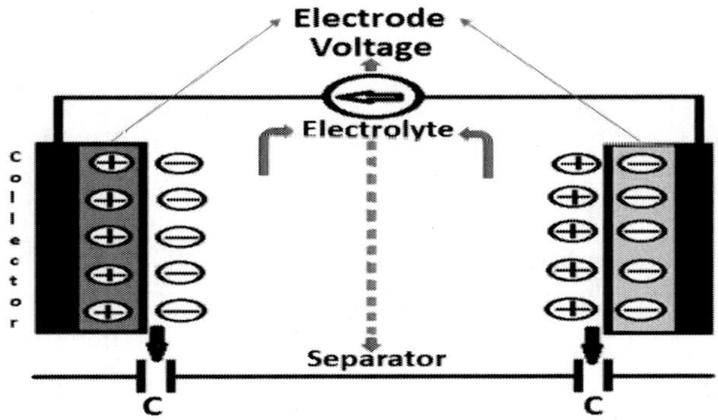

Figure 1. The inner structure of a supercapacitor.

The energy density of the supercapacitor is measured with the help of the equation given below.

$$E = \frac{1}{2} CV^2 \tag{2}$$

where E, C, and V denote the energy density, capacitance, and voltage of the supercapacitor, respectively. The equation (2) shows the energy density of the supercapacitor is proportional to capacitance [32-34].

Another significant parameter is called power density, which is measured by equation (3).

$$P_{max} = \frac{V^2}{4R} \tag{3}$$

where R denotes the equivalent series resistance (ESR) in Ω [35]. The equation (3) shows the power density is inversely proportional to internal resistance. So, the power density of a supercapacitor will be higher when internal resistance is lower [32-34]. The capacitance of a supercapacitor is measured by the equation (4).

$$C = \frac{\epsilon_0 \epsilon_r A}{d} \tag{4}$$

where ε_o and ε_r represent the dielectric constants of free space and the used material, respectively, A represents the electrode surface area, and d represents the space between two electrodes. Equation (4) indicates that the capacitance of the supercapacitor will be larger when the surface area of the supercapacitor electrode is larger and the distance between the two electrodes is lower [34].

3. Types of Supercapacitors

Usually, two types of supercapacitors serve as their energy storage mechanisms: the electrical double layer capacitor (EDLC) and the pseudocapacitor (Figure 2). If both mechanisms are used in a single electrode or two such diverse types of electrodes are used in a supercapacitor, that type of supercapacitor is called a hybrid supercapacitor [8].

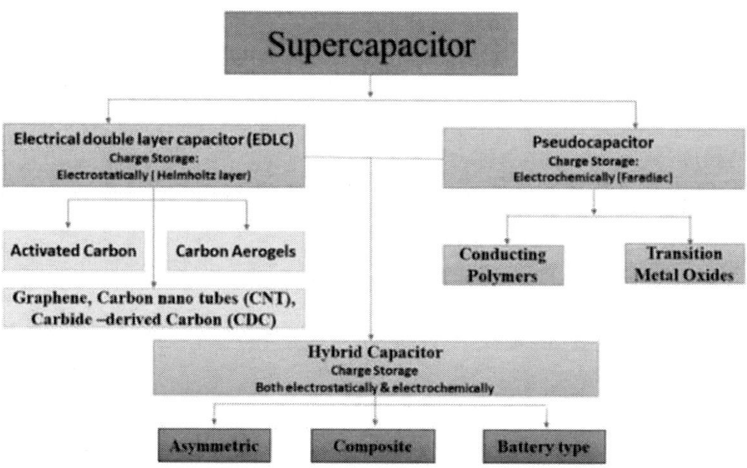

Figure 2. Types of supercapacitors as energy storage mechanisms.

3.1. Electrical Double Layer Capacitor (EDLC)

In this type of supercapacitor, two layers are grown on the electrode-electrolyte interface, such that one layer creates the outer surface of the electrode and another oppositely charged layer is in the electrolyte, and both layers are isolated by the solvent molecules that have dielectric properties such that in this region they behave like a capacitor. The same phenomenon occurs on another electrode of the EDLC supercapacitor. As a result, two capacitors are formed on adjacent electrodes of a single EDLC supercapacitor. Here carbon-like materials are used, for example, graphene, activated carbon, carbon nanotubes (CNT), etc., (Figure 2) for their large surface area, chemical stability, availability, and low cost. It accumulates energy completely electrostatically [35-40].

3.2. Pseudocapacitor

Pseudocapacitors are generated by the transfer of electron charge within the electrolyte to the electrode, which comes from desolvated and adsorption ions. This phenomenon is due to only the charge transfer between the electrolyte and electrode, called the Faradic redox reaction, and not the chemical reaction of adsorbed ions with the electrode atoms. Naturally, during the charging period, one electrode undergoes an oxidation reaction while another undergoes a reduction reaction; the opposite phenomenon occurs during the discharging period. In this electrode, transition metal oxides are used such as manganese dioxide (MnO_2), nickel oxide (NiO), and conducting polymers such as polyaniline, etc. These types of supercapacitors store energy by electrochemically, and their specific capacitance is higher than that of an EDLC [36-38, 40].

3.3. Hybrid Capacitor

Hybrid capacitors are those that combine both mechanisms, such as an electric double-layer capacitor (EDLC) and a pseudocapacitor in a supercapacitor as shown in Figure 3. It may occur when one electrode of a supercapacitor works on the electrical double layer capacitor (EDLC) mechanism and another electrode works as a pseudocapacitor, or when both the mechanisms are included in a single electrode of a supercapacitor. These types of supercapacitors are more prominent because they have a large specific capacitance for a pseudocapacitor as well as excellent cyclic stability and power execution due to the electrical double layer capacitor (EDLC). These are also three types, such as composite, asymmetric, and battery-type, as their electrode variations [14, 35].

3.4. Composite Electrode

Composite electrodes are built with a combination of carbon-like elements such as activated carbon, graphene, carbon nanotubes (CNTs), transition metal oxides such as manganese dioxide (MnO_2), nickel oxide (NiO), etc., or conducting polymers like polyaniline. So here both the energy storage mechanism of the EDLC as well as the pseudocapacitor are involved. This type of electrode has excellent features that make it different from the others. Here carbon-based materials display capacitance due to electrical double layer capacitor behavior, and that is huge for their large surface area. At the same time, pseudocapacitors are generated from transition metal oxides or conducting polymers, and that capacitance is also huge for the Faradaic redox reaction due to electron charge transfer from electrolyte to electrode [14].

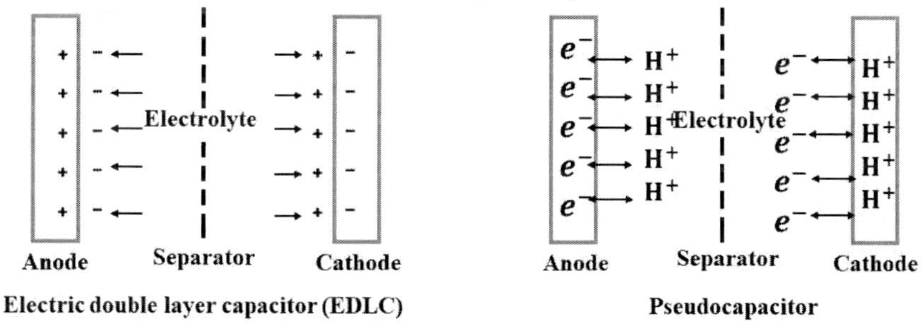

Figure 3. Schematic diagram of three different types of capacitance generation mechanism [41]. Reproduced after modification.

4. Nanosheets

Like graphene, nanosheets are two-dimensional polymers whose thickness is measured in nanoscale units (1-100 nm). Nanosheet has several valuable properties like large and stable carrier mobilities, vast areas of light absorption, an effective spin injector, excellent thermal conductivity, remarkable mechanical strength, and tunable functionality, which is why it has several applications such as in electronics, supercapacitors, solar cells, filter elements, etc. Most of the work has been done on transition metal dichalcogenide (TMD). Recently, studies of some other types of nanosheets, such as metal oxide, metal hydroxide, boron nitrides, and clay types, which are generated from multi-layered crystalline producer materials, have been called "top-down" nanosheets. Nanosheets are divided into two types based on their synthesis process: top-down nanosheets and bottom-up nanosheets. The bottom-up nanosheets are directly generated from atomic, molecular, or ionic components. It was a very old concept and is used now mainly for synthetic and analytical purposes [42].

4.1. Characteristics of Nanosheets

Nanosheets are excellent anisotropic two-dimensional crystallites in mono- or multilayers as shown in Figure 4. Table 3 represents several types of nanosheets displaying remarkable electrochemical properties, like layered metal oxides, layered metal chalcogenides, layered metal hydroxides, metal carbides, and metal-complex polymers. Inorganic 2D nanosheets display a large surface area, which is why the contact area with the electrolyte becomes larger and more suitable as an electrode for a supercapacitor due to their effective electrical double layer capacitance (EDLC). In the future, the execution of this nanosheet can be improved by producing large ions at the electrode-electrolyte interface via larger ion diffusion. Another important property of 2D inorganic nanosheets is the surface exposure of massive redox ions with a tunable construction, which is helpful for the pseudocapacitive mechanism. Despite their numerous advantages, 2D inorganic nanosheets are not widely used as supercapacitor electrodes due to their low electrical conductivity [42].

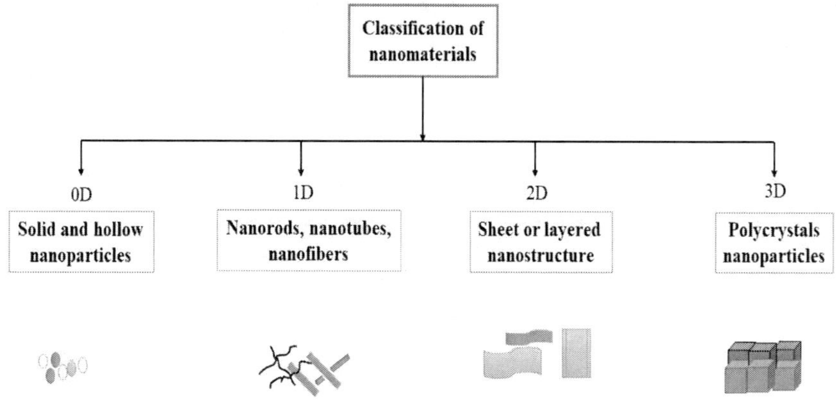

Figure 4 Several inorganic 2D nanosheets emerged as composite elements [42]. Reproduced after modification.

Table 3. Different types of nanosheets usage in supercapacitor electrodes [42]

element	Components	Reference
Metal oxides	MnO_2, RuO_2, h-MoO_3, V_2O_5, Bi_2MoO_6, NiO, Co_3O_4, Fe_2O_3, $Na_2Ti_3O_7$, $Ni_xCo_{3-x}O_4$, Cu_xO, $ZnCo_2O_4$, $NiCo_2O_4$, $MgCo_2O_4$	[42]
Metal hydroxide	Ni–Mn–LDH, Ni–Co–LDH, NiCl(OH), $Ni(OH)_2$	[42]
Metal chalcogenide	MoS_2, $MoSe_2$, VS_2, $Cu_{1-x}Ni_xS$, Bi_2Se_3	[42]
Metal carbide	Ti_3C_2Tx	[42]
Metal complex polymer	CONASH	[40]

5. Synthesis Process of Various Nanosheets for Supercapacitor Electrodes

The mechanisms of various types of nanosheets, such as carbon-related nanosheets, metal oxide nanosheets (NCO), metal sulphide nanosheets (CuS), and others, are discussed in this section.

5.1. Preparation of Carbon Based Nanosheets

To make oxidized asphaltene, start by oxidizing 20 g of asphaltene in 200 ml (10 mol/L) of HNO_3 for 12 hours at 80 °C. The prepared yellow suspension was filtered through a microporous membrane and then leached with water until a neutral sample was formed. Then, after 12 hours of drying at 100 °C, it becomes ASP(O). Carbon nanosheet synthesis process: dispersion of prepared ASP(O) in urea solution with the ratio of H_2O: ASP(O): urea = 40:20:40 wt% and stirring it at 25 °C for 1 hour, then drying it overnight in an oven with temperature at 75 °C, heating it again at 540 °C for 3 hours under nitrogen atmosphere after cooling the collected black powder sample indicates as ASP(O)-U. Finally, the prepared sample was mixed with KOH in the proportions of 1:1, 1:1.2, and 1:2 w/w and heated at 800 °C at a rate of 5 °C per minute before being carbonized in a nitrogen environment for 1 hour. After completing the heating process, wash it with 1 mol/L HCl and deionized water to eliminate the alkali metal residue. Finally, ACNs-x is prepared, where x denotes the KOH/ASP(O)-U ratio [43]. The stepwise process is depicted in Figure 5 [43].

5.2. Synthesis of Porous Nickel–Cobalt Oxide (NCO) Nanosheets

The entire chemical is analytically graded before being used directly in impure conditions. Nanosheets of NCO are synthesized with substrate independent electrochemical layout in a general three electrode system. Among the three electrodes, an F-doped SnO_2 coated glass sheet with a resistance of 14 Ω is introduced as the working electrode, a 4 cm^2 graphite rod works as the counter electrode, and the reference electrode is a saturated calomel electrode (SCE). Anode electrodeposition was carried out with a constant current of 1.67 mA/cm^2 in solutions containing xM $Co(NO_3)_2$, (0.02 -x) $Ni(NO_3)_2$ and 0.1 M NH_4NO_3. The generated NCO sample indicates NCO-Z, where Z denotes the Co/Ni ratio in the solution. Following the electrodeposition, the FTO substrate is sanctified with ultrasonication while still immersed in water. Ethyl alcohol, ethanol, acetone, and finally distilled water are all used. Whole electrodepositions are run at 70 °C for 20 minutes before being annealed at 200 °C for 3 hours. Finally, the NCO nanosheet is generated. Schematic description of species transportation on nanosheets of porous NCO in a solution of 1 M KOH is shown in Figure 6 [44].

Figure 5. Schematic diagram of synthesis process of as phaltene related carbon porous nanosheet [43]. Reproduced after modification.

5.3 Synthesis of CuS Nanosheets

These types of nanosheets are synthesized by chemical de-alloying methods of $Ti_{70}Cu_{30}$ an amorphous alloy kept in H_2SO_4 solution. Melt spinning processes are used for the synthesis of the Ti-Cu amorphous alloy. This alloy ribbon has dimensions of 20 mm × 1.5 mm × 25 m and is dissolved in various H_2SO_4 solutions. After a 3-hour reaction at 363 K, the samples are purified several times with water and alcohol before being dried in a vacuum oven for 10 hours. In this way, CuS materials are generated. A schematic description of the mechanism of growth on CuS materials is shown in Figure 7 [45].

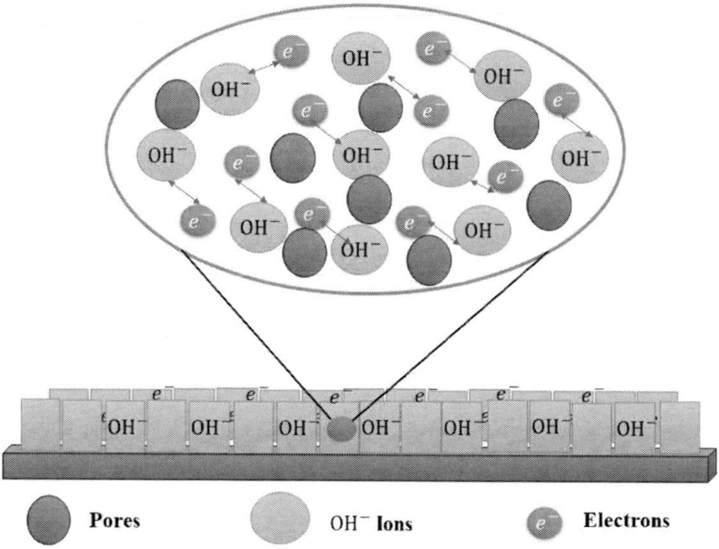

Figure 6. Schematic description of species transportation on nanosheets of porous NCO in a solution of 1 M KOH [44]. Reproduced after modification.

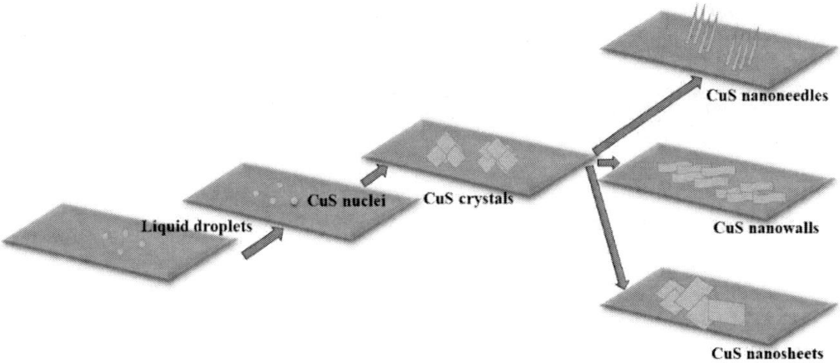

Figure 7. Schematic description of the mechanism of growth on CuS materials [45]. Reproduced after modification.

5.4. MoS_2 Nanosheet Synthesis

A simple hydrothermal method is used for the synthesis of MoS_2 nanosheets. Initially, Na_2MoO_4, $2H_2O$ is dissolved in distilled water and ultrasonicated, followed by the addition of 0.1 M NaOH to achieve a pH of 6.5, followed by the addition of L-cysteine by Na_2MoO_4, 2 H_2O according to the L-cysteine:1:3 ratio. After 5 minutes of stirring, place this mixture in a 100 ml teflon-lined autoclave for 48 hours at 180 °C. The collected sample is washed several times with water and ethanol before drying for 12 hours in a vacuum oven at 80 °C, shown in Figure 8 [46].

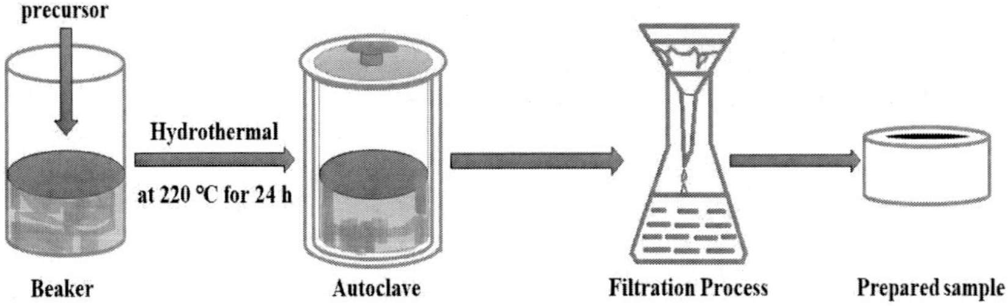

Figure 8. Schematic description of the synthesis process of MoS2 nanosheets [46]. Reproduced after modification.

6. Comparative Study of the Surface Morphologies of Prepared Nanosheets

SEM and TEM were used to examine the surface morphologies of ACNs-x, ASP(O)-U, ASP(O)-KOH, and ASP-KOH. The images of ASP(O)-U, ACNs-1, ACNs-1.2 [43], and ACNs-2, with mean thicknesses of about 40 to 60 nm. The images of NCO-2; we find a smaller thickness of 20 to 40 nm and almost the same surface morphology even after heating it at 200 °C for 3 hours while keeping it in air [44]. To calculate the surface morphology of CuS nanosheet samples, use SEM. The morphology depends on the H_2SO_4 concentration added in the synthesis process [45]. CuS has a surface morphology of about 50 nm. SEM instruments are used to recognize the surface morphology. The images of nanosheets of MoS_2 are a flower-like, identical structures were also found with an average thickness below 200 nm [46].

7. Results and Discussions

Discuss the various types of nanosheets separately according to CV, and GCD data, the characteristics information is elaborated below.

7.1. Electrochemical Features of ACNs-x with KOH Dissolution

In this study, ACNs-x has been successfully synthesized with a large pore structure and also successfully synthesized into graphite-like ribbons using the precursor asphaltene and the sheet structured agency of urea. ACNs-x displays prior electrochemical characteristics such as highly specific capacitance with 316 F/g at 1 A/g and 283 F/g at 100 A/g using a three-electrode scheme. When it is calculated using the two-electrode scheme in KOH electrolyte, the specific capacitance becomes 217 F/g at 1 A/g, and it also retains approximately 90% of the beginning capacitance in later cycles of 10000 with 4 A/g, indicating excellent cycle stability [43].

7.2. Electrochemical Property of NCO and NCO-2

Analyzing all the curves of electrochemical measurements of NCO developed on FTO substrates with natural template-free electrochemical deposit it was observed to be high-performing supercapacitor electrodes with specific capacitance 453 F/g at the scan rate of 5 mV/s, 506 F/g at 1 A/g and long-life cycle for their large surface area. On the other side when use both electrodes in a single supercapacitor like symmetric supercapacitor (NCO-2) displays a specific capacitance 89.2 F/g at 0.7A/g. Therefore, the compassion of two supercapacitor electrodes showed that the NCO was better than NCO-2 [44].

7.3. Electrochemical Property of CuS Nanosheet

CuS material was found to show that the prepared electrode has controlled a good morphology also excellent electrochemical performance for using in a supercapacitor. The spherical cluster displayed a high electrochemical execution with capacitance of 276 F/g with scan rate of 5mV/s and the retention of initial capacitance was 73% and 61% after 1000 and 2000 cycle respectively with current density 2 A/g [45].

7.4. Electrochemical Property of MoS_2 Nanosheet

In this paper MoS_2nanosheet wasgenerated by a common hydrothermal process. The structure of prepared materialswas net-like internally connected by nanoflakes and small thickness like nanometers. It displayed a high capacitance with 129.2 F/g with longer life cycle, capacitance retention 85.1% after 500 cycles. MoS_2 nanosheet has a large surface area and for this reason it displays good electrochemical performance [46].

Conclusion and Future Prospects

The energy density and the power density of supercapacitors are comparatively higher than normal capacitors and batteries. The supercapacitor has also got the advantages of long life cycle, fast charging and discharging with long range working temperature (-40 °C to 70 °C), low maintenance cost and environmentally friendliness. Hybrid supercapacitors consisting of electrical double layer capacitor and pseudocapacitors have got superiority over other supercapacitors. In this chapter, analyzing all the aspects it can be concluded that all the nanosheets are successfully synthesized and all are dimensionally near about similar range such as from 20nm to 200 nm approximately. The surface morphology depends upon some parameter like concentration synthesis agents, synthesis process, etc. ACNs-x nanosheet related electrode displays specific capacitance with 316 F/g at 1 A/g, excellent lifecycle with 90% retain after 10000 cycles. NCO nanosheet electrode exhibits high specific capacitance with 453 F/g at 5 mV/s and display long lifecycle for its large surface area. CuS nanosheet electrodes exhibit large capacitance with 276 F/g at scan rate 5 mV/s, 73% retention after 1000 cycles. MoS_2 nanosheet related electrode displays specific capacitance with 129.2 F/g at 5 mV/s and long cycle life, 85.1% retention after 500 cycles. So, after observing electrochemical execution of all the electrodes says that NCO shows highest specific capacitance among others and ACNs-x shows excellent cycle stability. The nanosheet electrode can be another better option for high performing supercapacitor electrode.

References

[1] J. Xu, J. Li, Q. Yang, Y. Xiong, and C. Chen, "*In-situ* Synthesis of MnO2@Graphdiyne Oxides Nanocomposite with Enhanced Performance of Supercapacitors," *Electrochim. Acta*, vol. 251, pp. 672–680, Oct. 2017. https://doi.org/10.1021/acs.chemrev.8b00288.

[2] Z. Bai, Hongji Li, Mingji Li, Cuiping Li, Xufei Wang, Changqing Qu, Baohe Yang, "Flexible carbon nanotubes-MnO2/reduced graphene oxide-polyvinylidene fluoride films for supercapacitor electrodes," *Int. J. Hydrogen Energy*, vol. 40, no. 46, pp. 16306–16315, Dec. 2015. https://doi.org/10.1016/j.ijhydene.2015.09.065.

[3] Z. Li, W. Zhang, H. Wang, and B. Yang, "Two-dimensional perovskite LaNiO3 nanosheets with hierarchical porous structure for high-rate capacitive energy storage," *Electrochim. Acta*, vol. 258, pp. 561–570, Dec. 2017. http://dx.doi.org/10.1016/j.electacta.2017.11.099.

[4] T. C. Nirmale, B. B. Kale, and A. J. Varma, "A review on cellulose and lignin based binders and electrodes: Small steps towards a sustainable lithium ion battery," *Int. J. Biol. Macromol.*, vol. 103, pp. 1032–1043, 2017, doi: https://doi.org/10.1016/j.ijbiomac.2017.05.155.

[5] B. P. Prasanna, Avadhani D N, Raghu M S, Yogesh Kumar K, "Synthesis of polyaniline/α-Fe2O3 nanocomposite electrode material for supercapacitor

applications," *Mater. Today Commun.*, vol. 12, pp. 72–78, 2017, http://dx.doi.org/doi:10.1016/j.mtcomm.2017.07.002.

[6] Q. Meng, K. Cai, Y. Chen, and L. Chen, "Research progress on conducting polymer based supercapacitor electrode materials," *Nano Energy*, vol. 36, pp. 268–285, 2017, doi: https://doi.org/10.1016/j.nanoen.2017.04.040.

[7] K. Song, H. Ni, and L. Z. Fan, "Flexible Graphene-Based Composite Films for Supercapacitors with Tunable Areal Capacitance," *Electrochim. Acta*, vol. 235, pp. 233–241, May 2017, doi: https://doi.org/10.1016/j.electacta.2017.03.065.

[8] A. González, E. Goikolea, J. A. Barrena, and R. Mysyk, "Review on supercapacitors: Technologies and materials," *Renewable and Sustainable Energy Reviews*, vol. 58. Elsevier Ltd, pp. 1189–1206, May 01, 2016. doi: https://doi.org/10.1016/j.rser.2015.12.249.

[9] N. R. Chodankar, S. H. Ji, and D. H. Kim, "Low-cost superior symmetric solid-state supercapacitors based on MWCNTs/MnO2 nanocomposite thin film," *J. Taiwan Inst. Chem. Eng.*, vol. 80, pp. 503–510, Nov. 2017, doi: https://doi.org/10.1016/j.jtice.2017.08.014.

[10] R. Rajagopal and K. S. Ryu, "Influence of rare earth elements on porosity controlled synthesis of MnO2 nanostructures for supercapacitor applications," *Electrochim. Acta*, vol. 265, pp. 532–546, Mar. 2018, doi: https://doi.org/10.1016/j.electacta.2018.01.161.

[11] A. Jain and S. K. Tripathi, "Experimental studies on high-performance supercapacitor based on nanogel polymer electrolyte with treated activated charcoal," *Ionics (Kiel).*, vol. 19, no. 3, pp. 549–557, 2013, doi: https://doi.org/10.1007/s11581-012-0782-0.

[12] K. Xu, J. Yang, and J. Hu, "Synthesis of hollow NiCo2O4 nanospheres with large specific surface area for asymmetric supercapacitors," *J. Colloid Interface Sci.*, vol. 511, pp. 456–462, 2018, doi: https://doi.org/10.1016/j.jcis.2017.09.113.

[13] R. Jose, S. G. Krishnan, B. Vidyadharan, I. I. Misnon, M. Harilal, R. A. Aziz, J. Ismail, M. M. Yusoff, "Supercapacitor Electrodes Delivering High Energy and Power Densities," in *Materials Today: Proceedings*, Elsevier Ltd, 2016, pp. S48–S56. doi: https://doi.org/10.1016/j.matpr.2016.01.007.

[14] Z. S. Iro, C. Subramani, and S. S. Dash, "A brief review on electrode materials for supercapacitor," *International Journal of Electrochemical Science*, vol. 11, no. 12. Electrochemical Science Group, pp. 10628–10643, 2016. doi: https://doi.org/10.20964/2016.12.50.

[15] A. Olad and H. Gharekhani, "Preparation and electrochemical investigation of the polyaniline/activated carbon nanocomposite for supercapacitor applications," *Prog. Org. Coatings*, vol. 81, pp. 19–26, 2015, doi: https://doi.org/10.1016/j.porgcoat.2014.12.009.

[16] M. S. Yadav and S. K. Tripathi, "Synthesis and characterization of nanocomposite NiO/activated charcoal electrodes for supercapacitor application," *Ionics (Kiel).*, vol. 23, no. 10, pp. 2919–2930, Oct. 2017, doi: https://doi.org/10.1007/s11581-017-2026-9.

[17] S. Faraji and F. N. Ani, "The development supercapacitor from activated carbon by electroless plating - A review," *Renew. Sustain. Energy Rev.*, vol. 42, pp. 823–834, 2015, doi: https://doi.org/10.1016/j.rser.2014.10.068.

[18] T. M. Masikhwa, J. K. Dangbegnon, A. Bello, M. J. Madito, D. Momodu, and N. Manyala, "Preparation and electrochemical investigation of the cobalt hydroxide carbonate/activated carbon nanocomposite for supercapacitor applications," *J. Phys. Chem. Solids*, vol. 88, pp. 60–67, Oct. 2015, doi: https://doi.org/10.1016/j.jpcs.2015.09.015.

[19] Z. A. Elsiddig, Hui Xu, Dan Wang, Wei Zhang, Xinli Guo, Yao Zhang, Zhengming Sun, Jian Chen., "Modulating Mn4+ Ions and Oxygen Vacancies in Nonstoichiometric LaMnO3 Perovskite by a Facile Sol-Gel Method as High-Performance Supercapacitor Electrodes," *Electrochim. Acta*, vol. 253, pp. 422–429, Nov. 2017, doi: https://doi.org/10.1016/j.electacta.2017.09.076.

[20] Y. Li, Bing Guan, Aimee Maclennan, Yongfeng Hu, Dandan Li, Jing Zhao, Yaqiong Wang, Huaihao Zhang, "Porous waxberry-like MnO2/La2O3 microspheres for high performance asymmetric supercapacitor," *Electrochim. Acta*, vol. 241, pp. 395–405, Jul. 2017, doi: https://doi.org/10.1016/j.electacta.2017.04.175.

[21] Y. Xiong Min Zhou Hao Chen Lei Feng Zhao Wang Xinzhu Yan Shiyou Guan, "Synthesis of honeycomb MnO 2 nanospheres/carbon nanoparticles/graphene composites as electrode materials for supercapacitors," *Appl. Surf. Sci.*, vol. 357, pp. 1024–1030, 2015, doi: https://doi.org/10.1016/j.apsusc.2015.09.111.

[22] J. Dong, Z. Wang, and X. Kang, "The synthesis of graphene/PVDF composite binder and its application in high performance MnO2 supercapacitors," *Colloids Surfaces A Physicochem. Eng. Asp.*, vol. 489, pp. 282–288, Jan. 2016, doi: https://doi.org/10.1016/j.colsurfa.2015.10.060.

[23] X. Li, Z. Wang, Y. Qiu, Q. Pan, and P. Hu, "3D graphene/ZnO nanorods composite networks as supercapacitor electrodes," *J. Alloys Compd.*, vol. 620, pp. 31–37, Jan. 2015, doi: https://doi.org/10.1016/j.jallcom.2014.09.105.

[24] F. Wang, G. Li, Q. Zhou, J. Zheng, C. Yang, and Q. Wang, "One-step hydrothermal synthesis of sandwich-type NiCo 2 S 4 @reduced graphene oxide composite as active electrode material for supercapacitors," *Appl. Surf. Sci.*, vol. 425, pp. 180–187, Dec. 2017, doi: https://doi.org/10.1016/j.apsusc.2017.07.016.

[25] X. Duan, J. Deng, X. Wang, and P. Liu, "Preparation of rGO/G/PANI ternary nanocomposites as high performance electrode materials for supercapacitors with spent battery powder as raw material," *Mater. Des.*, vol. 129, pp. 135–142, 2017, doi: https://doi.org/10.1016/j.matdes.2017.05.031.

[26] B. K. Saikia, S. M. Benoy, M. Bora, J. Tamuly, M. Pandey, and D. Bhattacharya, "A brief review on supercapacitor energy storage devices and utilization of natural carbon resources as their electrode materials," *Fuel*, vol. 282, no. July, p. 118796, 2020, doi: https://doi.org/10.1016/j.fuel.2020.118796.

[27] P. Liu, Jiang Liu, Shuang Cheng, Weizi Cai, Fangyong Yu, Yapeng Zhang, Peng Wu, Meilin Liu, "A high-performance electrode for supercapacitors: Silver nanoparticles grown on a porous perovskite-type material La0.7Sr0.3CoO3−Δ substrate," *Chem. Eng. J.*, vol. 328, pp. 1–10, 2017, doi: https://doi.org/10.1016/j.cej.2017.06.150.

[28] Y. Bi, A. Nautiyal, H. Zhang, J. Luo, and X. Zhang, "One-pot microwave synthesis of NiO/MnO2 composite as a high-performance electrode material for supercapacitors," *Electrochim. Acta*, vol. 260, pp. 952–958, Jan. 2018, doi: https://doi.org/10.1016/j.electacta.2017.12.074.

[29] L. Zhang, Tiehu Li, Xianglin Ji, Zhiyong Zhang, Wenbo Yang, Junjie Gao, Hao Li, Chuanyin Xiong, Alei Dang, "Freestanding three-dimensional reduced graphene oxide/MnO2 on porous carbon/nickel foam as a designed hierarchical multihole supercapacitor electrode," *Electrochim. Acta*, vol. 252, pp. 306–314, Oct. 2017, doi: https://doi.org/10.1016/j.electacta.2017.08.087.

[30] X. Liu, G. Du, J. Zhu, Z. Zeng, and X. Zhu, "NiO/LaNiO 3 film electrode with binder-free for high performance supercapacitor," *Appl. Surf. Sci.*, vol. 384, pp. 92–98, Oct. 2016, doi: https://doi.org/10.1016/j.apsusc.2016.05.005.

[31] Q. Tang, M. Sun, S. Yu, and G. Wang, "Preparation and supercapacitance performance of manganese oxide nanosheets/graphene/carbon nanotubes ternary composite film," *Electrochim. Acta*, vol. 125, pp. 488–496, Apr. 2014, doi: https://doi.org/10.1016/j.electacta.2014.01.139.

[32] A. Yadav, A., V. C. Lokhande, R. N. Bulakhe, and C. D. Lokhande, "Amperometric CO2 gas sensor based on interconnected web-like nanoparticles of La2O3 synthesized by ultrasonic spray pyrolysis," *Microchim. Acta*, vol. 184, no. 10, pp. 3713–3720, Oct. 2017, doi: https://doi.org/10.1007/s00604-017-2364-3.

[33] G. P. Pandey, Y. Kumar, and S. A. Hashmi, "Ionic liquid incorporated polymer electrolytes for supercapacitor application," *Indian J. Chem. - Sect. A Inorganic, Phys. Theor. Anal. Chem.*, vol. 49, no. 5–6, pp. 743–751, 2010 http://dx.doi.org/10.1021/acssuschemeng.6b00585.

[34] A. M. Abioye, Z. A. Noorden, and F. N. Ani, "Synthesis and Characterizations of Electroless Oil Palm Shell Based-Activated Carbon/Nickel Oxide Nanocomposite Electrodes for Supercapacitor Applications," *Electrochim. Acta*, vol. 225, pp. 493–502, Jan. 2017, doi: https://doi.org/10.1016/j.electacta.2016.12.101.

[35] R. S. Kate, S. A. Khalate, and R. J. Deokate, "Overview of nanostructured metal oxides and pure nickel oxide (NiO) electrodes for supercapacitors: A review," *Journal of Alloys and Compounds*, vol. 734. Elsevier Ltd, pp. 89–111, Feb.

[36] Q. Wang, J. L. Li, F. Gao, W. S. Li, K. Z. Wu, and X. D. Wang, "Activated carbon coated with polyaniline as an electrode material in supercapacitors," *Xinxing Tan Cailiao/New Carbon Mater.*, vol. 23, no. 3, pp. 275–280, 2008, doi: https://doi.org/10.1016/s1872-5805(08)60030-x.

[37] S. Khamlich, T. Mokrani, M. S. Dhlamini, B. M. Mothudi, and M. Maaza, "Microwave synthesis of simonkolleite nanoplatelets on 3D nickel foam-graphene for supercapacitor applications," *Energy Procedia*, vol. 88, pp. 614–618, 2016, doi: https://doi.org/10.1016/j.egypro.2016.06.087.

[38] A. Kafy, A. Akther, L. Zhai, H. C. Kim, and J. Kim, "Porous cellulose/graphene oxide nanocomposite as flexible and renewable electrode material for supercapacitor," *Synth. Met.*, vol. 223, pp. 94–100, Jan. 2017, doi: https://doi.org/10.1016/j.synthmet.2016.12.010.

[39] F. Markoulidis, C. Lei, C. Lekakou, D. Duff, S. Khalil, B. Martorana, I. Cannavaro, "A method to increase the energy density of supercapacitor cells by the addition of

multiwall carbon nanotubes into activated carbon electrodes," *Carbon N. Y.*, vol. 68, pp. 58–66, Mar. 2014, doi: https://doi.org/10.1016/j.carbon.2013.08.040.

[40] K. S. Lee, M. S. Park, and J. D. Kim, "Nitrogen doped activated carbon with nickel oxide for high specific capacitance as supercapacitor electrodes," *Colloids Surfaces A Physicochem. Eng. Asp.*, vol. 533, pp. 323–329, Nov. 2017, doi: https://doi.org/10.1016/j.colsurfa.2017.09.008.

[41] J. J. C. Mancera, R. Rengel, A. Z. Pérez, F. S. Manzano, E. López, and J. M. Andújar, "Hybrid Supercapacitors as a Promising Alternative for Hybrid Electric Vehicles Fueling," *Proc. Fom. la Cult. Científica, Tecnológica y Innovación en Ciudad. Intel. (ScienCity 2020), Huelva, Spain*, no. January 2021, pp. 17–19, 2020, https://www.researchgate.net/publication/348716472.

[42] S. M. Oh and S. J. Hwang, "Recent advances in two-dimensional inorganic nanosheet-based supercapacitor electrodes," *J. Korean Ceram. Soc.*, vol. 57, no. 2, pp. 119–134, 2020, doi: https://doi.org/10.1007/s43207-020-00023-2.

[43] F. Qin, X. Tian, Z. Guo, and W. Shen, "Asphaltene-Based Porous Carbon Nanosheet as Electrode for Supercapacitor," *ACS Sustain. Chem. Eng.*, vol. 6, no. 11, pp. 15708–15719, Nov. 2018, doi: https://doi.org/10.1021/acssuschemeng.8b04227.

[44] X. Lu, Xi Huang, Shilei Xie, Teng Zhai, Chengsheng Wang, Peng Zhang, Minghao Yu, Wei Li, Chaolun Liang and Yexiang Tong, "Controllable synthesis of porous nickel-cobalt oxide nanosheets for supercapacitors," *J. Mater. Chem.*, vol. 22, no. 26, pp. 13357–13364, 2012, doi: https://doi.org/10.1039/c2jm30927k.

[45] W. Xu Yanqin Liang Yungao Su Shengli Zhu Zhenduo Cui Xianjin Yang Akihisa Inoue Qiang Wei Chunyong Liang, "Synthesis and properties of morphology controllable copper sulphide nanosheets for supercapacitor application," *Electrochim. Acta*, vol. 211, pp. 891–899, 2016, doi: https://doi.org/10.1016/j.electacta.2016.06.118.

[46] K. J. Huang, J. Z. Zhang, G. W. Shi, and Y. M. Liu, "Hydrothermal synthesis of molybdenum disulfide nanosheets as supercapacitors electrode material," *Electrochim. Acta*, vol. 132, pp. 397–403, 2014, doi: https://doi.org/10.1016/j.electacta.2014.04.007.

Chapter 9

Nanosheets for Biomedical Applications

Suresh Kumar[1]
Megha Pant[1]
and Anirban Dandapat[2,*]

[1]Department of Biotechnology, Bhimtal, Kumaun University, Nainital, Uttarakhand, India
[2]University School of Automation and Robotics,
Guru Gobind Singh Indraprastha University, East Delhi Campus,
Surajmal Vihar, Delhi, India

Abstract

Nanosheets have drawn a great deal of interest to the scientific community due to their advanced physiochemical properties and their uses in different biomedical field, such as, targeted drug delivery, pharmaceutical, bioengineering, bio-imaging, biosensors, disease identification and diagnosis. Two-dimensional nanosheets have many promising features including anisotropic 2D nanocrystalline structures with high surface area, and atomic level thickness, which enable them to be used as unique materials suitable for various applications. The different types of nanosheets are used for the efficient delivery of various anticancer chemotherapeutics and biological drug carriers. Construction of 2D-nanomaterials acting as efficient and non-invasive biomedical applications is still an imperative matter. This chapter unveils the biomedical applications of different 2D nanosheets in various biomedical applications including anticancer, drug delivery, imaging, bone healing and antimicrobial potentials.

* Corresponding Author's Email: anirban.usar@ipu.ac.in.

In: Properties and Applications of Nanosheets
Editors: Nanda Gopal Sahoo and Anirban Dandapat
ISBN: 979-8-89113-381-5

Keywords: two-dimensional nanosheets, biosensing, antimicrobial agents, bioimaging, bone healing, drug delivery

1. Introduction

The advanced and valuable research in the field of two-dimensional (2D) materials as a "smart nanosheets" unlatched many beneficial approaches that would lead to the development of methodical products. Nanosheets are the emerging star in the field of two-dimensional (2D) materials. The synthesis of nanosheets from fabrication approaches leads to the development of large surface-area-to-thickness ratio sheets, which are known for their electronic structure with atomic thickness, creating two-dimensional (2D) nanosheets. The nanosheet provides very good mechanical strength, such as high adhesion- and flexibility properties making this material a first preference in the biomedical field. The nanosheets are used in the biomedical area to combat different illnesses [1-5].

The plethora of different nanosheets is an emerging and advanced area to develop various products in the field of catalysts, biomedicines, environmental sciences, physics, and renewable energy technologies. Some of basic nanosheets structures-based products are nitrides and carbonitrides (M Xenes) and hexagonal boron nitride (h-BN) and their derivatives exhibited the nanosheets structures, Black phosphorous (BP) and boron nanosheet (BNS), MoS_2 Nanosheets [6], electronic material graphite nanosheets, and graphene based-nanosheets [7]. The properties of these nanosheets depend upon weak interlayer bonding and strong covalent in-plane bonding. The mostly atoms of the nanosheets are exposed on the surface, the reason behind its exfoliation of the ultrathin formation. This is used to increase the surface areas of these sheets to a maximum level. Due to this, the material gains better efficiency in the form of chemical and physical reactivity. It also influences the 2D wave function from quantum confinement effects. The resultant product is ultrathin 2D features embedded with unique electronic, photonic, magnetic, and catalytic properties.

In the biomedical field, gel-based materials faced a major issue due to their low mechanical strength and unrestrained microstructures. Due to the development of nanosheets, it is used as a functionalized polymer gel material. These polymer gels are developed by cross-linked polymer chain networks which are treated chemically and physically and swollen by a solvent. The resultant product shows high biocompatibility, and malleability, and can stand strongly in external stimuli. All these characteristics made them a good choice to apply for soft actuators and biomedical materials [8].

Nobody will deny that the life expectancy of a human has been increasing rapidly. The other point of consideration is how we can early diagnose any disease; deal with stress through good management to provide a healthy lifestyle. The human body is endocrine-based and produces a number of specific chemicals at the time of stress, age, or during illness. With the advanced technology in the field of nanosciences, it becomes easy to sense those specific chemicals behind the threat, and how it is easily handled by the diagnosis of diseases and stress management. Although, the chemical released during these conditions are very minute in quantity, and normally detection is not possible. Therefore, it is a need to develop highly sensitive sensors; the nanosheets' structured-based sensor materials are the best option to detect these illnesses because of their well-defined high surface-to-volume ratio and their specific exposed crystal facets [9-13]. 2D nanosheets can be used as electrochemical biosensors for the detection of neurotransmitters, glucose detection in the body. These are also applicable as implantable materials such as chips, sensors, carriers for drug delivery and *in vivo* cell imaging, antimicrobial applications, etc. In addition, the advanced optical property of the 2D nanosheets is used to develop ultrathin polypyrrole (PPy) nanosheets with advanced features like biocompatibility and notable tumor ablation ability (both *in vitro* and *in vivo*) [14]. This chapter is focused on the potentials of different 2D nanosheets in various biomedical applications which include bone-healing, drug delivery, antimicrobial activities, hyperthermia, bioimaging, photothermal therapy, cancer therapy and biosensors.

2. Nanosheets as Bone Healing Material

The skeleton system of a human being has a tendency to regenerate its damaged part after injury. This revitalizing property is performed by the body's own organ, the liver, which is essential to support bones to heal at their previous shapes and strengths. Orthopaedic healing care is used to support this amazing natural healing property with some physiochemical support.

Unfortunately, all scenarios are not equal to getting these benefits, there are some glitches are still there that must be sorted out with advanced technology. There is a piece of evidence showing that poor healing is due to poor vascularity, poor nutrition, loss of substantial bone or soft tissue of that injured area, infection, and most important the efficacy of modern internal fixation techniques. These are all responsible for ineffective osteosynthesis. Modern techniques based on nanosheets are emerging as good bone substitutes and bioactive factors to facilitate the healing process in suboptimal conditions [15]. The field of nanosheet-based treatment in bone healing gain experienced unpredictable advancements over the last few years. These nanosheet based studies revealed an emphasis on their contribution to the advancement in the field of bone regeneration [16].

The various pathological and varied ranges of disorder of the synovial joint may lead to the failure of movement of the joints and causes arthritis [17]. An intra-articular injection is a promising application in this case but some rapid clearance of the drug in the target site is the drawback of this technique. Using some nanocarriers to scaffold the target drug is a proliferative approach. These drug nanosystems hold the therapeutic drug in the target site up to their completion of the mode of action and showed magnificent results. These 2D structures-based nanomaterials are used as a drug-loading carrier.

The intra-articular drug delivery system based on molybdenum disulfide (MoS_2) nanosheets scaffolding with chitosan (CS) (i.e., MoS_2@CS@Dex) works near-infrared (NIR) photo-responsive carriers to assemble dexamethasone anti-inflammatory drug [18]. Without the nanocarrier (MoS_2 nanosheets), the efficiency of this drug could not reach its maximum potential. Dexamethasone has emerged as a fighter against osteoarthritis but the viscous nature of the synovial fluid due to hyaluronic acid cleared the drug rapidly and minimizes its effect of the drug. Therefore, chitosan (CS) modified MoS_2 nanosheets were used as a scaffolding partner of dexamethasone to make them an efficient soldier to treat osteoarthritis with intra-articular injection. This stimulation reduced the clearance risk of the drug in the joint cavity and increases the residential time to prolonged work in the target site. Figure 1 showed the synthesis of the nanocarrier with the target drug, and how effectively it responds near the NIR light. This study showed here, the MoS_2@CS@Dex-based drug-positive results in in-vitro as well as in-vivo and the mechanism behind the drug release via photothermal conversion, by adjusting the behaviors of radiation of NIR light. The nanosheets regulate the efficient drug release in the target area.

The 2D materials MoS_2@CS@Dex containing dexamethasone at lower systematic doses, clears the pathway to reach at the target site to down-regulate the inflammatory factors TNF-α and IL-1β secretions through activated macrophages in the immune system. It provides protection to chondrocytes and improves the erosion of cartilage tissue.

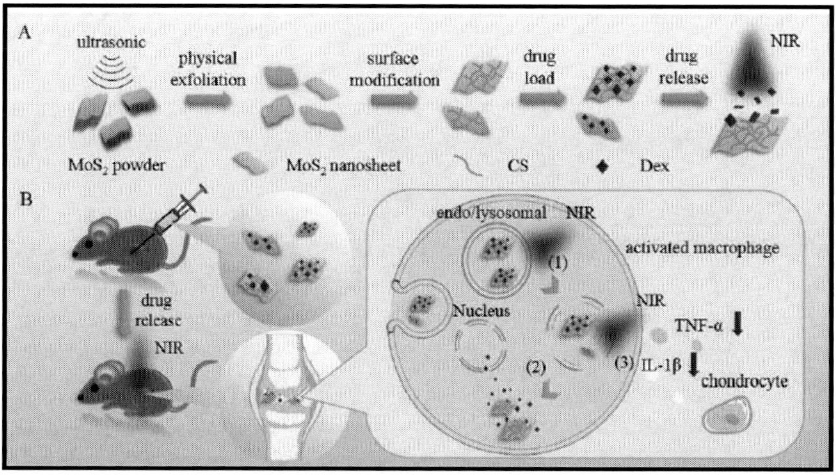

Figure 1. A. The mechanism of synthesis of in-vitro MoS_2@CS@Dex, scaffolding with chitosan (CS) modified molybdenum disulfide (MoS_2) nanosheets-based drug-loaded system in response to NIR light. B. In vivo application by using intra-articular injection and nano drug delivery system to treat osteoarthritis in response to NIR light, (reproduced with permission from Yingyu Zhao et al. [18], Copyright@ 2019, Americal Chemical Society).

3. Nanosheets as an Antimicrobial Agent

2D nanosheet-based nanostructures have shown profound biocompatibility and high dose-dependent anti-infective efficacy against infections. Bacterial resistance and the generation of new antibiotic-resistance genes are major dilemmas in the healthcare system. This problem is associated with artificial implants, clinical environment to agriculture areas also, due to regular misuse of antibiotics. In the regular use of implants (internal fixation in fractures and arthroplasty) in the field of orthopedics, this issue is just a common problem. It produces a high range of implant-related opportunistic infections, which result in catastrophic consequences [18-20]. The current antibiotics become resistant due to vertical as well horizontal gene transfer in bacterial species. There must be a need to develop efficient biomaterials with antibacterial activities to eradicate such issues. The development of 2D nanomaterials opens up new research areas to develop effective biomaterials to treat microorganisms-based infections. The nanotechnology approaches enlighten many pathways to develop useful biomaterials with great potential to treat difficult medical illnesses. These advanced biomaterials are known to kill bacteria by bactericidal polymers or release biocides bound to the biomaterials. Some useful bactericidal polymers are the derivatives of quaternary ammonium compounds (polycations) and polyethyleneimine showed membrane integrity disruption and the lysis of bacterial cell walls [21-24].

The traditional antibacterial agents such as metal-based nanoparticles like silver [25], zinc [26], magnesium [27], and copper [28] are scanty by bacterial drug resistance and are known to cause toxic side effects respectively. The present scenario demands biocompatible biomaterials which are efficient in their biomedical application with no harm. A Graphene-Like Silicon nanosheets-based study was conducted against the destruction of the cell membranes of both the species of Gram-positive and Gram-negative bacteria. The nanosheet-based biomaterials are an emerging source of bacteriocidal polymers or scaffolding agents that are used to transport the biocides materials inside the target site. Silicon-based 2D materials are emerged as prominent biomaterials due to their intrinsic biodegradability capability. The 2D graphene-based silicon nanosheets exhibit good topology that makes them better alternate strikingly from other 2D silicon-based nanomaterials. The previous silicon materials are porous, and silica based with distinct physiochemical properties like chiral superconductivity, quantum spin Hall effect, and giant magnetoresistance providing low-buckled topography [29-32].

Moreover, 2D nanomaterials are the most suitable material in cancer phototherapy, bone healing agents, biosensors, and antimicrobial assays. The antibacterial assays performed by different 2D nanomaterials are graphenes and their derivatives based, MXene [33], Colloidal $Ti_3C_2T_x$ MXene Nanosheets [34], MoS2 [35], are proven as the best antibacterial materials with remarkable substantially overcome the possibility to develop the antibacterial resistance.

4. Nanosheets in Drug Delivery and Photothermal Therapy

Drug delivery is one specialized biomedical application of nanosheets. The crucial problems of accessing the specific targets without hurting the normal cells can have an alternate solution of using nanosheets in drug-delivery applications. Cancer continues to be a major issue faced by humanity today and it's also one of the key causes of deaths worldwide. One of the most promising methods to effectively combat cancer cells without unfavourable severe side effects is nanotechnology. Silicon-based nano-platforms have so far been created for a number of medicinal uses. Due to its superior photo-thermal conversion ability and potent photo-absorption, silicene is typically used as a photo-thermal nanoagent to cause tumor hyperthermia ablation. Photothermal therapy (PTT), which exploits localized hyperthermia to achieve optimal antitumor activity, is indeed a potential therapeutic approach. Lin et al. developed few-layered silicene nanosheets using a wet chemical method and investigated thoroughly their potential use in photothermal therapy [36]. The single-/few-layered silicene nanosheets were displaying a freestanding structure with a thickness of 0.6 nm similar to graphene. According to optical physics studies, silicene has a photothermal conversion efficiency of 36.09% when exposed to NIR-II radiation which is significantly higher than the traditional 2D photothermal conversion nanomaterials like Ti_3C_2which has a 30.6% value [37]. Without damaging healthy cells, silicene-based nanomedicine can undoubtedly cause tumour overheating to eradicate cancerous cells. The formation of silicene in the tumor site could significantly raise the therapeutic temperature, by observing the thermal infrared fluctuations of tumor-bearing mice exposed to NIR-II radiation the study introduced silicene as a novel photothermal nano-agent that efficiently slows the growth of cancer cells in tumor-bearing mice and creatively developed silicene-based nanomaterials for biomedical field (Figure 2).

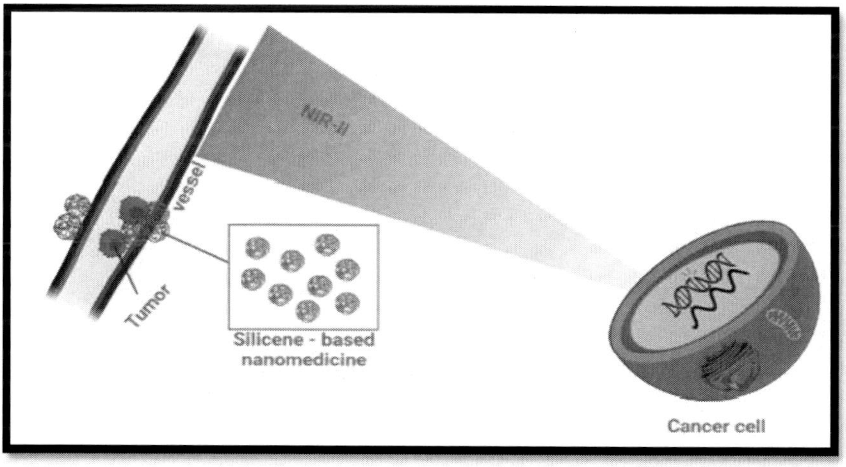

Figure 2. Scheme to illustrate the photothermal therapeutic process of silicene-based photothermal nano agents.

PTT and chemotherapy used together have significant benefits. PTT agents have the ability to convert light energy into localized high temperature under laser irradiation, directly destroying cancer cells [38-40]. The efficacy of synergistic therapy is considerably increased by the heat dissipated from PTT agents, which can both speed up drug release from their surfaces and improve drug uptake by enhancing cell membrane permeability [39-42].

He and his colleagues further developed a lysine demethylase KDM3A enzyme regulated PTT approach while examining the complicated mechanism of photonic hyperthermia in the BC model [43]. Following photo-thermal therapy treatment, RNA sequencing on the Mcf-7 breast cancer cell line was carried out to investigate the molecule-level alterations. The PTT treatment may demonstrate the presence of an antagonistic effect between both the thermal sensitivity-regulated KDM3A and anticancer function-related p53. This research provided fresh perspectives on the molecular causes of tumor cell apoptosis following thermal ablation and the further decrease of thermal resistance by the lysine demethylase KDM3A.

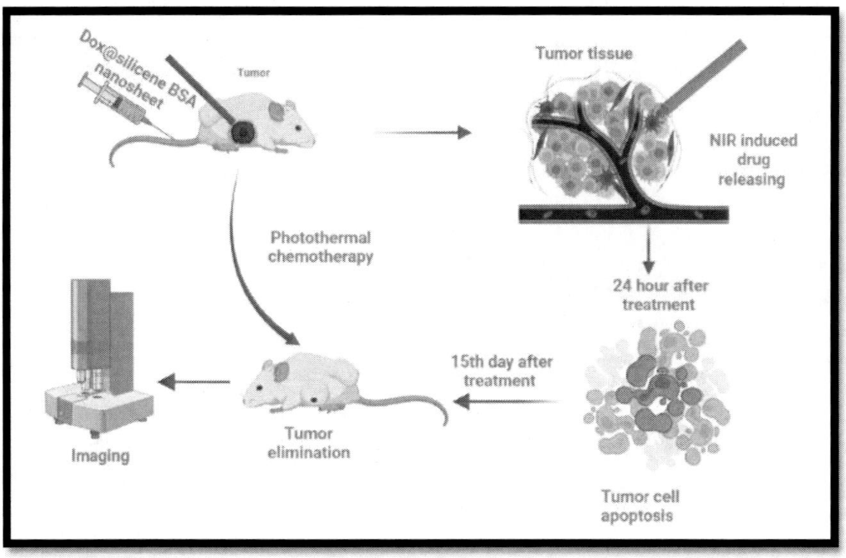

Figure 3. Scheme of DOX-loaded silicene nanosheets for PA imaging-guided chemo-/phototherapy.

To act as versatile nano-platforms in synergistic therapeutics, 2D silicene-based nano materials have now been created. Three groups of silicene's have synergistic therapeutic properties (Figure 3) of drug delivery systems for chemotherapeutics and Fenton-like reaction nano-catalysts for chemodynamic therapy radiosensitizers. Silicene nanosheets with high specific surface area and numerous surface functional groups, similar to normal 2D nanostructures, can be employed as drug delivery carriers [44]. Wang et al. developed Doxorubicin-loaded silicene nanosheets for the dual pH-/NIR-triggered drug-release performance, resulting in a nanocomposite-enabled DDS [45]. They discovered that silicene has a drug loading capacity of 137.0% (w/w %), which is higher than the vast majority of conventional Nano-carriers and ensures that significant dosages of anti-neoplastic medications can be administered to lesion locations.

Lately, Hu et al. reported that silicene-based silicene@Pt composite nanosheets were used to accomplish a dual-sensitized radiotherapy/ photothermal synergetic therapy by building the Pt nanoparticles onto the surface of the silicene nanosheets [46]. The development of tumour growth, proliferation, angiogenesis, and metastasis as well as the development of oxygen-dependent treatment resistance has all been linked to hypoxic tumor microenvironment [47, 48].

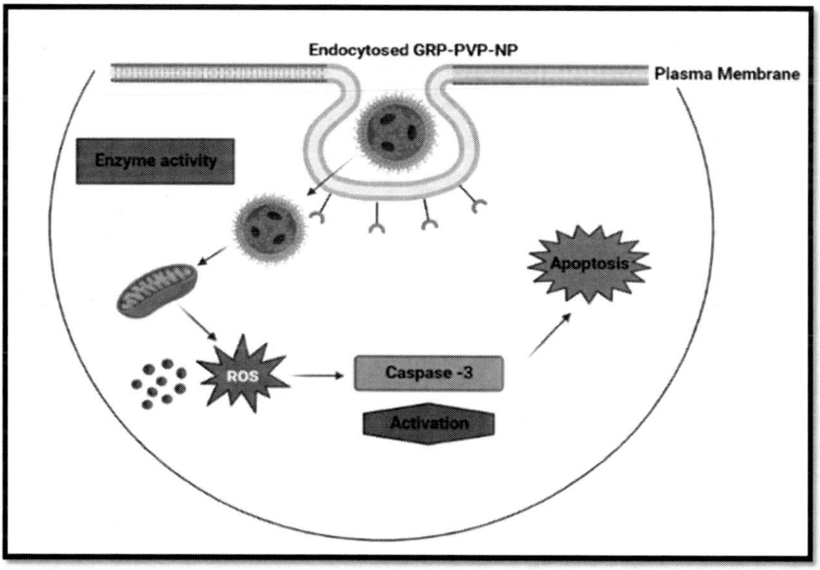

Figure 4. Diagram illustrating a putative mechanism demonstrating the cytotoxicity of the graphene-poly(vinylpyrrolidone) nanoparticle composite (GRP-PVP-NP) in cancer cells.

In the study poly(vinylpyrrolidone) nanoparticles (PVP-NP), which contain higher concentrations of loaded drug induced substantial toxicity both against mouse embryonic fibroblasts (NIH-3T3) and human colorectal cancer cells, were utilized to exfoliate graphene nanosheets. As the graphene nanosheets have a large surface area and the ability for the pharmaceuticals to stick to both sides of a single sheet, the composite graphene-poly(vinylpyrrolidone) nanoparticle composite (GRP-PVP-NP) nanosheets were used as nano-carriers for anti-cancer medications (Figure 4). According to the experimental results, larger concentrations of graphene nanosheets can reduce the capacity of cancer cells to form colonies and cause cell death by oxidative stress [49].

The graphene oxide (GO) sheet was proven to be a transporter for adriamycin (ADR), which has been shown to overcome drug resistance in breast carcinoma cells [50]. ADR was heavily drug-loaded (up to 93.6%) onto the surface of the GO sheet (ADR-GO). Drugs can be loaded efficiently because the GO sheet has intact graphitic areas interspersed with sp^3- and sp^2-hybridized carbon atoms on the sheet surface and aromatic network. A range of anticancer medications, including ADR, are pumped out of the cells by P-gp using the energy that is released by ATP hydrolysis. Adenosine triphosphate was released into the cytoplasm in response to low endosomal pH. According to cellular tests, GO sheets demonstrate a considerable increase in ADR accumulation in MCF-7/ADR cells with a significantly higher level of cytotoxicity than free ADR. GNS has the potential to have higher loading capacities than some other nano-carriers delivering biological and chemotherapeutic medicines.

Cheng and colleagues synthesized Ce-WS_2 NSs with improved colloidal stability and bio - compatibility that are functionalized via LA conjugated PEG [51]. In Balb/c mice with 4T1 tumors, the in vivo PTT performance of WS_2-PEG NSs was investigated. The temperature of the tumor surface increased from 30°C to 65°C under the irradiation of an 808 nm laser at a power level of 0.8 W cm^2 for five minutes after intratumorally (i.t.) administering (2 mg kg1) for 0.5 hours or intravenously (i.v.) injecting (20 mg kg^1) of WS_2-PEG NSs for 24 hours. When WS_2-PEG NSs were injected intravenously or intramuscularly into the tumors after NIR laser irradiation, additional trials revealed that the tumors were entirely eradicated without regrowth after 45 days, demonstrating the remarkable PTT efficacy.

Liu and colleagues used a hydrothermal technique to create ultrathin 1T-WS_2 NSs using NH_4^+ intercalation [52]. Results showed the 1T-WS_2 NSs with a size of 150 nm exhibit substantial optical absorption in the NIR range and good hydrophilicity. The temperature of Hela tumours in NOD/SCID mice significantly raised to 50°C under the irradiation of 808 nm laser at the power density of 0.6 W cm^2 within 4 min and then remained at this temperature for the next 6 min after injection of 1T-WS_2 NSs (1.2 mg mL1, 40 L). During the course of treatment, all tumors were removed 4 days following PTT without recurrence. PEGylated MoS_2 NSs were used as 2D nanocarriers to load several therapeutic compounds for cancer therapy, such as DOX, SN38, and Ce_6 [53]. They discovered that MoS_2 NSs had high drug loading ratios, significantly greater than those of graphene of 239% for DOX 118% for SN38 and 39% for Ce6.

A unique and simple method for creating multifunctional nanocomposites with silica-coated Fe_2O_3 particle cores and $NaYF_4$-Yb/Er shells were developed on top of which MnO_2 NSs were later formed for the delivery and release of a model drug, Congo red (CR). The MnO_2 NSs in this nanosystem worked as effective quenchers for the UC luminescence to track intracellular GSH content in addition to serving as transporters to load and release of CR in vitro [54]. When GSH reduced MnO_2 to Mn^{2+}, the drug was released, and the UC luminescence increased at the same time. Owing to its own GSH-stimulated efficient drug delivery and UC luminescence monitoring, the constructed nanocomposite is a potential platform.

5. Nanosheets for Biosensing

Biological responses can be transformed into quantifiable signals by analytical tools called biosensors [55, 56]. Biosensors typically have two components: a bioreceptor for interacting with the target analytes and a transducer for turning recognition events into audible signals. [57]. Different biosensors have been created for biological, food security, pollution management, healthcare, and clinical diagnostics because of their high sensitivity, selectivity, low cost, and simplicity [58, 59]. Numerous nanostructures have been effectively used as transducers to capture the physical alterations brought on by biological interactions and convert them into quantifiable signals [60]. For the detection of diverse biomolecules, such nucleic acids, proteins, enzymes, etc. High sensitivity, selectivity-based sensors utilizing nanomaterials have been thoroughly explored during the past two decades [61-64].

Multiple Two-dimensional transition metal dichalcogenides (TMDC) NS-based biosensors have been developed in recent years. Further electrochemical biosensors based on TMDC NSs have been thoroughly investigated, taking advantage of special electrical features, convenience for fabrication, and surface functionalization. Furthermore, TMDC NSs offer excellent prospects for fluorescent bio-sensing due to their robust fluorescence quenching capacity. Specifically, certain TMDC NSs feature high carrier mobility, low dangling bonds, and configurable bandgaps, which make them great contenders for field-effect transistor (FET)-based biosensors.

The fluorescent sensor based on TMDC NSs was first demonstrated by Zhang et al. Their research shown that ssDNA had a greater affinity for MoS_2 NSs than dsDNA, which can be used to create fluorescent biosensors that can be turned on or off. The labeled-dye reporting ssDNA appears as single strand form in the absence of the targeted ssDNA and binds to the surface of MoS_2 NSs via van der Waals forces, effectively quenching the dyes fluorescence. Due to the significantly weaker binding between dsDNA and MoS_2 NSs than that of ssDNA, dye labelled ssDNA hybridizes with the target ssDNA to create a DNA duplex. This allows the fluorescence to be preserved. This study not only provides a superb illustration of a straightforward, quick, and inexpensive DNA biosensor, but also demonstrates the intrinsic high fluorescence quenching efficiency and exceptional ability to distinguish between ssDNA and dsDNA, which greatly expands the potential applications of TMDC NSs for fluorescent biosensing [64].

Recently a dual target aptasensor based on decorated MoS_2 NSs with Au NPs for the simultaneous determination of thrombin and adenosine triphosphate (ATP) was developed [65]. These were developed on Au NPs@MoS_2 NSs, that were tagged with methylene blue (MB) and ferrocene (Fc) two distinct aptamers were immobilized. With both thrombin and ATP, the two aptamer structures would switch, resulting in MB being far from the electrode surface and Fc being close to the electrode surface. The dual target aptasensor for thrombin and ATP have a limit of detection (LOD) of 0.74 nM and 0.0012 nM respectively.

A DNA electrochemical biosensor was constructed by Huang and colleagues using WS_2-graphene composites surface modified with Au NPs and chitosan [66]. Au NPs were used to increase the composites conductivity in addition to linking thiolated DNA. This biosensor was capable of detecting the target DNA down to 2.3 fM. Au NPs and GOD we further combined to MoS_2-multi-walled carbon nanotubes (MoS_2-MWCNTs) to further increase the detection sensitivity [66]. A biosensor of this kind might effectively detect several signal amplitudes, lowering the LOD to 0.79 fM. In order to measure immunoglobulin E (IgE), Huang et al. developed WS_2-graphene composites functionalized with aptamers and Au NPs [67]. Signal amplification was effectively accomplished by WS_2-graphene films and Au NPs, and the LOD of this biosensor may go as low as 0.12 pM. Additionally, Xia et al. developed an electrochemical tryptophan biosensor based on MoS_2 NSs embellished with silver nanoflakes (Ag NFs) With a LOD of 0.05 M and a broad linear range from 0.5 to 120 M, this biosensor displayed an outstanding approach for the catalytic oxidation of tryptophan [68].

A very robust microRNA (miRNA) biosensor based on WS_2 NSs and a duplex-specific nuclease (DSN) signal amplification method (Figure 5) was created by Xi et al. [69]. According to the findings, shorter ssDNA oligonucleotide fragments had a weaker affinity for WS_2 NSs than longer ssDNA oligonucleotide fragments (more than 10 bases) (less than 10 bases). The cycle of cleaving, releasing, and hybridizing occurs when a ssDNA probe binds to a target miRNA. DSN can then precisely break the DNA strand in the DNA/RNA complex, liberating the target miRNA, which will then hybridize with another ssDNA again. The fluorescence signal of the short ssDNA probe will be preserved following the addition of WS_2, whereas the fluorescence of long ssDNA probes will be muted, since the cleaved short ssDNA probe won't be adsorbed on the surface of WS_2 NSs. A biosensor of this type might offer a LOD of miRNA as low as 300 fM.

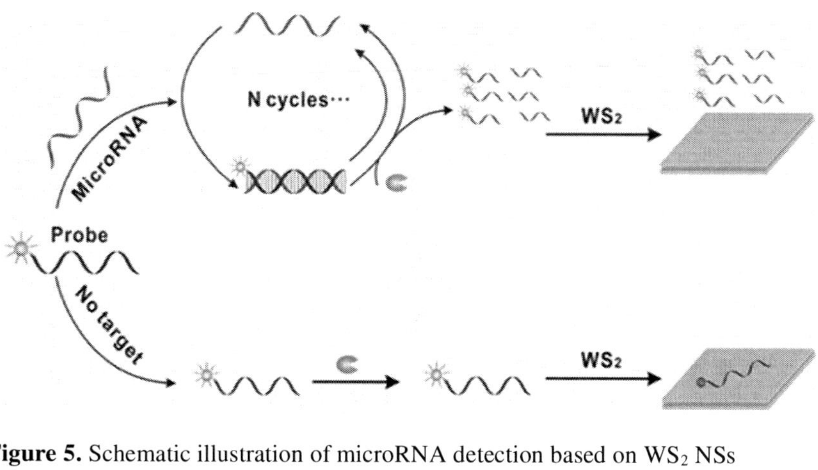

Figure 5. Schematic illustration of microRNA detection based on WS_2 NSs coupled with duplex specific nuclease signal amplification strategy (Reproduced with permission from Qiang Xi et al. [69], Copyright@ 2014, Americal Chemical Society).

The FET type gas sensors were developed by Zhang et al. [70]. They created FET-type sensors for NO detection using mechanically exfoliated MoS_2 NSs with various layer counts (1 L–4 L). The results indicated that monolayer MoS_2 FET had a quick but unstable response, while FET-type sensors based on bilayer (2 L) and multilayer (3 L, 4 L) MoS_2 NSs showed a LOD down to 0.8 ppm for NO. Additionally, according to Late and colleagues, five-layer MoS_2 devices responded to NO_2, NH_3, and water vapor more sensitively than bilayer MoS_2 FETs [71]. Liu et al. employed monolayer MoS_2 NSs produced through CVD to create a Schottky-contacted FET-type sensor [72]. With a LOD as low as 20 ppb and 1 ppm, respectively, this FET-type sensor displayed exceptionally high sensitivity for NO_2 and NH_3, among other gases. A NH_3 sensor with a LOD shows highly efficient as low as 300 ppb was built by Lee et al. using MoS_2 NSs [73].

Ji et al. developed the CaO_2/MnO_2@polydopamine-methylene blue (MB) nanosheets (CMP-MB), a multifunctional nanosystem in which the MnO_2 NS inhibits the fluorescence of MB. The MnO_2 NSs may break down into Mn^{2+} when exposed to a tumor microenvironment, which results in MB fluorescence. Switch-controlled tumor cell imaging was thus made possible [74]. The degradation of MnO_2 NS by ascorbic acid (AA) allows for the reversal of the MnO_2 NS-mediated quenching effect. Based on this, a carbon dot (CD)-MnO_2 nanocomposite has been developed for the hydrolysis of 2-phosphate (AAP) into AA, which aids in the detection of alkaline phosphatase (ALP). In this way, a label-free fluorescence switching method for detecting ALP activity can be established with a limit of detection (LOD) of 0.4 U/L using the CD-MnO_2 nanocomposite as a sensing probe. Wang and colleagues used fluorescent CDs and MnO_2 NSs as an energy donor-acceptor combination [75] in order to detect GSH. Yan et al. created an optical sensing platform based on graphene quantum dots (GQD) and MnO_2 NS for the detection of GSH in addition to using MB and CDs as fluorescence donors [76].

A MnO_2 NS-modified upon version (UC) nanosystem was created by Chu and colleagues for sensitive switchable fluorescence detection of H_2O_2 and glucose in blood be According to the equation: $MnO_2 + H_2O_2 + 2H^+ = Mn^{2+} + 2H_2O + O_2$.

The enzymatic breakage and unification of glucose-by-glucose oxidase (GOx) produced H_2O_2, which was subsequently used to convert MnO_2 to Mn^{2+}, analogous to GSH.

6. Nanosheets in Bioimaging Applications

Different contrast agents have been created to increase the clarity and resolution of bioimaging for tissues or vessels. These agents also promote functional molecular bioimaging and significantly increase the sensitivity and selectivity for identifying significant biomolecules or signal pathways [77]. Novel contrast agents based on nanoparticles have gained significant interest and demonstrated considerable potential in several bioimaging modalities with the rapid advancement of nanotechnology. Due to transition metal atoms great capacity to attenuate X-rays (e.g., Mo, W), several 2D TMDC NSs can be employed as contrast agents. MoS_2 and WS_2 NSs have both been studied as contrast agents for CT imaging up to this point and have produced results that are even better than those conventional contrast agents [78, 79].

To serve as contrast agents for in vivo dual-modal CT/PA imaging, Cheng et al. produced PEGylated WS_2 NSs [78]. The Hounsfield unit (HU) value of PEGylated WS_2 NSs is around 22.01 HU Lg^{-1}, which is significantly higher than that of the commercial CT contrast agent iopromide (15.9 HU Lg^{-1}). Furthermore, once Balb/c mice with 4T1 tumours were injected with PEG-WS_2 NSs intravenously or intraperitoneally, noticeably improved CT and PA signals were seen. Dual-modal CT/PA imaging not only enables whole-body imaging with great spatial resolution and sensitivity, but it also offers perceptible data on the intratumoral distribution of nanoagents, which can aid in our understanding of the microstructures of tumours.

MoS_2/Bi_2S_3-PEG (MBP) NSs with high photothermal effect, colloidal stability, and biocompatibility were synthesized by Wang et al. using a simple one-pot approach [80]. In this two-component theranostic system, MoS_2 NSs served as PTT agents with outstanding PA imaging performance, while Bi atoms might improve the CT imaging and RT effect.

With a high spatial resolution and multiple dimensions, magnetic resonance imaging (MRI) is a non-invasive clinical imaging method. Typically, MRI contrast agents modify the longitudinal (T1) and transverse (T2) relaxation rates of water protons in the targeted biological tissues to produce imaging contrasts [81, 82]. Mn^{2+} ions have the potential to be a good contrast agent for MRI, benefiting from the low toxicity and high r_1 relaxation (1/T1, longitudinal T1 relaxation rate), when compared to the commercial T1 contrast agent Gd3 [83]. The mentioned nanocomposite was made by growing MnOx nanoparticles in situ on a silicene substrate (referred to as MS-BSA) [84]. Mn^{2+} ions were released after extra GSH was reduced, and they displayed r_1 relativity of 2.22 $mM^{-1}s^{-1}$, demonstrating the effective TME-responsive performance of MS-BSA. In a very recent study, silicene-based nanocomposites were used as a new type of CT contrast agent to provide an ideal balance between the effectiveness of targeting and the toxicity of surrounding tissues by incorporating a significant amount of strong X-ray attenuation into the silicene structures.

Conclusion

Two-dimensional nanosheets have attracted due to their biocompatibility, biodegradability in vivo, and less cell toxicity compared to the other available materials in the market. 2D nanosheets served as one of the best biomedical materials, sensors, gas sensors, aids in drug release systems, and several other biological applications. 2D nanosheets are being used in bone tissue repairing, healing engineering, and also used as scaffolding and coating materials applied on the surface of the prosthetics to enhance the osteoinductive and antibiotic proficiency of that biomedical equipment. In spite of everything, the future novelty in 2D materials through research may open up several exciting challenges, particularly with regard to theranostics applications, and tissue engineering, in the field of oncology, advanced anatomy, and physiology. The mechanism behind the antibacterial activity of 2D nanomaterials is based on the direct interaction with the cells, the development of reactive oxygen species, photothermal therapy in the presence of light, photocatalytic ablation, the release of polysulfane compounds, and generation of piezoelectric effect. By following these multiple biological multiple pathways, these 2D nanomaterials show effective activities without facing any resistant causes by the bacterial species. The nanosheets disrupt the cellular components, lipids contents of the cell wall, protein structures, and genetic material ultimately leading to the death of the bacterial cell. The different functional groups $-NH_2$, -SH, and -COOH are used to attain by tuning or by alternating the shape, size, and orientation of the 2D materials. This is achieved by the incorporation of metal nanoparticles, metal oxide nanoparticles, halogens, polymers, or quaternary ammonium/phosphonium salts assembled into these 2D-based nanomaterials. Based on these all previous reports, there is need to address more challenges and future development trends of 2D-nanomaterials as effective agents used in biomedical applications.

References

[1] Guo S., Dong S. (2011). Graphene nanosheet: Synthesis, molecular engineering, thin film, hybrids, and energy and analytical applications. *Chem. Soc. Rev. 40* (5), 2644–2672.

[2] Mallwitz F., Laschewsky A. (2005). Direct access to stable, freestanding polymer membranes by layer-by-layer assembly of polyelectrolytes. *Adv. Mater. 17*(10), 1296–1299.

[3] Geim A. K. (2009) Graphene: Status and prospects. *Science*. 324, 1530–1534.

[4] Umar, A., Alshahrani, A. A., Algarni, H., Kumar, R., (2017). CuO nanosheets as potential scaffolds for gas sensing applications. *Sensors and Actuators B: Chemical*, *250*, 24-31.

[5] You, Y., Yang, C., Zhang, X., Lin, H., Shi, J. (2021). Emerging two-dimensional silicene nanosheets for biomedical applications. *Materials Today Nano*, *16*, 100132.

[6] Song, H., Wang, Q., Zhao, W. M. (2020). A novel SPR sensor sensitivity-enhancing method for immunoassay by inserting MoS2 nanosheets between metal film and fiber. *Optics and Lasers in Engineering*, *132*, 106135.

[7] Lee, H. P., Gaharwar, A. K. (2020). Light-responsive inorganic biomaterials for biomedical applications. *Advanced Science*, *7*(17), 2000863.

[8] Miyamoto, N., Yamamoto, S. (2017). Functional layered compounds for nanoarchitectonics. In *Supra-Materials Nanoarchitectonics* (pp. 173-192). William Andrew Publishing.

[9] Wongchoosuk, C., Lutz, M., Kerdcharoen, T. (2009). Detection and classification of human body odor using an electronic nose. *Sensors*, *9*(9), 7234-7249.

[10] Bernier, U. R., Kline, D. L., Barnard, D. R., Schreck, C. E., Yost, R. A. (2000). Analysis of human skin emanations by gas chromatography/mass spectrometry. 2. Identification of volatile compounds that are candidate attractants for the yellow fever mosquito (Aedes aegypti). *Analytical chemistry*, *72*(4), 747-756.

[11] Fuchs, P., Loeseken, C., Schubert, J. K., Miekisch, W. (2010). Breath gas aldehydes as biomarkers of lung cancer. *International Journal of Cancer*, *126*(11), 2663-2670.

[12] Zhu, C., Zeng, Z., Li, H., Li, F., Fan, C., Zhang, H. (2013). Single-layer MoS2-based nanoprobes for homogeneous detection of biomolecules. *Journal of the American Chemical Society*, *135*(16), 5998-6001.

[13] Haze, S., Gozu, Y., Nakamura, S., Kohno, Y., Sawano, K., Ohta, H., Yamazaki, K. (2001). 2-Nonenal newly found in human body odor tends to increase with aging. *Journal of Investigative Dermatology*, *116*(4), 520-524.

[14] Wang, X., Ma, Y., Sheng, X., Wang, Y., Xu, H. (2018). Ultrathin polypyrrole nanosheets via space-confined synthesis for efficient photothermal therapy in the second near-infrared window. *Nano letters*, *18*(4), 2217-2225.

[15] Dang, W., Li, T., Li, B., Ma, H., Zhai, D., Wang, X., Chang, J., Xiao, Y., Wang, J. Wu, C. 2018. A bifunctional scaffold with CuFeSe2 nanocrystals for tumor therapy and bone reconstruction. *Biomaterials*, *160*, 92-106.

[16] Wang, Z., Zhao, J., Tang, W., Hu, L., Chen, X., Su, Y., Zou, C., Wang, J., Lu, W. W., Zhen, W., Zhang, R., 2019. Multifunctional nanoengineered hydrogels

consisting of black phosphorus nanosheets upregulate bone formation. *Small, 15*(41), 1901560.

[17] Nuki, G. (1999). Osteoarthritis: a problem of joint failure. *Zeitschrift für Rheumatologie*, *58*(3), 142-147.

[18] Zhao, Y., Wei, C., Chen, X., Liu, J., Yu, Q., Liu, Y., Liu, J. (2019). Drug delivery system based on near-infrared light-responsive molybdenum disulfide nanosheets controls the high-efficiency release of dexamethasone to inhibit inflammation and treat osteoarthritis. *ACS Applied Materials & Interfaces*, *11*(12), 11587-11601.

[19] Arciola, C. R., Campoccia, D., Montanaro, L. (2018). Implant infections: adhesion, biofilm formation and immune evasion. *Nature Reviews Microbiology*, *16*(7), 397-409.

[20] Luo, Y., Ge, M., Lin, H., He, R., Yuan, X., Yang, C., Wang, W., Zhang, X. (2020). Anti-infective application of graphene-like silicon nanosheets via membrane destruction. *Advanced Healthcare Materials*, *9*(3), 1901375.

[21] Ding, X., Duan, S., Ding, X., Liu, R., Xu, F. J. (2018). Versatile antibacterial materials: an emerging arsenal for combatting bacterial pathogens. *Advanced Functional Materials*, *28*(40), 1802140.

[22] Hofstee, M. I., Muthukrishnan, G., Atkins, G. J., Riool, M., Thompson, K., Morgenstern, M., Stoddart, M. J., Richards, R. G., Zaat, S. A., Moriarty, T. F., (2020). Current concepts of osteomyelitis: from pathologic mechanisms to advanced research methods. *The American Journal of Pathology*, *190*(6), 1151-1163.

[23] Zhang, J., Chen, Y. P., Miller, K. P., Ganewatta, M. S., Bam, M., Yan, Y., Nagarkatti, M., Decho, A. W., Tang, C. (2014). Antimicrobial metallopolymers and their bioconjugates with conventional antibiotics against multidrug-resistant bacteria. *Journal of the American Chemical Society*, *136*(13), 4873-4876.

[24] Liu, Y., Ding, S., Shen, J., Zhu, K. (2019). Nonribosomal antibacterial peptides that target multidrug-resistant bacteria. *Natural Product Reports*, *36*(4), 573-592.

[25] Kumar, A., Kumar, S., Kiran, K., Banerjee, S., Pande, V., Dandapat, A. (2021). Myco-nanotechnological approach to synthesize silver oxide nanocuboids using endophytic fungus isolated from Citrus pseudolimon plant. *Colloids and Surfaces B: Biointerfaces*, *206*, 111948.

[26] Xie, Y., He, Y., Irwin, P. L., Jin, T., Shi, X. (2011). Antibacterial activity and mechanism of action of zinc oxide nanoparticles against Campylobacter jejuni. *Applied and Environmental Microbiology*, *77*(7), 2325-2331.

[27] Bhattacharya, P., Dey, A., Neogi, S. (2021). An insight into the mechanism of antibacterial activity by magnesium oxide nanoparticles. *Journal of Materials Chemistry B*, *9*(26), 5329-5339.

[28] Bezza, F. A., Tichapondwa, S. M., Chirwa, E. (2020). Fabrication of monodispersed copper oxide nanoparticles with potential application as antimicrobial agents. *Scientific Reports*, *10*(1), 1-18.

[29] Molle, A., Grazianetti, C., Tao, L., Taneja, D., Alam, M. H., Akinwande, D. (2018). Silicene, silicene derivatives, and their device applications. *Chemical Society Reviews*, *47*(16), 6370-6387.

[30] Lin, H., Qiu, W., Liu, J., Yu, L., Gao, S., Yao, H., Chen, Y., Shi, J. (2019). Silicene: wet-chemical exfoliation synthesis and biodegradable tumor nanomedicine. *Advanced Materials*. *31*(37), 1903013.

[31] Balendhran, S., Walia, S., Nili, H., Sriram, S., Bhaskaran, M. (2015). Graphene Analogues: Elemental Analogues of Graphene: Silicene, Germanene, Stanene, and Phosphorene (Sm all 6/2015). *Small*, *11*(6), 633-633.

[32] Zhang, F., Wang, M., Wang, Z., Han, K., Liu, X., Xu, X. (2018). Nonlinear absorption properties of silicene nanosheets. *Nanotechnology*, *29*(22), 225701.

[33] Rizi, K. S. (2022). MXene nanosheets as a novel nanomaterial with antimicrobial applications: A literature review. *Journal of Molecular Structure*, *1262*.

[34] Arabi Shamsabadi, A., Sharifian Gh, M., Anasori, B., Soroush, M. (2018). Antimicrobial mode-of-action of colloidal Ti3C2T x MXene nanosheets. *ACS Sustainable Chemistry & Engineering*, *6*(12), 16586-16596.

[35] Mei, L., Zhu, S., Yin, W., Chen, C., Nie, G., Gu, Z., Zhao, Y. (2020). Two-dimensional nanomaterials beyond graphene for antibacterial applications: current progress and future perspectives. *Theranostics*, *10*(2), 757.

[36] Lin, H., Qiu, W., Liu, J., Yu, L., Gao, S., Yao, H., Chen, Y., Shi, J., 2019. Silicene: wet-chemical exfoliation synthesis and biodegradable tumor nanomedicine. *Advanced Materials*, *31*(37), 1903013.

[37] Lin, H., Chen, Y., Shi, J., 2018. Insights into 2D MXenes for versatile biomedical applications: current advances and challenges ahead. Advanced Science, *5*(10), 1800518.

[38] Zeng, J., Goldfeld, D., Xia, Y. (2013). A Plasmon-Assisted optofluidic (PAOF) system for measuring the photothermal conversion efficiencies of gold nanostructures and controlling an electrical switch. *Angewandte Chemie International Edition, 52*(15), 4169-4173.

[39] Menon, J. U., Jadeja, P., Tambe, P., Vu, K., Yuan, B., Nguyen, K. T., 2013. Nanomaterials for photo-based diagnostic and therapeutic applications. *Theranostics*, *3*(3), 152.

[40] Zhang, Z., Wang, J., Chen, C. (2013). Near-infrared light-mediated nanoplatforms for cancer thermo-chemotherapy and optical imaging. *Advanced Materials*, *25*(28), 3869-3880.

[41] Yin, W., Yan, L., Yu, J., Tian, G., Zhou, L., Zheng, X., Zhang, X., Yong, Y., Li, J., Gu, Z., Zhao, Y. (2014). High-throughput synthesis of single-layer MoS_2 nanosheets as a near-infrared photothermal-triggered drug delivery for effective cancer therapy. *ACS Nano*, *8*(7), 6922-6933.

[42] Liu, T., Wang, C., Cui, W., Gong, H., Liang, C., Shi, X., Li, Z., Sun, B., Liu, Z. (2014). Combined photothermal and photodynamic therapy delivered by PEGylated MoS_2 nanosheets. *Nanoscale*, *6*(19), 11219-11225.

[43] He, C., Yu, L., Ding, L., Yao, H., Chen, Y., Hao, Y. (2020). Lysine demethylase KDM3A regulates nanophotonic hyperthermia resistance generated by 2D silicene in breast cancer. *Biomaterials*, *255*, 120181.

[44] Han, X., Huang, J., Lin, H., Wang, Z., Li, P., Chen, Y. (2018). 2D ultrathin MXene-based drug-delivery nanoplatform for synergistic photothermal ablation and chemotherapy of cancer. *Advanced Healthcare Materials*, *7*(9), 1701394.

[45] Wang, F., Duan, H., Zhang, R., Guo, H., Lin, H., & Chen, Y. (2020). Potentiated cytosolic drug delivery and photonic hyperthermia by 2D free-standing silicene nanosheets for tumor nanomedicine. *Nanoscale*, *12*(34), 17931-17946.

[46] Hu, R., Chen, Z., Dai, C., Guo, X., Feng, W., Liu, Z., Lin, H., Chen, Y., Wu, R. (2021). Engineering two-dimensional silicene composite nanosheets for dual-sensitized and photonic hyperthermia-augmented cancer radiotherapy. *Biomaterials*, *269*, 120455.

[47] Zhong, H., De Marzo, A. M., Laughner, E., Lim, M., Hilton, D. A., Zagzag, D., Buechler, P., Isaacs, W. B., Semenza, G. L., Simons, J. W. (1999). Overexpression of hypoxia-inducible factor 1α in common human cancers and their metastases. *Cancer Research*, *59*(22), 5830-5835.

[48] Bristow, R. G., Hill, R. P. (2008). Hypoxia and metabolism: Hypoxia, DNA repair and genetic instability. *Nature Reviews Cancer*, *8*(3).

[49] Tyagi, N., Attia, N. F., Geckeler, K. E. (2017). Exfoliated graphene nanosheets: pH-sensitive drug carrier and anti-cancer activity. *Journal of Colloid and Interface Science*, *498*, 364-377.

[50] Wu, J., Wang, Y. S., Yang, X. Y., Liu, Y. Y., Yang, J. R., Yang, R., & Zhang, N. (2012). Graphene oxide used as a carrier for adriamycin can reverse drug resistance in breast cancer cells. *Nanotechnology*, *23*(35), 355101.

[51] Cheng, L., Liu, J., Gu, X., Gong, H., Shi, X., Liu, T., Wang, C., Wang, X., Liu, G., Xing, H., Bu, W. (2014). PEGylated WS2 nanosheets as a multifunctional theranostic agent for in vivo dual-modal CT/photoacoustic imaging guided photothermal therapy. *Advanced Materials*, *26*(12), 1886-1893.

[52] Liu, Q., Sun, C., He, Q., Khalil, A., Xiang, T., Liu, D., Zhou, Y., Wang, J., Song, L. (2015). Stable metallic 1T-WS2 ultrathin nanosheets as a promising agent for near-infrared photothermal ablation cancer therapy. *Nano Research*, *8*(12), 3982-3991.

[53] Liu, T., Wang, C., Cui, W., Gong, H., Liang, C., Shi, X., Li, Z., Sun, B., Liu, Z. (2014). Combined photothermal and photodynamic therapy delivered by PEGylated MoS_2 nanosheets. *Nanoscale*, *6*(19), 11219-11225.

[54] Zhao, P., Zhu, Y., Yang, X., Shen, J., Jiang, X., Zong, J., Li, C. (2014). Multifunctional MnO 2 nanosheet-modified Fe_3O_4@ SiO_2/NaYF 4: Yb, Er nanocomposites as novel drug carriers. *Dalton Transactions*, *43*(2), 451-457.

[55] Turner, A., Karube, I., Wilson, G. S. (1987). *Biosensors: fundamentals and applications*. Oxford university press.

[56] Banica, F. G. (2012). *Chemical sensors and biosensors: fundamentals and applications*. John Wiley & Sons.

[57] Zhang, J., Chen, M., Peng, Y., Li, S., Han, D., Ren, S., Qin, K., Li, S., Han, T., Wang, Y., Gao, Z., 2023. Wearable biosensors for human fatigue diagnosis: A review. *Bioengineering & Translational Medicine*, *8*(1), e10318.

[58] Lee, S. H., Sung, J. H., Park, T. H. (2012). Nanomaterial-based biosensor as an emerging tool for biomedical applications. *Annals of Biomedical Engineering*, *40*(6), 1384-1397.

[59] Morales-Narváez, E., Merkoçi, A. (2012). Graphene oxide as an optical biosensing platform. *Advanced Materials*, *24*(25), 3298-3308.

[60] Liu, Y., Dong, X., Chen, P. (2012). Biological and chemical sensors based on graphene materials. *Chemical Society Reviews*, *41*(6), 2283-2307.

[61] Zhong, W. (2009). Nanomaterials in fluorescence-based biosensing. *Analytical and Bioanalytical Chemistry*, *394*(1), 47-59.

[62] Su, S., Chao, J., Pan, D., Wang, L., Fan, C. (2015). Electrochemical Sensors Using Two-Dimensional Layered Nanomaterials. *Electroanalysis*, *27*(5), 1062-1072.

[63] Luo, X., Morrin, A., Killard, A. J., Smyth, M. R. (2006). Application of nanoparticles in electrochemical sensors and biosensors. *Electroanalysis: An International Journal Devoted to Fundamental and Practical Aspects of Electroanalysis*, *18*(4), 319-326.

[64] Zhu, C., Zeng, Z., Li, H., Li, F., Fan, C., & Zhang, H. (2013). Single-layer MoS2-based nanoprobes for homogeneous detection of biomolecules. *Journal of the American Chemical Society*, *135*(16), 5998-6001.

[65] Su, S., Sun, H., Cao, W., Chao, J., Peng, H., Zuo, X., Yuwen, L., Fan, C., Wang, L. (2016). Dual-target electrochemical biosensing based on DNA structural switching on gold nanoparticle-decorated MoS2 nanosheets. *ACS Applied Materials & Interfaces*, *8*(11), 6826-6833.

[66] Huang, K. J., Liu, Y. J., Wang, H. B., Gan, T., Liu, Y. M., Wang, L. L. (2014). Signal amplification for electrochemical DNA biosensor based on two-dimensional graphene analogue tungsten sulfide–graphene composites and gold nanoparticles. *Sensors and Actuators B: Chemical*, *191*, 828-836.

[67] Huang, K. J., Liu, Y. J., Cao, J. T., Wang, H. B. (2014). An aptamer electrochemical assay for sensitive detection of immunoglobulin E based on tungsten disulfide–graphene composites and gold nanoparticles. *RSC Advances*, *4*(69), 36742-36748.

[68] Xia, X., Zheng, Z., Zhang, Y., Zhao, X., Wang, C. (2014). Synthesis of Ag-MoS2/chitosan nanocomposite and its application for catalytic oxidation of tryptophan. *Sensors and Actuators B: Chemical*, *192*, 42-50.

[69] Xi, Q., Zhou, D. M., Kan, Y. Y., Ge, J., Wu, Z. K., Yu, R. Q., Jiang, J. H. (2014). Highly sensitive and selective strategy for microRNA detection based on WS2 nanosheet mediated fluorescence quenching and duplex-specific nuclease signal amplification. *Analytical chemistry*, *86*(3), 1361-1365.

[70] He, Q., Wu, S., Yin, Z., Zhang, H. (2012). Graphene-based electronic sensors. *Chemical Science*, *3*(6), 1764-1772.

[71] Late, D. J., Huang, Y. K., Liu, B., Acharya, J., Shirodkar, S. N., Luo, J., Yan, A., Charles, D., Waghmare, U. V., Dravid, V. P., Rao, C. N. R. (2013). Sensing behavior of atomically thin-layered MoS2 transistors. *ACS Nano*, *7*(6), 4879-4891.

[72] Liu, B., Chen, L., Liu, G., Abbas, A. N., Fathi, M., Zhou, C. (2014). High-performance chemical sensing using Schottky-contacted chemical vapor deposition grown monolayer MoS2 transistors. *ACS Nano*, *8*(5), 5304-5314.

[73] Gatensby, R., McEvoy, N., Lee, K., Hallam, T., Berner, N. C., Rezvani, E., Winters, S., O'Brien, M., Duesberg, G. S. (2014). Controlled synthesis of transition metal dichalcogenide thin films for electronic applications. *Applied Surface Science*, *297*, 139-146.

[74] Ji, C., Lu, Z., Xu, Y., Shen, B., Yu, S., Shi, D. (2018). Self-production of oxygen system CaO2/MnO2@ PDA-MB for the photodynamic therapy research and switch-

control tumor cell imaging. *Journal of Biomedical Materials Research Part B: Applied Biomaterials*, *106*(7), 2544-2552.

[75] Wang, Y., Jiang, K., Zhu, J., Zhang, L., Lin, H. (2015). A FRET-based carbon dot–MnO2 nanosheet architecture for glutathione sensing in human whole blood samples. *Chemical Communications*, *51*(64), 12748-12751.

[76] Yan, X., Song, Y., Zhu, C., Song, J., Du, D., Su, X., Lin, Y. (2016). Graphene quantum dot–MnO2 nanosheet based optical sensing platform: a sensitive fluorescence "turn off–on" nanosensor for glutathione detection and intracellular imaging. *ACS Applied Materials & Interfaces*, *8*(34), 21990-21996.

[77] Cho, E. C., Glaus, C., Chen, J., Welch, M. J., Xia, Y. (2010). Inorganic nanoparticle-based contrast agents for molecular imaging. *Trends in Molecular Medicine*, *16*(12), 561-573.

[78] Yang, G., Gong, H., Liu, T., Sun, X., Cheng, L., Liu, Z. (2015). Two-dimensional magnetic WS2@ Fe3O4 nanocomposite with mesoporous silica coating for drug delivery and imaging-guided therapy of cancer. *Biomaterials*, *60*, 62-71.

[79] Liu, J., Zheng, X., Yan, L., Zhou, L., Tian, G., Yin, W., Wang, L., Liu, Y., Hu, Z., Gu, Z., Chen, C., Zhao, Y. (2015). Bismuth sulfide nanorods as a precision nanomedicine for in vivo multimodal imaging-guided photothermal therapy of tumor. *ACS Nano*, *9*(1), 696-707.

[80] Deng, R., Yi, H., Fan, F., Fu, L., Zeng, Y., Wang, Y., Li, Y., Liu, Y., Ji, S., Su, Y. (2016). Facile exfoliation of MoS2 nanosheets by protein as a photothermal-triggered drug delivery system for synergistic tumor therapy. *RSC Advances*, *6*(80), 77083-77092.

[81] Terreno, E., Castelli, D. D., Viale, A., Aime, S. (2010). Challenges for molecular magnetic resonance imaging. *Chemical Reviews*, *110*(5), 3019-3042.

[82] Caravan, P., Ellison, J. J., McMurry, T. J., Lauffer, R. B. (1999). Gadolinium (III) chelates as MRI contrast agents: structure, dynamics, and applications. *Chemical Reviews*, *99*(9), 2293-2352.

[83] Park, M., Lee, N., Choi, S. H., An, K., Yu, S. H., Kim, J. H., Kwon, S. H., Kim, D., Kim, H., Baek, S. I., Ahn, T. Y., Park, O. K., Son, J. S., Kim, Y. W., Wang, Z., Pinna, N, Hyeon, T. (2011). Large-scale synthesis of ultrathin manganese oxide nanoplates and their applications to T1 MRI contrast agents. *Chemistry of Materials*, *23*(14), 3318-3324.

[84] Duan, H., Guo, H., Zhang, R., Wang, F., Liu, Z., Ge, M., Yu, L., Lin, H. Chen, Y. (2020). Two-dimensional silicene composite nanosheets enable exogenous/endogenous-responsive and synergistic hyperthermia-augmented catalytic tumor theranostics. *Biomaterials*, *256*, 120206.

Chapter 10

Two Dimensional Nanosheets for Sensing Applications

Arnab Maity[1,*]
Vajinder Kumar[2]
Partha Pratim Bag[1]
and Atul Kumar[2]
[1]Department of Chemistry, SRM University Sikkim, 5th Mile, Tadong, Sikkim, India
[2]Department of Chemistry, Akal University, Talwandi Sabo, Bathinda, Punjab, India

Abstract

Two dimensional nanosheet materials as sensor become suitable for flexible chemical resistance, exceptional physicochemical, electrical, and mechanical properties. The electrical properties of these nanosheet sensors have been customized via suitable doping, surface modification and manufacture composite materials. Instead of pristine graphene, two dimensional nanosheets; phosphorous-doped, boron-doped or fluorine-doped graphene nanosheet sensors show evidence of better sensing response with either P-type or N-type sensing characteristics. Apart from graphene nanosheet sensors, several metal oxides, transition metal di-chalcogenides sensors and composites of metal oxide and metal di-chalcogenides also exhibit excellent sensing performance. These sensors are recognized for prompt response-recover behavior, low limit of detection and easy to fabricate on light weight electrode surface. Elucidation of various electrochemical sensor devices improvised the use of several nanosheets as active sensing materials have been discussed in simple and lucid manner in this article through citation of literature examples.

* Corresponding Author's Email: arnab.maity09@gmail.com.

In: Properties and Applications of Nanosheets
Editors: Nanda Gopal Sahoo and Anirban Dandapat
ISBN: 979-8-89113-381-5

Keywords: Pristine graphene, graphene nanosheet (GNS), P doped GNS, transition metal dichalcogenide (TMD), humidity sensors, P-type semiconductor, N-type semiconductor, exfoliated gas sensor

1. Introduction

Owing to their distinctive features nanomaterials are widely utilized in modern industrial science and technology. Two-dimensional (2D) nanomaterials possess exceptional qualities, which include a high surface-to-volume ratio, flexible chemical resistance, and remarkable physicochemical, electrical, and mechanical capabilities that are not seen in their bulk counterparts. In sensing, the interaction of foreign molecule with two dimensional materials have attracted great interested due to the two dimensional arrangement that allow the entire volume to be utilized for adsorption purpose where optical or electrical properties of 2D materials have altered [1, 2]. With the help of surface modification including defect engineering, fictionalization via chemical doping, and incorporation of metal nanoparticles; the sensitivity of chemical sensors can be enhanced many fold [3-5]. After the discovery of graphene in 2004 researchers found great interest in exploring the essential role of dimensionality in determining the intrinsic properties of nanomaterials and applied them as potential candidates for a extensive range of applications including energy conversion [6], storage device [7], catalysis [8] and sensors [9, 10]. Other types of 2D nanomaterials, such as transition metal dichalcogenide (TMD), transition metal oxide/sulfide, and boron nitride nanosheets, are also attracted a great deal of scientific attention for sensing property.

Most of the report published till date is based on surface modification of nanosheet where defects are purposefully introduced utilizing reactive ion etching plasma treatment, ultraviolet or wet chemical treatment to improve the low limit of detection, higher adsorption energy and ultrahigh performance and enhance charge transfer [11-14]. In the above Figure 1, various metal oxide, mixed metal oxide and graphene and doped graphene nanosheet sensors for selective detection of a variety of molecules are depicted. It is reported that by employing opposite category of substrate or divulging sensors to opposite dopant gases before sensing, the sensitivity to a particular type of dopant gases are improved [15]. Additionally, since the surface-to-volume ratios rise as the thicknesses decrease, the sensing performance of 2D materials was improved by adjusting their thicknesses [16]. In this chapter various sensors based on metal, metal oxide, composite, non-metal 2D nanosheet are discussed in lucid manner with various examples cited from different authors.

Most of the report published till date is based on surface modification of nanosheet where defects are purposefully introduced utilizing reactive ion etching plasma treatment, ultraviolet or wet chemical treatment to improve the low limit of detection, higher adsorption energy and ultrahigh performance and enhance charge transfer [11-14]. In the above Figure 1, various metal oxide, mixed metal oxide and graphene and doped graphene nanosheet sensors for selective detection of a variety of molecules are depicted. It is reported that by employing opposite category of substrate or divulging sensors to opposite dopant gases before sensing, the sensitivity to a particular type of dopant gases are improved [15]. Additionally, since the surface-to-volume ratios rise as the thicknesses decrease, the sensing performance of 2D materials was improved by adjusting their thicknesses [16]. In this chapter various sensors based on metal, metal oxide, composite, non-metal 2D nanosheet are discussed in lucid manner with various examples cited from different authors.

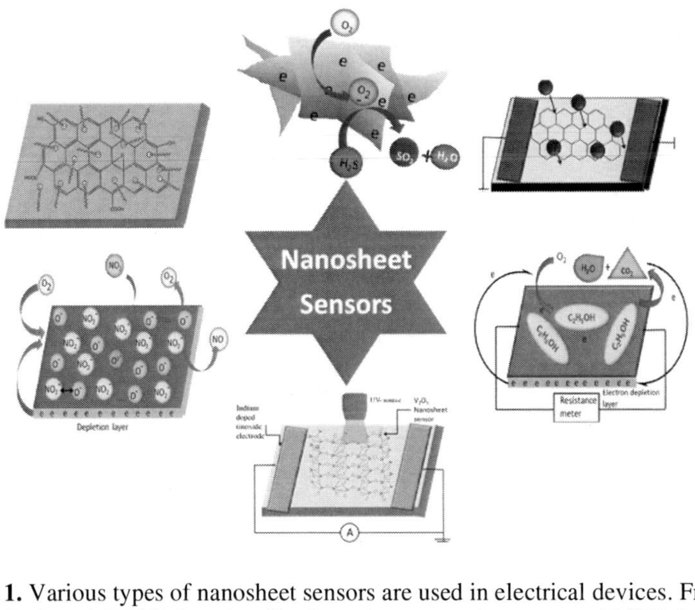

Figure 1. Various types of nanosheet sensors are used in electrical devices. From top left clockwise: a) CTAB functionalized graphene nanosheet sensor, b) WS_2 based selective NO_2 and H_2S sensor, c) N and Si doped graphene based NO_2 sensor, d) CeO_2-SnO_2 selective ethanol sensor, e) V_2O_5 coated selective humidity sensor, f) Mo-doped CeO_2-SnO_2 selective sensor for NO_2.

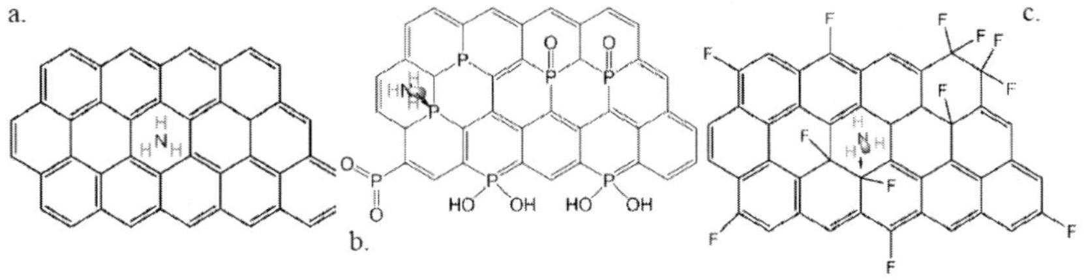

Figure 2. a) Schematic representation of adsorption mechanism of NH_3 on pure graphene nanosheet, b) phosphorous doped graphene nanosheet with adsorbed NH_3 molecule, and c) fluorine doped graphene nanosheet with adsorbed NH_3 molecule.

2. Graphene-based Gas Sensors

Graphene from the beginning of discovery attracts attention for its superior electrical conductivity, large surface area, very thin two dimensional structures etc. However, in comparison to pristine graphene the sensitivity of graphene derivative is promising [17]. There are several techniques through which the sensitivity of graphene can be improved, including doping with hetero atoms [4], incorporation of defects [18, 19], and functionalization of surface etc. [20].

2.1. Phosphorous Doped Graphene Nanosheet Sensor

The electronic characteristics of graphene might get changed due to the insertion of heteroatom that modulates the band gaps. Several doped graphene based sensor has been validated. By doping graphene nanosheet with phosphorous (P doped GNS) atoms and annealing mixtures of graphene oxide and triphenyl phosphine at high temperatures, researchers have created a highly sensitive NH_3 gas sensor [21] (Figure 2b). The P doped GNS based NH_3 sensors demonstrated significantly increased sensor responsiveness, short response-recovery time at room temperature, due to the adsorption of NH_3 onto the extra phosphorus atoms. The P doped GNS-400 for example confirmed a response value of 5.4% in presence of 100 ppm of NH_3. This was in contrast to thermally reduced graphene. Duan et al., developed a site-specific, fluorinated graphene, ion beam-induced functionalized approach that give graphene-based sensors with high response for NH_3 sensing as shown in Figure 2c. Fluorinated graphene demonstrates an eight-fold increase in sensing response to NH_3 as compared to pure graphene with a limit of detection value as low as 64 ppb [20].

2.2. Boron Doped Graphene Sensor

According to Srivastava et al., boron-doped graphene has been used to create a potential aspirant of NH_3 gas sensor. The dopant boron atom creates an electron shortage in the graphene sheet. Both pure and boron-doped few layer graphene demonstrate P-type sensing mechanisms, which are identified by changing the electric signal of graphene upon adsorbing NH_3 onto the sensors [4] (Figure 3). This sensor also has the advantages of being relatively stable for up to 25 days and being easily recoverable by exposing it to ambient air as shown in Figure 4.

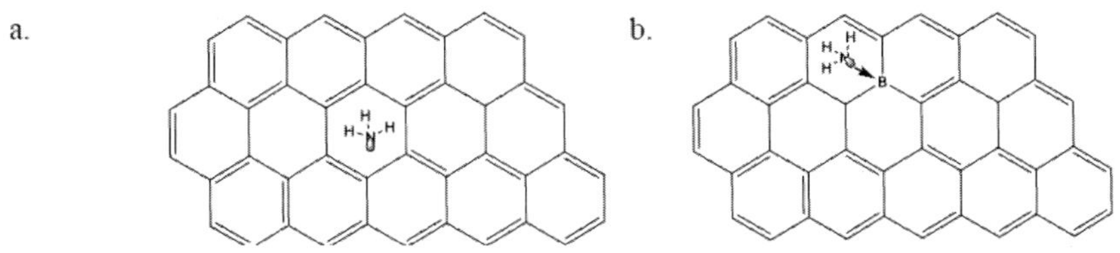

Figure 3. Schematic representation of adsorption of NH_3 molecule in a) un-doped graphene and b) boron doped graphene (reference [4]).

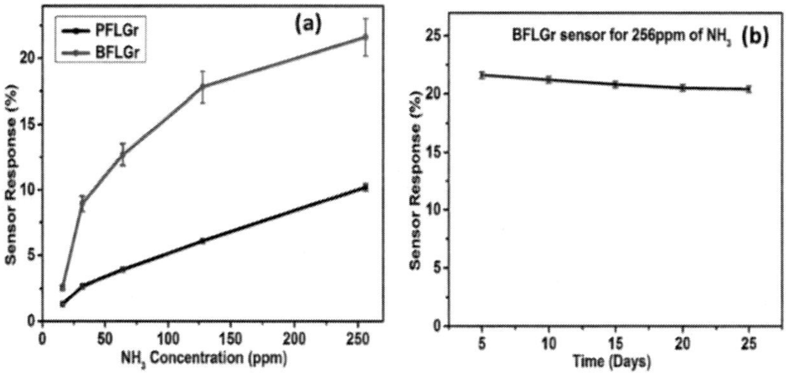

Figure 4. a) Sensor response plotted against NH_3 concentration in ppm with error bars, b) A stability study of the boron doped few layer graphene sensors for 256 ppm of NH_3 as a function of time in days with error bars (reprinted with the permission from ref [4]).

2.3. CTAB Functionalized Graphene Sensor

Graphene nanosheets can potentially be employed for the detection of hazardous dyes used in the industry. Mao et al., developed an electrochemical sensor of graphene for Sudan-1 dye where the surface of the electrode is covered with cetyl trimethyl ammonium bromide (CTAB) modified graphene nanosheet which eventually enhances the adsorption of Sudan-1 to the electrode surface as shown in Figure 5.

Since the hydrophobic tail of CTAB accumulates more Sudan-1, electrochemical signal has been improved significantly by coating the electrode surface with modified graphene nanosheet and a low limit of detection has been reported to be 0.7 $nmolL^{-1}$ [3].

2.4. Graphene Supported NO_2 Sensors

Nitrogen dioxide (NO_2) molecules can be preferentially sensed by graphene in a single sheet or sheets of few layers at ambient conditions. Fast, reversible, and selective NO_2 gas sensors were created by Ko et al., based on charge transfer processes. The repeatability of this sensor as air is substituted for NO_2 is one of its advantages. A single sheet of graphene can be used or the configuration of the sensor can be altered to increase the sensitivity. A sensitivity of 9% has been reported in presence of 100 ppm of NO_2 gas [22].

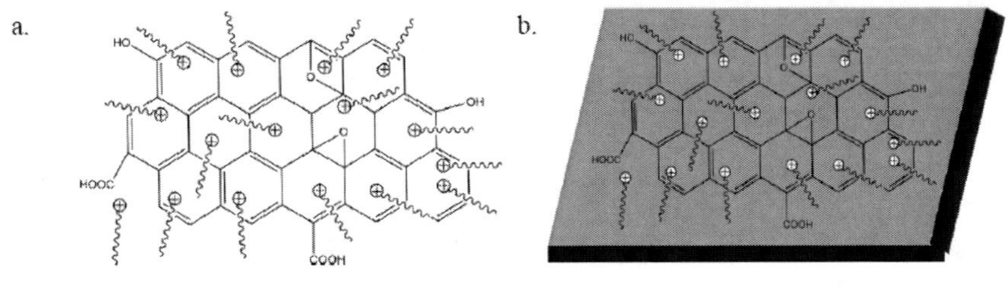

Figure 5. a) Graphic representation of CTAB functionalized graphene nanosheet and b) electrode modified with CTAB functionalized graphene nanosheet (reference [3]).

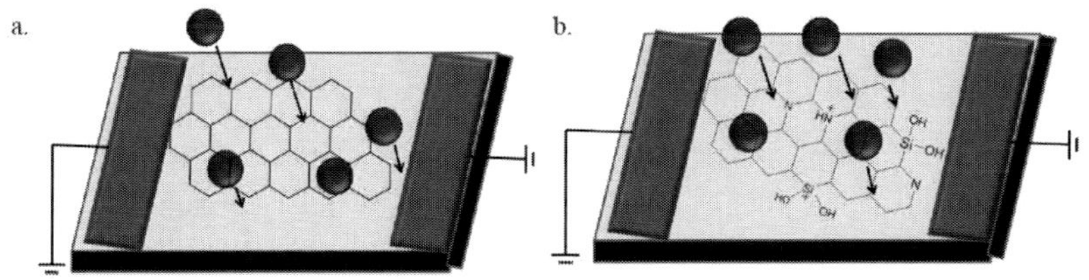

Figure 6. Graphical representation of NO_2 sensing on a) only graphene coated electrode and b) N, Si co-doped graphene coated electrode (Adapted from ref [23]).

In a related study, the co-doping of nitrogen (N) and silica (Si) in graphene sheets using high temperature annealing of N and Si-containing graphene oxide-ionic liquid composite improved the capacity to detect NO_2 gas as shown in Figure 6a and 6b [23]. When N and Si atoms are included in the graphene plane, the sensitivity to NO_2 gas improves because the N atom serves as the NO_2 absorption site and the Si atom improves the electrical behavior.

2.5. Graphene Oxide Sensor for Heavy Metal

Graphene oxide (GO) can also be used in the field of heavy metal detection with the aid of Forster Resonance Energy Transfer (FRET) mechanism between a dye-labeled T-rich ss-DNA and GO nanosheet. A dye-labeled ss-DNA can readily adsorbed to the surface of GO due to hydrophobic and π-π^* interaction and eventually extinguish its fluorescence via FRET mechanism. Kong et al., developed a heavy metal sensor depends on the concept of fluorescence quenching and recovery phenomena based on FRET-shell insulated metal nanoparticles surface enhanced fluorescence spectroscopy (SIMNSEF) [24]. In light of the above factors a combination of graphene oxide (GO), shell insulated anisotropic gold nanoparticles has been utilized for Hg^{2+} detection with the aid of Thiamine-riched single strand DNA functionalization with fluorescein probe. The ss-DNA encapsulated metal nanoparticles easily adsorbed on the GO surface and dumped the excited state energy of fluorescein to GO without further emission via FRET mechanism but in presence of Hg^{2+}, formation of DNA hairpin structure is stimulated due to T-Hg^{2+}-T interaction that simultaneously weakens the interaction between GO and DNA capped metal nanoparticles. From Figure 7 it is clear that the above consequence help to revive the fluorescence due to the surface enhanced fluorescence method and amplify the signal strength of fluorescein; signify the detection of Hg^{2+} with a limit of detection value of 0.3 nM [24].

3. Metal Oxide Nanosheet Based Sensors

3.1. Cobalt Oxide Humidity Sensor

Research on occurrence and detection of humidity is essential for the agricultural, culinary, cotton, pharmaceutical, and medical industries as well as human activities, therefore humidity sensors are required to detect and continually monitor environmental humidity. A humidity sensor built on complementary metal oxide semiconductor technology was described by Dai et al., in order to construct a micro electro mechanical system on chip that can detect humidity through changes in capacitance [25]. The sensitivity of this sensor has been reported to be 14.5 kHz/% RH at 25°C. In order to improve the sensitivity of the humidity sensor Yang et al., employed cobalt oxide as the active sensing unit and follow the same complementary metal oxide semiconductor, micro electro mechanical synthesis strategy [26]. The cobalt oxide sensor is connected with a seven stage ring oscillator circuit fabricated on chip that appropriately convert the small capacitance change of sensor to the frequency output upon adsorbing humidity. Porous architectures of cobalt oxide nanostructure help to adsorb humidity efficiently due to large contact area that clinches sensitivity result of 310 kHz/% relative humidity [26].

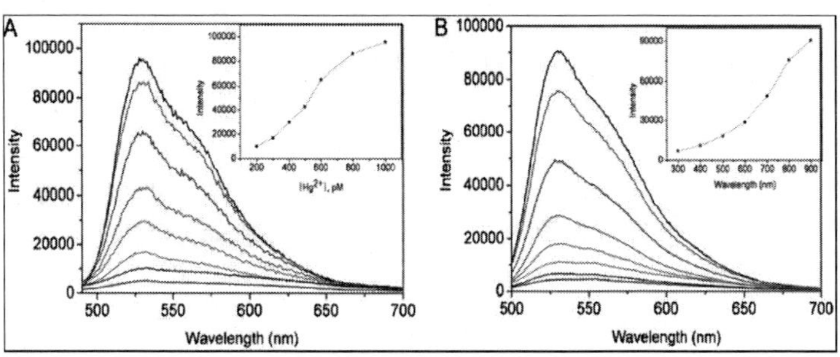

Figure 7. Fluorescence spectra of the GO/triangular gold nanoplate@SiO2/mercury specific oligonucleotide at different Hg^{2+} concentrations, (A) using the first sensing approach (Hg^{2+}: 0, 0.2, 0.3, 0.4, 0.5, 0.6, 0.8, and 1 nM, respectively), with the calibration curve for Hg^{2+} detection inset; (B) using the second technique (Hg^{2+}: 0, 0.3, 0.4, 0.5, 0.6, 0.8) Reprinted with the permission of reference [24].

3.2. Vanadium Pentaoxide Humidity Sensor

The sensitivity of humidity sensor has been enhanced by means of 2D transition metal oxide of highest oxidation state. Pawar et al., developed a V_2O_5 nanosheet humidity sensor photo detector via hydrothermal method with a bandgap of 2.2 eV to 2.7eV that correspond to the visible light [27]. Figure 8 describes the optical properties of V_2O_5 nanosheet sensor with a superior response and recovery time and a maximum photoresponsitivity of 6.2%. The sensitivity of the V_2O_5 photodetector depends on the number of water molecule adsorbed on the surface since the current decreases with increasing relative humidity value and a maximum sensitivity value of 45.3% has been reported.

3.3. Mixed CeO_2-SnO_2 Ethanol Vapour Sensor

Instead of using a single metal oxide as a gas sensor, a composite or doping with rare earth metal oxide can improve the sensing system's ability to identify a specific gas. By integrating a little amount of CeO_2 into the SnO_2 matrix, Yan et al., have created a very effective ethanol vapour sensor that may enhance the sensing performance of SnO_2 in terms of excellent responsiveness, quick reaction and recovery time, and an extraordinary selectivity towards ethanol vapour [28].

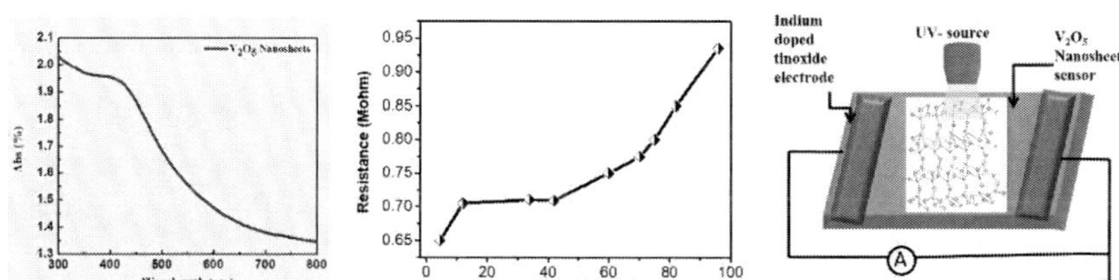

(a) (b) (c)

Figure 8. a) UV-Vis absorbance spectra of V_2O_5 nanosheet based sensor, b) change of resistance of V_2O_5 sensor with gradually increasing relative humidity value and c) schematics of electronics set up for humidity detection using V_2O_5 nanosheet sensor. Reprinted with permission of reference [27].

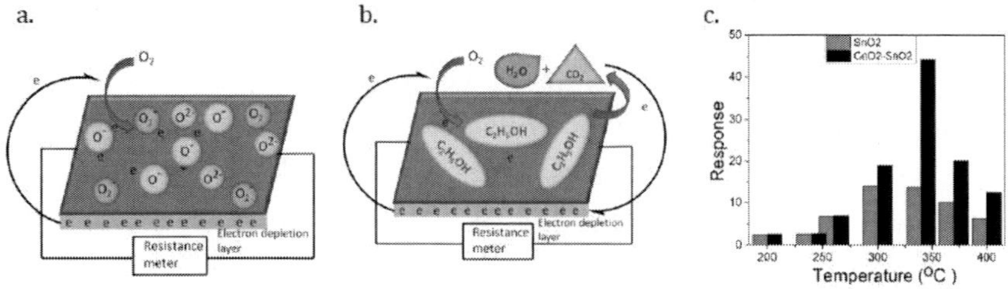

Figure 9. Schematic of gas sensing mechanism of CeO_2-SnO_2 nanosheet in presence of a) air, b) ethanol vapour and c) Sensor responses to 100 ppm ethanol at various operating temperatures (200-400°C) for SnO_2, CeO_2-SnO_2. Adapted from ref [28].

At a particular temperature (340°C) CeO_2- SnO_2 sensor exhibit brilliant selectivity for ethanol in presence of other vapour like glacial acetic acid, N, N-dimethyl formamide, ammonia, acetone etc. The presence of Ce^{4+}/Ce^{3+} species in SnO_2 has given rise to heterostructure, with the two distinct oxidation states and acid-base characteristics that regulating the electrical signal of the sensor. Figure 9 reveals that in presence of air, O_2 molecule getting converted into different surface adsorbed species such as O_2^-, O^- and O^{2-} with the help of electron from the conduction band, resulting into surface electron deficiency layer that increase the resistance of the surface. Whereas the ethanol vapor converted into CO_2 and water vapor under the influence of various oxygen species and simultaneously returning back the electrons to the conduction band that eventually decreases the resistance. Apart from that mixed CeO_2-SnO_2 sensor exhibit superior response and recovery time in comparison to SnO_2 alone (Figure 9c) [28].

4. Exfoliated Gas Sensors: Phosphorous Exfoliated NO_2 Sensor

In addition to conventional two-dimensional (2D) materials like graphene, transition metal oxides, or sulphides, 2D black phosphorus made of exfoliated black phosphorous can be extremely important in the selective sensing of gas. Flakes are supported on other materials in the majority of the research previously published that were based on 2D materials, and the interference effects of coulomb scattering at the interface or charge trapping cannot be ignored. Anisotropy is created in the matrix from the odd non-planar structures like armchairs or zigzags, which ultimately aid in the sensing properties. In contrast to supported flake sensors, Lee et al., utilized SiO_2/Si's supported black phosphorous (BP) suspended sensors, which were constructed on Ti/Au electrode surfaces, have the advantage that both of the BP surfaces are open for gas adsorption [29]. In comparison to a supported one, the suspended BP sensor likewise exhibits good sensitivity and repeatability, with an improved response value of 23% at 200 ppm of NO_2 gas.

5. Metal Di-Chalcogenide Exfoliated Gas Sensor

5.1. Tungsten Diselenide NO_2 Gas Sensor

Instead of using bulk transition metal di-chalcogenide, exfoliated one can works well due to large surface area, high affinity for NO_2 gas, improved energy of adsorption and quick response-recovery cycle. Guo et al., reported a room temperature WSe_2 nanosheet sensor synthesized via sonication mediated liquid phase exfoliation method with a short response time as 5.06 at 50ppb of NO_2 [30]. The exfoliated WSe_2 nanoheet provide a large surface area as well as plenty of active adsorption site for NO_2 in comparison to bulk WSe_2 that confirms an enhancement in response [31]. Similar to other transition metal di-chalcogenides, NO_2 molecule has been physisorbed on WSe_2 surface. Due to high electron acceptor character of NO_2 molecule extract electron from valence band of WSe_2 resulting in to the increase in the hole concentration. Thus resistance of WSe_2 sensor increases with NO_2 concentration and a P-type character is demonstrated.

5.2. Lithium Chloride Doped Tin Selenide Humidity Sensors

A semiconductor layer material with higher sensitivity, better linear response, prompt response and recover time is leading area of sensor delve into. Semiconductor tin selenide (SnSe) is best in class for its large natural resources, well tuned physical as well as chemical properties, innocuous to environment and broad range of humidity sensing capability which covers 0-97% of relative humidity (RH). Although semiconductor SnSe humidity sensors have exceptional repeatability and long-lasting stability, their use and further development are constrained by their low sensitivity caused by the slow desorption of water molecules. Instead of using SnSe alone, a composite material consists of SnSe nanosheet doped with LiCl nanoparticle proved to be a better humidity sensor in terms of very short response (12s) and recover time (4s) as suggested by Zhu et al., [5]. First water molecule chemisorbed on the LiCl:SnSe nanosheet surface and after two layer of thickness physisorption starts as the RH increases. Proton being the active current carrying species, the conduction of LiCl:SnSe composite show evidence of gradual enhancement of conductivity with increasing value of RH. The change of phase transition of LiCl to hydrated LiCl actually activate the LiCl:SnSe nanosheet surface for faster adsorption and desorption of water molecule [5].

5.3. Molybdenum Doped Tin Disulphide Gas Sensors

Two-dimensional metal dichalcogenides, like MoS_2, WS_2, SnS_2, etc., have the ability to detect NO_2 molecules especially at low temperatures [32, 33]. Transition metal oxide-based sensors, namely for NO_2, have been described, but the issue with these sensors is that they must run at higher temperatures to obtain good gas detecting kinetics, which restricts their practical utility [34]. Zhou et al., have solved this issue by solvothermal molybdenum (Mo) doping in layered SnS_2 nanosheets with selective and sensitive NO_2 detection [35]. Depending on the temperature and NO_2 concentration, a 3%Mo doped SnS_2 nanosheet, was claimed to demonstrate a switch in the sensing mechanism from P to N-type detection. At NO_2 concentration 25 ppm or below the sensor exhibits a P-type sensing mechanism with decreasing resistance while the same sensor clinch a N-type semiconductor behavior at NO_2 concentration of 50 ppm or above. The probable mechanism of P to N type sensing swift has been illustrated in accordance to DFT calculation [35]. Figure 10 demonstrated that a thin layer of oxygen is formed on the SnS_2 nanosheet upon exposure to O_2, due to chemisorbed O_{ads}^- and very few surface active for $NO_{2(ads.)}$ is available. At low NO_2 concentration, $NO_{2(ads.)}$ will occupy the surface and eventually $NO_2^-{}_{(ads.)}$ try to replace $O_{(ads)}-$ from the surface but as the distance between $NO_2^-{}_{(ads.)}$ and $O_{(ads)}-$ is sufficiently large substitution only play the crucial role resulting in the P-type sensing mechanism. (Reaction -1) [35].

$$NO_2(g) + 2O_{(ads.)}^- \rightarrow NO_{2(ads.)}^- + O_2(g) + e^- \tag{1}$$

At high NO_2 concentration, not only more $NO_2^-{}_{(ads.)}$ will substitute $O_{(ads)}-$ at the same time the relative distance between $NO_2^-{}_{(ads.)}$ and $O_{(ads)}-$ undergo further reaction. (Reaction-2) [35].

$$NO_{2(ads.)}^{-} + 2O(ads.)^{-} + e^{-} \rightarrow NO(g) + 1/2O_2(g) + 2O_2(ads.)^{-} \quad (2)$$

The following reaction scheme explains in detail how, at low NO_2 concentrations, electrons will be donated into the SnS_2 nanosheet, (Reaction - 1) decreasing the resistance of the sensor and resulting in the observation of P-type mechanisms, while, at high NO_2 concentrations, electrons will be donated from the SnS_2 nanosheet, (Reaction -2) increasing the resistance and exhibiting N-type behavior.

5.4. Tungsten Disulphide NO_2 and H_2S Gas Sensor

It should be mentioned that the oxygen molecule is important for the sensing mechanism of NO_2 gas on Mo doped-SnS_2 sensor in the previous study, although it is unclear whether O_2 has any effect on the selectivity of NO_2 gas. Liu et al., clarify the effect of local oxygen on the WS_2 nanosheet sensor's detecting mechanism with regard to the selectivity of NO_2 and H_2S gas [36]. Since NO_2 has a high affinity for electrons, it will physically adsorb on the surface of the WS_2 nanosheet sensor when it is exposed to NO_2 gas, collecting electrons from the surface and turning them into $NO_2^{-}{}_{(ads.)}$. The adsorbed O_2 molecule also converted to $O_{2(ads.)}^{-}$. (Reaction-3) The number of holes on the surface of the WS_2 nanosheet rises with charge transfer from WS_2 to NO_2, which lowers resistance and results in the appearance of a P-type sensing mechanism.

$$WS_2 \text{ (nanosheet)} + NO_2(g) + O_2 \rightarrow NO_{2(ads.)}^{-} + O_{2(ads.)}^{-} \quad (3)$$

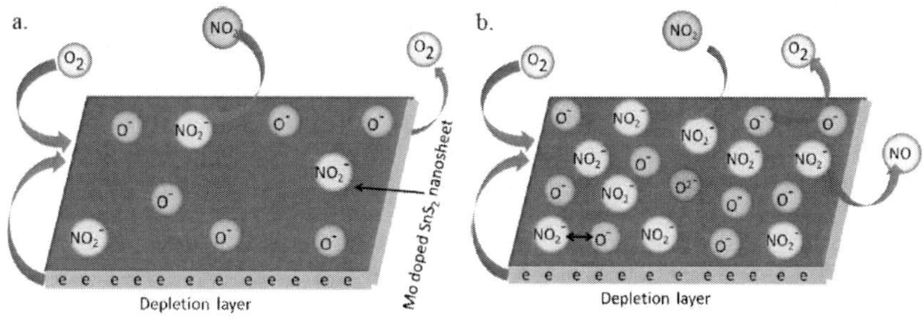

Figure 10. Schematic illustration of NO_2 sensing mechanism at low pressure a) and at high pressure b) by Mo doped CeO_2-SnO_2.

The sensing responsiveness of the WS_2 nanosheet decreased as O_2 concentration was raised while NO_2 concentration remained constant, and the temperature remained constant. Thus, any potential critical involvement of O_2 in the NO_2 sensing mechanism was eliminated. For H_2S sensing study first O_2 molecule are adsorbed on the surface of WS_2 nanosheet and converted to $O_2(ads.)^-$ in presence of trapped electron as shown in Figure 11. After that when this sensor is exposed to H_2S environment, it will be adsorbed and interact with $O_2(ads.)^-$ and eventually converted to SO_2 and H_2O. (Reaction-4) [36].

$$WS_2 \text{ (trapped electron)} + O_2 \rightarrow O_{2(ads.)}^- + WS_2 \text{ (untrapped electron)}$$

$$WS_2 \text{ (untrapped electron)} + 3O_{2(ads.)}^- + H_2S \rightarrow 2SO_2 + 2H_2O + WS_2 \text{ (trapped electron)} \quad (4)$$

The trapped electrons are eventually liberated from this proccss and returned to the WS_2 sensor's conduction band, where they recombine with the holes to raise the sensor's resistance. This is a hallmark feature of N-type sensing mechanisms, and the role of the O_2 molecule has also been shown.

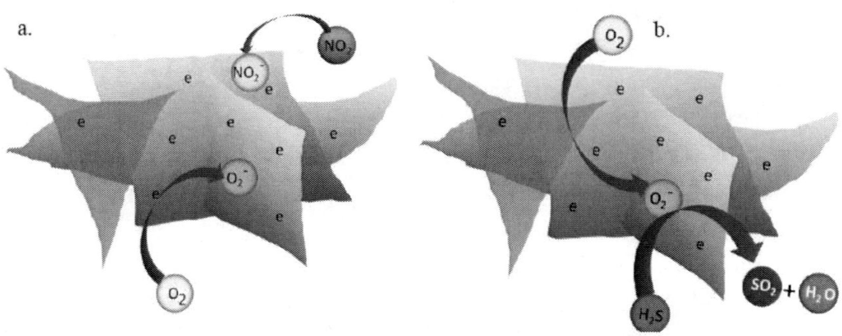

Figure 11. Schematic representation of NO_2 a) and H_2S b) sensing mechanism of WS_2 nanosheet, the scheme is illustrated depends on the mechanism described above.

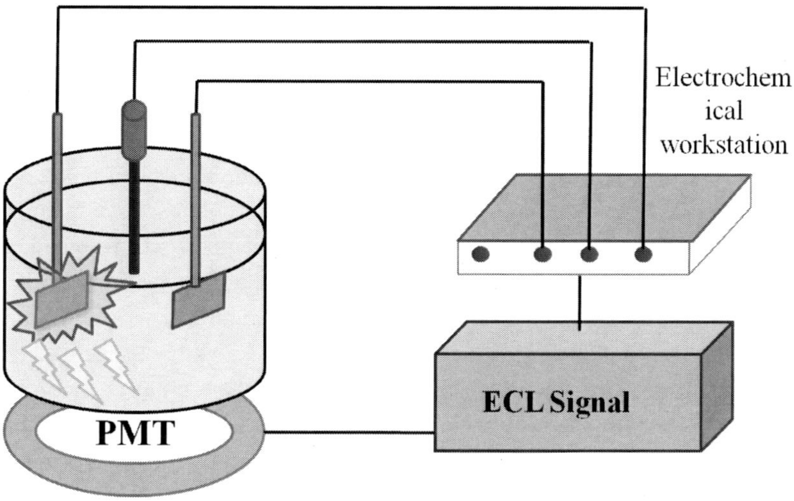

Figure 12. Schematic representation of a g-CN nanosheet based ECL sensor.

6. Electro Chemiluminescence Carbon Nitride Nanosheet Sensors

Electro chemiluminescence is a phenomenon where high energy electron transfer reaction undergone by the products generated at the electrode to form excited state species which emits photons [37]. Technically in presence of electrical energy in terms of voltage, a suitable precursor in close proximity of electrode will generate an intermediate species which subsequently interacts with co-reactants to produce an excited state that eventually emits lights. (Figure 12) Greatly tunable electronic structure of graphitic carbon nitride (g-CN) nanosheet owing to the presence of nitrogen in carbon network pave the way to be an efficient material for electro chemical luminescence (ECL) material.

Wang et al., developed a sensor for DNA and Hg^{2+} ion, derived from g-CN nanosheet for its strong affinity towards single stranded DNA (ss-DNA) over double stranded DNA (ds-DNA) and the fluorescence quenching of g-CN in presence of DNA or the fluorophore labeled to DNA. Within the medium, upon hybridization of probe analyte with target DNA via the formation of ds-DNA or depends on specific recognition of bases of the target analyte towards the probe DNA via intermolecular hybridization mechanism, the quenching efficiency is weakened which leads to the enhancement of fluorescence intensity. Shang et al., developed a dual-ECL signal dependent graphitic phase polymeric carbon nitride (GPPCN) luminophore sensor for selective detection of Ni^{2+} ion in presence of other ions [38]. Needless to mention that dual-ECL signal sensors has already been reported by Zhang et al., [39]. But most of them utilize multiple luminophore or label techniques. Apart from that the interference from other metal ions presented within the system can also be eliminated comparing the ECL signal both from cathodic potential as well as anodic potential [39]. Thin graphite like carbon nitride (g-C3N4) shows excellent sensitivity toward tryptophan moiety with a minimum detection limit of 0.024μM [40]. The synthesized g-C3N4 nanosheet has been effectively attached to the glassy carbon electrode that actually performs as sensors and demonstrates excellent linear relationship of Tryptophan concentration from 0.1 to 110μM.

7. Boron Nitride Nanosheet Sensors

Analogous to graphene, inorganic hexagonal boron nitride nanosheet (h-BNNSs) can also act as sensor. Two dimensional natures allow total exposure of all atoms of the nanaosheet to the adsorbing gas molecule that ultimately contribute to the sensitivity of the sensor materials. Few atomic layer thickness and chemical alteration of B and N atom within the layer generates ionic nature within the crystal that might enhance sensitivity to various gas molecules. Chemically inert and high thermal stability allows this material to be used in tough environment where use of other material is limited. Sajjad et al., developed a resistance based BNNSs on silicon substrate (Si) for efficiently detecting methane gas [41]. Upon adsorption of CH_4 gas on the surface of BNNSs nanosheet it undergo reduction that removes free electrons from the base materials and generate potential barrier at the grain boundaries of BNNSs results in the increases in the resistance of the materials. Lin et al., utlilized hexagonal boron nitride nanosheets (h-BNNSs) for the efficient detection of reduced organic gases [42]. Like metal oxide nanosheets molecular oxygen adsorbed on BNNSs oxidized to O_{ads}^{2-} via capturing of free electron from the conduction band resulting in the increase in resistance of the sensor which is typical behavior of n-type semiconductor. Ethanol vapour after adsorption on the surface of BNNSs converted to CO_2 and H_2O in presence of O_{ads}^{2-} resulting in the release of electron to the conductance band of BNNS leading to the sudden decrease in resistance.

$$C_2H_5OH\ (ads) + O_{ads}^{2-} \rightarrow CO_2(g) + H_2O\ (g) + 2e^- \quad (5)$$

It is also important to mention that although the sensitivity of h-BNNSs towards ethanol vapour is inferior to many other metal oxides based nanosheet sensors, such as ZnO, CdO/ZnO/Yb_2O_3 etc. [43-45] but due to short response time and excellent repeatability makes it a potential candidate for gas sensors.

Conclusion

Two dimensional nanomaterials as sensors is an excellent choice for their outstanding qualities like high surface to volume ratio, prompt physicochemical, electrical, mechanical characteristics, tunable chemical resistance etc that generally not found in their bulk materials. Graphene nanosheet, being one atomic thickness with high electron density can interact with several analyte molecules with simultaneous change in resistance. However, the sensitivity of pristine graphene can be enhanced via doping with suitable materials, creating surface defects as well as making suitable composite material where the charge transfer behavior has been altered many fold. Sometime the nanosheet sensor has been doped with opposite charge type species just to amplify the attraction for particular gas molecule. Modifying graphene nanosheet with surfactant molecule also enhances the interaction of dye molecule with nanosheet sensor in solution phase. Not only graphene nanosheet sensor many transition metal di-chalcogenides sensors proven to be appropriate for their response and recovery time, high temperature appliance and zig-zag like structure. Metal oxide nanosheet doping a with suitable redox type metal can serve as a better sensor for specific gas molecule and show evidence of distinctly different sensing mechanism for two gas molecules. Apart from all these benefits nanosheet sensors have inherent drawbacks which debars them using for ready applications. For example CeO_2-SnO_2 nanosheet sensor can achieve maximum sensitivity only at 340°C and sensitivity falls if temperature decreases from it. In presence of a mixture of gases specificity for a particular one is not always shown by a number of sensors and at the same time their bigger size, complicated fabrication technique and limited response-recovery cycle clearly demands for further improvement in this segment.

References

[1] Yuan, W. and G. Shi, Graphene-based gas sensors. *Journal of Materials Chemistry A*, 2013. 1(35) : p. 10078-10091.

[2] Schedin, F., Geim, A. K., Morozov, S. V., Hill, E. W., Blake, P., Katsnelson, M. I. and Novoselov, K. S., Detection of individual gas molecules adsorbed on graphene. *Nature Materials,* 2007. 6(9) : p. 652-655.

[3] Mao, Y., Fan, Q., Li, J., Yu, L. and Qu, L. B., A novel and green CTAB-functionalized graphene nanosheets electrochemical sensor for Sudan I determination. *Sensors and Actuators B : Chemical,* 2014. 203 : p. 759-765.

[4] Srivastava, S., Jain, S. K., Gupta, G., Senguttuvan, T. D. and Gupta, B. K., Boron-doped few-layer graphene nanosheet gas sensor for enhanced ammonia sensing at room temperature. *RSC Advances,* 2020. 10(2) : p. 1007-1014.

[5] Zhu, K., Tang, Y., Zhong, X., Xiong, L., Zhang, Y., Tan, C., Song, H. and Wang, J., Improved Response/Recovery Time and Sensitivity of SnSe Nanosheet Humidity Sensor by LiCl Incorporation. *Advanced Electronic Materials,* 2020. 6(5) : p. 1901330.

[6] Aizudin, M., Goei, R., Ong, A. J., Tan, Y. Z., Lua, S. K., Pottammel, R. P., Geng, H., Wu, X. L., Tok, A. L. Y. and Ang, E. H., Sustainable development of graphitic carbon nanosheets from plastic wastes with efficient photothermal energy conversion for enhanced solar evaporation. *Journal of Materials Chemistry A*, 2022. 10(37) : p. 19612-19617.

[7] He, P. and S. Chen, Vertically Oriented Graphene Nanosheets for Electrochemical Energy Storage. *ChemElectroChem,* 2021. 8(5) : p. 783-797.

[8] Wang, X., Ma, Y., Wu, Q., Wen, Y. and Xiao, F. S., Zeolite nanosheets for catalysis. *Chemical Society Reviews*, 2022. 51(7) : p. 2431-2443.

[9] Liu, C., Zhang, B., Chen, W., Liu, W. and Zhang, S., Current development of wearable sensors based on nanosheets and applications. *TrAC Trends in Analytical Chemistry*, 2021. 143 : p. 116334.

[10] Kumar, A., Zhao, Y., Mohammadi, M. M., Liu, J., Thundat, T. and Swihart, M. T., Palladium Nanosheet-Based Dual Gas Sensors for Sensitive Room-Temperature Hydrogen and Carbon Monoxide Detection. *ACS Sensors,* 2022. 7(1) : p. 225-234.

[11] Zhang, Y. H., Chen, Y. B., Zhou, K. G., Liu, C. H., Zeng, J., Zhang, H. L. and Peng, Y., Improving gas sensing properties of graphene by introducing dopants and defects: à first-principles study. *Nanotechnology,* 2009. 20(18) : p. 185504.

[12] Lee, G., Kim, J., Kim, K. and Han, J. W., Precise control of defects in graphene using oxygen plasma. *Journal of Vacuum Science & Technology A*, 2015. 33(6) : p. 060602.

[13] Li, F., Peng, H., Xia, D., Yang, J., Yang, K., Yin, F. and Yuan, W., Highly Sensitive, Selective, and Flexible NO_2 Chemiresistors Based on Multilevel Structured Three-Dimensional Reduced Graphene Oxide Fiber Scaffold Modified with Aminoanthroquinone Moieties and Ag Nanoparticles. *ACS Applied Materials & Interfaces*, 2019. 11(9) : p. 9309-9316.

[14] Kaniyoor, A., Jafri, R. I., Arockiadoss, T. and Ramaprabhu, S., Nanostructured Pt decorated graphene and multi walled carbon nanotube based room temperature hydrogen gas sensor. *Nanoscale,* 2009. 1(3) : p. 382-386.

[15] Gupta, R. K., Alqahtani, F. H., Dawood, O. M., Carini, M., Criado, A., Prato, M., Garlapati, S. K., Jones, G., Sexton, J., Persaud, K. C. and Dang, C., Suspended graphene arrays for gas sensing applications. *2D Materials*, 2021. 8(2) : p. 025006.

[16] Cui, S., Pu, H., Wells, S. A., Wen, Z., Mao, S., Chang, J., Hersam, M. C. and Chen, J., Ultrahigh sensitivity and layer-dependent sensing performance of phosphorene-based gas sensors. *Nature Communications*, 2015. 6(1) : p. 8632.

[17] Fowler, J. D., Allen, M. J., Tung, V. C., Yang, Y., Kaner, R. B. and Weiller, B. H., Practical Chemical Sensors from Chemically Derived Graphene. *ACS Nano,* 2009. 3(2) : p. 301-306.

[18] Cagliani, A., Mackenzie, D. M. A., Tschammer, L. K., Pizzocchero, F., Almdal, K. and Bøggild, P., Large-area nanopatterned graphene for ultrasensitive gas sensing. *Nano Research,* 2014. 7(5) : p. 743-754.

[19] Hajati, Y., Blom, T., Jafri, S. H. M., Haldar, S., Bhandary, S., Shoushtari, M. Z., Eriksson, O., Sanyal, B. and Leifer, K., Improved gas sensing activity in structurally defected bilayer graphene. *Nanotechnology*, 2012. 23(50) : p. 505501.

[20] T Duan, T., Li, H., Daukiya, L., Simon, L. and Leifer, K., Enhanced Ammonia Gas Adsorption through Site-Selective Fluorination of Graphene. *Crystals,* 2022. 12 : p. 1117.

[21] Niu, F., Tao, L. M., Deng, Y. C., Wang, Q. H. and Song, W. G., Phosphorus doped graphene nanosheets for room temperature NH3 sensing. *New Journal of Chemistry*, 2014. 38(6) : p. 2269-2272.

[22] Ko, G., Kim, H. Y., Ahn, J., Park, Y. M., Lee, K. Y. and Kim, J., Graphene-based nitrogen dioxide gas sensors. *Current Applied Physics,* 2010. 10(4) : p. 1002-1004.

[23] Niu, F., Liu, J. M., Tao, L. M., Wang, W. and Song, W. G., Nitrogen and silica co-doped graphene nanosheets for NO2 gas sensing. *Journal of Materials Chemistry A,* 2013. 1(20) : p. 6130-6133.

[24] Kong, L., Wang, J., Zheng, G. and Liu, J., A highly sensitive protocol (FRET/SIMNSEF) for the determination of mercury ions : à unity of fluorescence quenching of graphene and enhancement of nanogold. *Chemical Communications,* 2011. 47(37) : p. 10389-10391.

[25] Dai, C.-L., A capacitive humidity sensor integrated with micro heater and ring oscillator circuit fabricated by CMOS–MEMS technique. *Sensors and Actuators B : Chemical,* 2007. 122(2) : p. 375-380.

[26] Yang, M. Z., Dai, C. L., Shih, P. J. and Chen, Y. C., Cobalt oxide nanosheet humidity sensor integrated with circuit on chip. *Microelectronic Engineering*, 2011. 88(8) : p. 1742-1744.

[27] Pawar, M. S., Bankar, P. K., More, M. A. and Late, D. J., Ultra-thin V 2 O 5 nanosheet based humidity sensor, photodetector and its enhanced field emission properties. *RSC advances,* 2015. 5(108) : p. 88796-88804.

[28] Yan, S., Liang, X., Song, H., Ma, S. and Lu, Y., Synthesis of porous CeO2-SnO2 nanosheets gas sensors with enhanced sensitivity. *Ceramics International*, 2018. 44(1) : p. 358-363.

[29] Lee, G., Kim, S., Jung, S., Jang, S. and Kim, J., Suspended black phosphorus nanosheet gas sensors. *Sensors and Actuators B : Chemical,* 2017. 250 : p. 569-573.

[30] Guo, R., Han, Y., Su, C., Chen, X., Zeng, M., Hu, N., Su, Y., Zhou, Z., Wei, H. and Yang, Z., Ultrasensitive room temperature NO2 sensors based on liquid phase exfoliated WSe2 nanosheets. *Sensors and Actuators B : Chemical,* 2019. 300 : p. 127013.

[31] Hu, J., Zou, C., Su, Y., Li, M., Yang, Z., Ge, M. and Zhang, Y., One-step synthesis of 2D C3N4-tin oxide gas sensors for enhanced acetone vapor detection. *Sensors and Actuators B : Chemical*, 2017. 253 : p. 641-651.

[32] Xia, Y., Hu, C., Guo, S., Zhang, L., Wang, M., Peng, J., Xu, L. and Wang, J., Sulfur-Vacancy-Enriched MoS_2 Nanosheets Based Heterostructures for Near-Infrared Optoelectronic NO_2 Sensing. *ACS Applied Nano Materials*, 2020. 3(1) : p. 665-673.

[33] Huang, Y., Jiao, W., Chu, Z., Wang, S., Chen, L., Nie, X., Wang, R. and He, X., High Sensitivity, Humidity-Independent, Flexible NO2 and NH3 Gas Sensors Based on SnS2 Hybrid Functional Graphene Ink. *ACS Applied Materials & Interfaces*, 2020. 12(1) : p. 997-1004.

[34] Yan, W., Lv, C., Zhang, D., Chen, Y., Zhang, L., Ó Coileáin, C., Wang, Z., Jiang, Z., Hung, K. M., Chang, C. R. and Wu, H. C., Enhanced NO2 Sensitivity in Schottky-Contacted n-Type SnS_2 Gas Sensors. *ACS Applied Materials & Interfaces*, 2020. 12(23) : p. 26746-26754.

[35] Zhou, J., Xue, K., Liu, Y., Liang, T., Zhang, P., Zhang, X., Zhang, W. and Dai, Z., Highly sensitive NO2 response and abnormal P-N sensing transition with ultrathin Mo-doped SnS_2 nanosheets. *Chemical Engineering Journal,* 2021. 420 : p. 127572.

[36] Liu, D., Tang, Z. and Zhang, Z., Comparative study on NO2 and H2S sensing mechanisms of gas sensors based on WS2 nanosheets. *Sensors and Actuators B : Chemical,* 2020. 303 : p. 127114.

[37] Richter, M. M., Electrochemiluminescence (ECL). *Chemical Reviews,* 2004. 104(6) : p. 3003-3036.

[38] Shang, Q., Zhou, Z., Shen, Y., Zhang, Y., Li, Y., Liu, S. and Zhang, Y., Potential-Modulated Electrochemiluminescence of Carbon Nitride Nanosheets for Dual-Signal Sensing of Metal Ions. *ACS Applied Materials & Interfaces,* 2015. 7(42) : p. 23672-23678.

[39] Zhang, H. R., Xu, J. J. and Chen, H. Y., Electrochemiluminescence Ratiometry: A New Approach to DNA Biosensing. *Analytical Chemistry,* 2013. 85(11) : p. 5321-5325.

[40] Liu, X., Zhang, J., Di, J., Long, Y., Li, W. and Tu, Y., Graphene-like carbon nitride nanosheet as a novel sensing platform for electrochemical determination of tryptophan. *Journal of Colloid and Interface Science*, 2017. 505 : p. 964-972.

[41] Sajjad, M. and Feng, P., Study the gas sensing properties of boron nitride nanosheets. *Materials Research Bulletin,* 2014. 49 : p. 35-38.

[42] Zhan, Y., Yan, J., Wu, M., Guo, L., Lin, Z., Qiu, B., Chen, G. and Wong, K. Y., Boron nitride nanosheets as a platform for fluorescence sensing. *Talanta,* 2017. 174 : p. 365-371.

[43] Gai, L. Y., Lai, R. P., Dong, X. H., Wu, X., Luan, Q. T., Wang, J., Lin, H. F., Ding, W. H., Wu, G. L. and Xie, W. F., Recent advances in ethanol gas sensors based on

metal oxide semiconductor heterojunctions. *Rare Metals*, 2022. 41(6) : p. 1818-1842.

[44] Bhardwaj, R. and K. G. Sharma, Semiconductor metal oxide based ethanol gas sensor using ZnO : A short review. *AIP Conference Proceedings*, 2020. 2297(1) : p. 020027.

[45] Rahman, M. M., Alam, M. M., Asiri, A. M. and Islam, M. A., Ethanol sensor development based on ternary-doped metal oxides ($CdO/ZnO/Yb_2O_3$) nanosheets for environmental safety. *RSC advances*, 2017. 7(37) : p. 22627-22639.

Chapter 11

Wastewater Treatment on Carbon Supported Nanosheets

Bholanath Panda
Julekha Khatun
and Debasis Dhak*
Nanomaterials Research Laboratory, Department of Chemistry,
Sidho-Kanho-Birsha University, Purulia, WB, India

Abstract

Water sources have become contaminated with heavy metals and dangerous toxins as a consequence of the rapid development. Researchers have developed a variety of methods to get dangerous toxins out of the aquatic environment. Nowadays, there is growing interest in using composite materials based on graphene oxide as effective adsorbents for wastewater treatment, desalination, separation, and purification. High mechanical strength, hydrophilicity, amazing flexibility, simplicity of synthesis, atomic thickness, and compatibility with other materials are a few of their notable characteristics. High separation performance and stable graphene-based laminar structures have been the primary objectives in water treatment. The best performance in water filtration has been achieved using nanocomposites treated with graphene oxide. This chapter focusses on the fabrication of GO/rGO based nanocomposites as well as the fundamental characterization procedures required to assess the physiochemical and structural characteristics of the graphene-based nanocomposites. The possible research difficulties for remediating water contaminants such

* Corresponding Author's Email: debasis.chem@skbu.ac.in.

In: Properties and Applications of Nanosheets
Editors: Nanda Gopal Sahoo and Anirban Dandapat
ISBN: 979-8-89113-381-5

harmful metals, dyes using synthesized GO/RGO based nanocomposite as an adsorbent are also addressed.

Keyword: graphene oxide/reduced graphene oxide, water pollution, adsorption, photodegradation, mitigation

1. Introduction

Water plays a vital role in daily life, due to the necessity of water for all human activities, and the fact that there are several causes of pollution such as organic dyes, pesticides, fluoride, arsenic, heavy metals, etc. [1]. Membrane separation is one of the most important techniques to remove those pollutants. Modification of membrane structural properties such as hydrophilicity, charge and charge density, thermal and mechanical stability, porosity, and roughness, nanosheets offer the chance to create customized membranes. They may be able to improve the balance between permeability and selectivity while reducing the tendency for fouling, which can be a major barrier to membrane separations. The invention of new membrane materials, adjustment of operating conditions, management of membrane structure, and the implementation of pretreatment have all contributed to significant advancements in membrane efficiency and quality. Advancements in nanosheet technology including layered oxide, reduced graphene oxide, graphene oxide, layered dichalcogenides, molybdenum sulfide, exfoliated transition metal carbides, metal-organic framework in two dimensions, nitrides, covalent organic framework or organic framework were chosen as a potential choice for constructing advance membrane. Characteristics of all these nanosheets such as their nano porous structure, nanochannel, and nanoscale thickness provide low transport resistance, therefore allowing for rapid water transportation and filtration [2–7]. This study mainly focused on the preparation and exploitation of graphene oxide and reduced graphene oxide based nanosheets and their behavior of the different toxic elements present in wastewater. GO nanosheet is typically produced using processes such as crystal growth, chemical vapor deposition, mechanical exfoliation from bulk graphite, and others. Their nanosheets are obtained via sonication exfoliation, solvent-assisted exfoliation, mechanical exfoliation, chemical exfoliation, and so on [8, 9]. The modification of nanosheets is implemented to improve the characteristics and performance of nanosheets [10–12]. Typical modification methods include crosslinking nanosheets with other organic matter with rich functional groups [13], self-crosslinking [14], and introducing inorganic particles or multivalent M^{n+} into nanosheet layers [15].

2. Causes of water pollution and their effects

Freshwater pollution originates from many sources, including municipal, industrial, and agricultural waste, wastewater and nutrient run-off, power generation, heavy industry, automobiles, and others. Around the world, some 2 billion tons of human waste are disposed of in water courses every day. The main types of water pollution include groundwater pollution, surface water pollution, suspended matter, oil spillages, microbiological pollution, chemical water pollution, thermal pollution, and oxygen-depletion pollution. A schematic representation of different pollution is shown in Figure 1. All of these can cause pollutants and contaminants to enter our waterways. Consumption of polluted water is a major cause of ill health in the world. Surface water is mainly polluted by industrial waste and groundwater pollution mainly comes from different rocks and leaching out of surface water pollutants. Dyes pollutant mainly comes from the textile industry and most of the dyes are very much carcinogenic. Fluoride and arsenic pollution come from the leaching out of different fluoride and arsenic containing rocks which cause skeletal fluorosis and backfoot diseases and ultimately cancer. Besides that, heavy metals nowadays are a major concern because long-term heavy metal exposure causes several health issues like Alzheimer disease, multiple sclerosis, muscular dystrophy, Perkinson's disease, hypertension, diabattius mellitus, and ismechic heart disease, etc. For the above reason, World Health Organization (WHO) set a permissible limit on the above chemical pollutant in drinking water and these data are listed in Table 1.

Table 1. A list of chemical pollutant and their permissible limit in drinking water

Chemical Pollutant	Permissible limit
Fluoride	1.5 mg/L
Arsenic	10 μg /L
Lead	10 μg /L
Mercury	06 μg /L
Uranium	30 μg /L
Cadmium	03 μg /L
Copper	2000 μg /L
Chromium	50 μg /L
Nitrate	10 mg/L

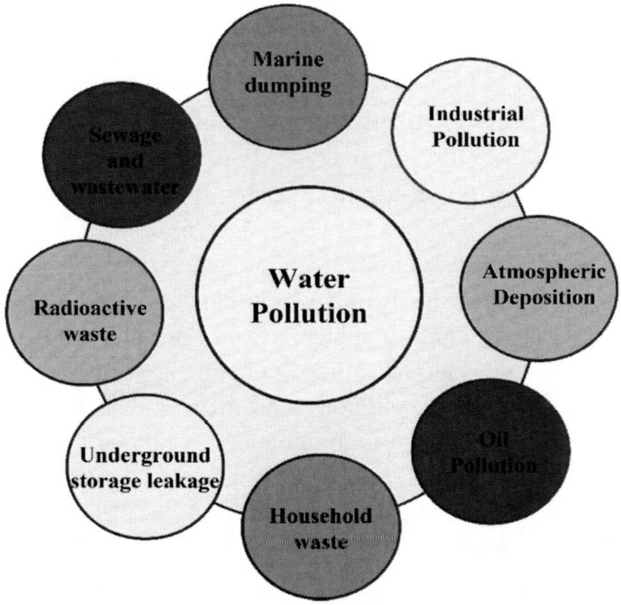

Figure 1. Schematic diagram depicting the various sources of water pollution.

3. Preparation of Graphene Oxide/Reduced Graphene Oxide nanosheet

There are two main carbons supported nanosheets one is a graphene based nanosheet and the other is a reduced graphene oxide based nanosheet.

Brodie (1859), Staudenmaier (1898) and Hummers and Offeman (1958), and others have developed methods to produce a significant amount of GO. According to Brodie (1859), graphite oxide (GrO) can be produced by oxidizing graphite flakes with a solution of potassium perchlorate and strong nitric acid [16]. Staudenmaier (1898) produced GrO by oxidation of graphite using a solution of sulfuric acid, nitric acid, and potassium perchlorate [17]. Further, Hummers and Offeman (1958) synthesized GrO by graphite oxidation with a mixture of potassium permanganate, sodium nitrate, and concentrated sulfuric acid [18].

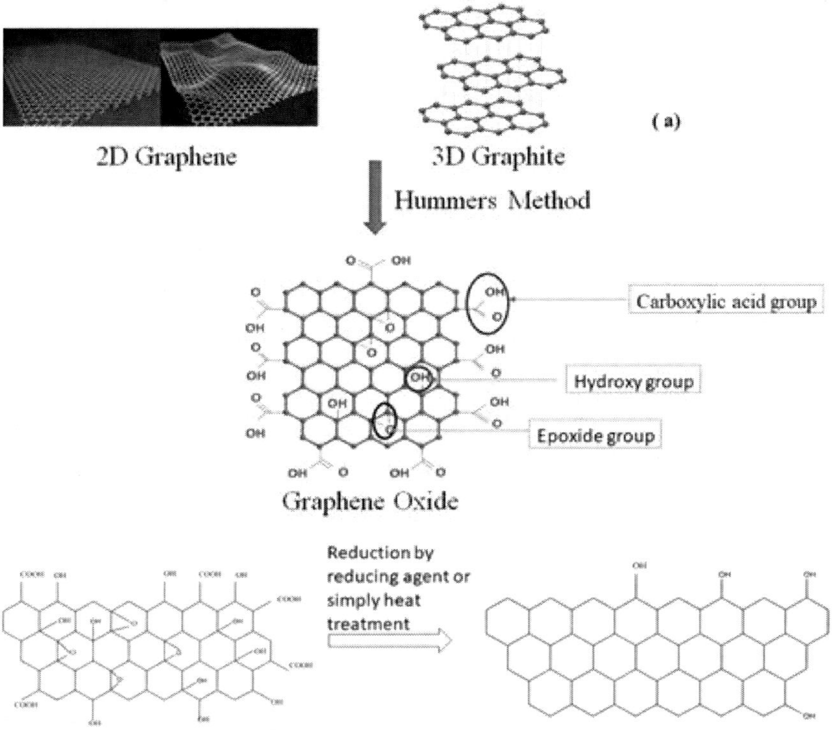

Figure 2. Preparation of graphite to graphene oxide by Hammer's method (a). Graphene oxide to reduced graphene oxide (b).

GrO has a structure of graphitic layers with variable interlayer distances [19–25]. Increased interlayer spacing reduces physical interaction between layers and increases the probability of GrO which will produce from the exfoliation from GO via mechanical stirring or ultrasonic procedure [20]. The increased oxygenous functional groups that cause the synthesis of exfoliated GO also help the increased interlayer distance [19, 21–23, 26–32]. The concentration of exfoliated GO is therefore substantially higher in water than in organic solvent [33]. Most of the researchers later, stated an improved method of Hummers' approach for the direct synthesis of GO. Nowadays GO preparation typically uses the modified Hummers method [20, 22, 23, 25, 26, 34–38]. Two methods have commonly been used. A concentrated H_2SO_4:H_3PO_4 solution (360:40 mL), 3g of graphite powder, and a definite amount of $KMnO_4$ were slowly added under continuous stirring for 36 hours at 50 °C. After obtaining a reddish solution, it was chilled to room temperature and mixed with around 400 ml of ice water that contained 3 mL of 30% H_2O_2. After that, the resulting suspension was centrifuged and repeatedly washed with HCl, water, and methanol. Khatamian et al. used a different method in which they combined 5 g of graphite with 2.5 g of $NaNO_3$ and 130 mL of 98% H_2SO_4 and stirred the mixture continuously for 1 hour at 0 °C. After that, 15 g $KMnO_4$ was added to the mixture slowly under continuously stirring and the mixture was kept for 12 h in a reflux condition. Next, the suspension was mixed with 350 ml of water, and the reaction temperature rose quickly to 98 °C. When a yellow precipitate was formed, 200 mL of 10% HCl, 200 mL of water, and 60 mL of 30% H_2O_2 were gently added to the suspension. After filtering, the mixture was washed with HCl and water [39]. The reduced graphene oxide can be produced by dissolving 400 mg graphene oxide in 800 mL distilled water and while stirring and heating at 95 °C [40] or by reduction of graphene oxide with a reducing agent such as ascorbic acid [41], substituted sodium borohydride [42]. Preparation of graphene oxide from 2D graphene and 3D graphite by modified Hummer's method (Figure 2a) and its reduction to reduced graphene oxide is graphically represented in Figure 2b.

4. Modification of GO and RGO by different nanocomposite

The use of GO for water filtration is quite appealing, however it has the drawback of recovering after adsorption. The development of magneto responsive GO can solve this problem [43]. By adding magnetic particles to the GO matrix, which are easily retrieved by the application of an external magnetic field, magneto responsive GO can be synthesized. There have been many magneto-responsive GO synthesized. In large-scale water treatment, this method enables the use of "virgin" GO as an effective adsorbent. Different magneto-responsive GO have been synthesized using the conventional co precipitation method, such as, GO-$MnFe_2O_4$ [44], FeO_x-GO [45], α-FeOOH@GCA [46], and among others. The precursors ferric chloride ($FeCl_3$. H_2O), ferrous chloride ($FeCl_2$), or ferrous sulphate ($FeSO_4$), for the preparation of FeO_x-GO [45], and ferric chloride ($FeCl_3.6H_2O$), and manganese sulphate ($MnSO_4.H_2O$), for the preparation of GO-$MnFe_2O_4$. The pH of the solution was then adjusted to 10–10.5 by adding an alkaline solution such NaOH solution or ammonia solution, and the temperature of the solution was then elevated up to 80 °C while being continuously stirred. After a predetermined time interval, the reaction was stopped and cooled to room temperature. The resultant precipitate was separated magnetically, repeatedly washed with methanol and water, and then allowed to dry overnight [46]. Modified their method to create a graphene oxide-carbon nanotube aerogel that is coated with α-FeOOH (α-FeOOH@GCA). Briefly, 100 mL of deionized water was dissolved in GO:CNT (7:3) and 1.81 g of ferrous sulphate heptahydrate ($FeSO_4.7H_2O$) (0.065 M), and the mixture was then sonicated for 30 minutes at room temperature. The resultant homogeneous mixture was further hydrolyzed in an oil bath for 6 hours at 95 °C to produce the α-FeOOH@GCA hydrogel. The synthesized hydrogel was finally collected after being dialyzed with deionized water and lyophilized using a freeze drier. There are many oxygen atoms on the graphitic backbone of GO, which extend from its surface in the form of epoxy, hydroxyl, and carboxyl groups. These groups can easily connect to other functionalized materials, such as chitosan, via coordination, electrostatic interaction, and covalent interaction, and give a lot of active sites for charged ions [20].

Chitosan-functionalized GOs were synthesized by Kumar and Jiang in 2016. Each chitosan glucosamine monomer has one amino and two hydroxyl groups, which function as interaction sites. The electrostatic attraction between the chitosan's amino group (NH_2), primary and secondary hydroxyl groups (CH_2OH, CHOH), and the surface carboxyl (COOH) and hydroxyl (OH) groups on the GO results in synthesis of chitosan-functionalized GO. In this procedure, GO (1.0 g) was first combined with a 50 mL solution of thionyl chloride ($SOCl_2$)-dimethylformamide (DMF), and the mixture was then refluxed at 60 °C for 24 hours. The GO-COCl precipitate that was produced was washed with tetrahydrofuran (THF) by several times, and finally vacuum-dried. After GO-COCl (1.0 g) was obtained, it was slowly mixed with chitosan solutions (0.5 g of chitosan in 10 mL of 1% (v/v) acetic acid solutions) under continuous stirring for 10 min until a uniform mixture was obtained. The mixture was then gradually heated to 50 °C under continual stirring for 3 h, and the final product was then washed with acetone several times, filtered, and dried [47].

Kumar and Jiang synthesized$^{\beta}$-Cyclodextrins (CDs) functionalized GO ($^{\beta}$-CDs@GO hybrid), magnetic NPs decorated with $^{\beta}$-CDs-functionalized GO (Fe_3O_4-$^{\beta}$-CDs@GO hybrid), and polyelectrolyte of poly (diallyl dimethyl ammonium chloride) (pDADMAC) with paramagnetic anions based on $FeCl_4$-GO (Fe-poly DADMAC@GO) [48].

Khatamian et al. (2017) used the solid-state dispersion (SSD) approach to modify the GO with copper doped-Zeolite (Cu-ZEA) as Cu-Zeolite/GO, Fe_3O_4/RGO/Cu-ZEA, and Fe_3O_4/GO/Cu-ZEA. By dispersing 1 g of zeolite (ZEA) into 100 mL of a solution of 0.02 M CuSO4.5H2O while stirring continuously for 24 hours, the first precursor of Cu-doped zeolite (Cu-ZEA) was produced. The obtained suspension of Cu-ZEA was then filtered, washed, and dried for further studies. Cu-ZEA/GO was synthesized by dispersing GO:Cu-ZEA (1:1) in 15 mL of ethanol using ultrasound for 3 hours. After that, the solution was stirred until the solvent completely evaporated and finally dried at 60 °C. Similarly, the solid-state dispersion (SSD) technique was used to create the Fe_3O_4/RGO/Cu-ZEA composite [39]. The synthesized samples were characterized by various instrumental techniques and used for chemical water pollutant samples. A general method for the preparation of graphene oxide or reduced graphene oxide metal oxide /hydroxide composite are shown in Figure 3.

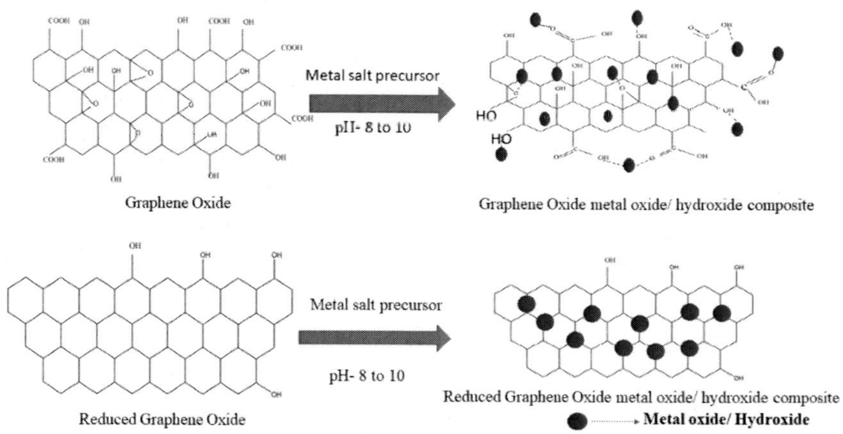

Figure 3. A general method for the preparation of graphene oxide/ reduced graphene oxide metal oxide/ hydroxide composite.

5. Characterization of GO, RGO, and its different composite

Generally, characterization of GO, RGO, and its composite was done by FTIR, XRD, XPS, and Raman Spectroscopy and the morphology by SEM.

The characteristics absorption peak from 3415-3210, 2840-2960, 1705-1735, 1615-1640, 1410-1545, 1225-1410, and 1015-1070 cm^{-1} appeared in the Fourier-transform infrared (FTIR) spectrum in the graphene oxide and its different composite [44, 45, 47]. The bands in the (O-H) stretching range were 3415-3210 cm^{-1} [47]. The presence of peaks around 1735-1705 cm^{-1} confirmed the presence of the C=O group, while the absorption peaks in the

2960-2840 cm^{-1} region confirmed the C-H stretching of the CH_3 and CH_2 groups [45, 47]. By presenting the peaks in the region at 1640–1615 cm^{-1}, the C=C stretching was identified [45]. The C-H deformation bond was confirmed by the absorption bands in the frequency range of 1445-1410 cm^{-1}, and for epoxy and alkoxy, a band range at 1410–1015 cm^{-1} was seen due to C-O stretching vibration [45]. The characteristics peaks of the oxygenated functional groups always appeared in FTIR spectrum confirmed the preparation of GO matrixes for further modification. The doping of metal and metal oxides in the GO matrix has been confirmed by identifying the absorption peaks of M-O between 750–400 cm^{-1} [45]. The FeOx-GO nanocomposite have the two additional vibration peaks at 552 and 442 cm^{-1} for Fe-O. Similarly, two additional absorption peaks at 490 and 577 cm^{-1} were observed in FTIR spectrum for M-O stretching vibrations of manganese ferrite of $MnFe_2O_4$-GO [44]. The FTIR spectra of GO/Cu-ZEA showed the characteristic vibration peaks at 1027, 704, 602 and 449 cm^{-1}, which were in conformity with the TO_4 (T= Si and Al) tetrahedral of zeolite (ZEA) in GO/Cu-ZEA [39]. Peak at 1027 cm^{-1} confirms the presence of ZEA in the composite structure, while vibration frequency at 704 cm^{-1} is associated with internal vibration of the TO_4 tetrahedral moiety of the parent ZEA [39]. Chitosan showed corresponding vibration bands at the following frequencies: 3345 cm^{-1} for OH and NH stretching vibrations, 2878 cm^{-1} for C-H stretching vibration in CH_3 and CH_2, 1655 cm^{-1} for NH bending vibration in NH_2, 1424 and 1380 cm^{-1} for NH deformation vibration in NH_2, and 1087 cm^{-1} for CO stretching vibration in C-OH. In contrast, the two significant peaks that appeared at 1630 and 1522 cm^{-1} were attributed to the C=O stretching vibration of NH-CO (amide I) and the N-H bending from NH_2, respectively. The FTIR spectrum of the chitosan-GO composite showed a strong peak around 3452 cm^{-1}, due to the stretching vibration of O-H, the extension vibration of N-H, and the intermolecular hydrogen bonds of the polysaccharide [47]. The peaks at 3452 (O-H) and (C-O) 1087 cm^{-1} had greater intensities than those of GO, due to covalent interactions between OH of GO with chitosan. The peak at 1728 cm^{-1}, which was assigned to carboxylic groups, changed to a lower frequency (1630 cm^{-1}), as a result of the reaction between the carboxylic groups and the NH_2 group of chitosan, which corresponded to stretching vibrations of the C-N bond [47]. Chandra et al. also observed the Similar result [34].

The transformation of graphite into GO is shown by the shift in the X-ray diffraction (XRD) peak from 2θ value, 26.6° to 11.20° [49]. Based on literature the XRD peaks occurred in the range 2θ = 23.0°-23.5°, which matched to the

(002) plane of RGO, whereas the XRD peaks appeared in the range 2ϴ = 9.0°-11.20°, which related to the (001) plane of GO [39]. Sometimes the partial reduction of GO occurs during the synthesis of its composites, the peak corresponding to the reflection plane of the substance is decreasing. In previous studies, the XRD pattern of Fe_3O_4/GO composite have shown the strong diffraction peak corresponded to the (001) plane of GO at near 2ϴ = 10.40°. The relatively weak peaks at 2ϴ = 18.57° (111), 30.34°(220), 35.46° (311), 43.38° (400), 53.03°(422), 56.88°(511) and 62.55°(550) attributed the presence of Fe_3O_4 in Fe_3O_4/GO composite [50]. This finding was similar to the previously reported pattern [44]. However, for Fe_3O_4/RGO nanocomposite the sharp peak of GO at 2ϴ= 10.43° (001) vanished in the case of Fe_3O_4/RGO composite, and a new, broad peak was seen at 2ϴ = 24.5° (002), indicating the coexistence of RGO and Fe_3O_4 [39]. Since the NPs have grown on the surface of the GO to prevent its agglomeration, the peak corresponding to the GO is significantly reduced in the XRD patterns of the different GO composites. The XRD patterns of rGO, GO, α-FeOOH, and α-FeOOH@GCA were described for the case of goethite anchoring on GO-CNTs aerogel [46]. This study showed that the sharp, narrow peak that the XRD pattern of GO attributed to it at 2ϴ = 10.6°(001) for GO had disappeared in the XRD pattern of RGO, and a new peak that corresponded to the (002) plane of RGO had appeared at 2ϴ=26.5°. This is because the reduction process turned GO into RGO. The orthorhombic phase of α-FeOOH in α-FeOOH@GCA was confirmed by the XRD pattern of α-FeOOH@GCA, which displayed relative diffraction peaks at 2ϴ = 21.2° (110), 3.2° (130), 35.5° (101), 36.1° (040), 39.1° (200), 41.2° (140), 53.2° (221), 57.4° (231) and 61.4° (002). Due to the complete inclusion of α-FeOOH NPs into the GO sheets, the diffraction peaks of GO in the α-FeOOH@GCA composite could not be observed [46]. This result was matched with Joint Committee on Powder Diffraction Standards (JCPDS) File No. 29-0713. JCPDS maintains the Powder Diffraction File (PDF) including the d spacing and relative intensities of observable diffraction peaks which is used to identify the substance on the basis of X-ray diffraction data. The diminishing of GO peaks in XRD pattern can also be understood by the XRD pattern of the GO and its modified form; FeOx–GO-36, FeOx–GO-80, and the iron oxide control sample given by [45]. Interlayer spacing between stacked GO sheets arise the strong XRD peaks at 2ϴ = 12° in GO XRD spectrum which was absent in the FeOx–GO-36 and FeOx–GO-80. This is due to the complete exfoliation of GO sheets by iron oxide. The broad peak at 43° corresponded to the plane (100) also observed in the spectrum of GO which contributed to the peak for graphitic structures. The control sample, iron oxide showed sharp and

strong diffraction peaks that also agreed with Fe_3O_4 (JCPDS File No. 75-0033) or α-Fe_2O_3 (JCPDS File No. 39-1346). XRD pattern of FeOx–GO-36 (low content iron oxide) and FeOx–GO-80 (high content of iron oxide) showed the diffraction peaks corresponded to the plane (104) and (110) which agreed the diffraction pattern of α-Fe_2O_3 (JCPDS File No. 80-2377), however, others peaks of diffraction pattern of α-Fe_2O_3 was largely invisible in FeOx–GO-36 than FeOx–GO-80, this might be due to the low iron oxide content in the FeOx–GO-36 [45]. These XRD results confirm the modification in GO.

X-ray photoelectron spectroscopy was also used to characterize GO and its composites (XPS). The only method for determining the elemental composition, empirical formula, chemical state, and electronic state of the elements present in a material is XPS [34, 45]. From the literature, XPS studies revealed that GO samples have a 2:1 ratio of graphitic carbon (36%) to oxidized carbon (64%) which implies that each GO sheet has substantial oxygenous groups along its borders, and basal planes such as carboxyl, epoxy, and hydroxyl group, etc. [45]. XPS spectrum of GO showed its C1s spectrum peaks at near 285.0, 287.0, 289.0, and 290.0 eV corresponding to the carbon sp^2 (C-C), epoxy/hydroxyls (C-O), carbonyl (CO), and carboxylates (O-C), respectively. Due to the O^{2-} and O-H bond present on the GO, the O 1s spectrum peaks appeared to be near the 529.0-533.0 eV [34, 45, 48]. GO are largely dispersible in water by ultrasonic treatment due to a large number of the functional group present on GO, and they may easily form a composite with other functionalized material. The XPS spectrum for FeOx-GO-36 and FeOx-GO-80 suggested that on the loading of iron oxide into the GO having 40.4 at. % of C and 55.0 at % of O, the counts of C 1s (34.5 and 13.1 at. % for FeOx–GO-36 and FeOx–GO-80, respectively) and O 1s (36 and 27.7 at. % for FeOx–GO- 36 and FeOx-GO-80, respectively) decrease and the counts of Fe (25.2 at. % for FeOx–GO-36 and 57.7 at. % FeOx-GO-80) increased dramatically [45]. The composite of Fe_{2p} XPS spectra revealed two photoelectron peaks with energies of 711.1 eV (Fe $2p_{3/2}$) and 724.6 eV (Fe $2p_{1/2}$), which were associated with the Fe^{3+} and Fe^{2+}, respectively. FeOx-GO-36 and FeOx-GO-80 were found to have molar ratios of Fe^{3+}/Fe^{2+} of 2.92 and 1.87, respectively, while iron oxide, the control sample, was calculated to have a molar ratio of 2.03 in both cases. These Fe^{3+}/Fe^{2+} molar ratios for FeOx-GO-80 and the control sample were close to the predicted theoretical value for Fe_3O_4, indicating that Fe_3O_4 was present. However, this molar ratio for FeOx-GO-36 was different from the predicted value, indicating that Fe^{3+} ions were more strongly complexed with oxygen functions on GO than Fe^{2+} ions. For oxygen atoms bound to O-Fe, the samples deconvoluted O 1s spectra show a peak at 531.6. On the contrary, the former one is absent in GO. The deconvoluted peak intensity associated to C atoms in C-O in the C 1s spectra of FeOx-GO-36 and FeOx-GO-80 was lower while that for C-C bonds was higher. This implied that some of the GO had been chemically changed into graphene (CCG). The GO-chitosan composite's wide-scan XPS spectra identified two peaks with binding energies of 285.1 eV and 530.9 eV, which were associated with the C 1s and O 1s, respectively. The sp^2-hybridized carbon, C-N, C-C, $C_{1/4}C$, $C_{1/4}N$, and C-O species each had five photoelectron peaks in the C 1s core level spectra at 283.2, 284.3, 284.8, 285.3, and 286.0 eV respectively [47]. So, the information regarding the elemental composition, empirical formula, chemical state, and electronic state of the elements present in a material was provided by the XPS analysis of GO and their composite.

Raman spectroscopy is a tool that is used to get valuable evidence for carbonaceous materials such as GO. It has been used frequently to describe GO and its compounds. The Raman spectrum for GO obtained from various studies showed two characteristics peaks from 1550 to 1630 and 1320–1370 cm^{-1} corresponding to the first order E_{2g} mode from sp_2 carbon domains (G-band) and disorder mode (D-band), respectively. The incorporation of nanoparticles like Fe_2O_3, $MnFe_2O_4$, Cu, and Zr into GO has produced the identifiable peaks for M-O-GO in the range of 450-650 cm^{-1}. The Raman spectrum for $MnFe_2O_4$-GO showed peaks associated to disorder mode (D-band) and first order E_2g mode from sp_2 carbon domains for GO, as well as manganese ferrite, $MnFe_2O_4$, at about 600 cm^{-1} [44]. The Raman spectrum of GO, β-CDs@GO, Fe_3O_4NPs, and β-CDsGO@Fe_3O_4 NPs indicated corresponding peaks at 1360 and 1603 cm^{-1} for GO related to D and G bands, respectively. However, the D band shifted slightly from 1360 to 1354 cm^{-1}, and the G band shifted slightly from 1603 to 1600 cm^{-1} upon functionalization of β-CD onto the GO [47]. The structural distortions caused by the inclusion of Fe_3O_4 NPs into β-CDs@GO, β-CDs-GO@Fe_2O_3 displayed a slightly higher intensity ratio of ID/IG than GO [48]. The chitosan modified GO spectra exhibited two significant peaks at 1610$cm^{-1}$and 2445 cm^{-1} in the Raman spectrum corresponding to the G band and D band, as well as one broad peak at 1360 cm^{-1}. Due to the interaction of the -COOH and -OH group of chitosan with GO, two additional peaks at 718 and 1047 cm^{-1} in chitosan modified-GO were also observed (Kumar and Jiang, 2016). Both Chandra et al. (2010) and Chen et al. (2010) reported similar results. Scanning electron microscopy (SEM) images of graphite, graphite flakes, GO, and their composites can be used to study their morphology. The Sem images that describe the surface morphology showed that pure graphite exhibits veins and that graphite flakes are composed of intrinsic graphite morphology. According to the surface morphology characterization of GO and RGO, GOs are very porous, amorphous solids, homogeneous, and interconnected structures. Moreover, there are lots of leaf-like veins that come from scrolling GO [47]. Additionally, RGO SEM images clearly indicate a wrinkled edge [46]. Large oxygenous functional groups that are present on the surface of GO allow for homogeneous decoration and easy interaction with other functionalized materials, NPs, or materials. Transmission electron microscopy (TEM) images collected from literature [46] showed that graphene oxide has a nanosheet structure and also a uniform distribution of nanoparticles on the surface of the graphene oxide.

6. Mitigation of water contaminant by carbon based nanosheets

Carbon based nanosheets like Graphene oxide, reduced graphene oxide, etc were often introduced into some substrates to prepare membranes for water treatment. Nanosheets with hydrophilic groups and nanochannels played a significant role in the selective sieving of chemical pollutants and improved the characteristics and efficiency of advanced membranes. The outstanding mechanical properties of the membranes allowed for the fabrication of ultrathin membranes which results in high water flux, low filtration resistance, and low energy consumption.

6.1. Graphene Oxide and Reduced Graphene Oxide based nanosheet for water treatment

GO is the most commonly used nanosheet in membranes for its easy preparation method and moderate performance. Different scientists use different Graphene oxide-based nanosheets for wastewater treatment. In this section, we discussed different graphene based and reduced graphene oxide based nanosheets for the treatment of different chemical pollutants.

Nowadays, several process industries, including those in the paper, textile, paint, plastic, and leather sectors, need pigments and dyes to color their products. Excess color in the outflow results in industrial effluent. These dyes are toxic molecules, and their presence in an aquatic environment influences photosynthesis for aquatic plants by preventing sunlight from penetrating and diminishing aesthetics. Overall, the danger posed by hazardous substances, which are resistant to a chemical reaction in wastewater and can cause cancer, mutagenesis, and other serious issues in people and aquatic life, is increased by the presence of dyes. Due to dyes' complex chemical structures, they are incredibly resistant to biodegradation [51, 52]. As a result, these colors must be effectively removed before the effluent is released into the neighboring aquatic environment. The greatest concern has prompted numerous researchers and scientists to create technologies and corrective procedures for its elimination. Methylene Blue, Methylene Orange, Malachite Green, Congo Red, BR-12, Azo Acid Red 14, Acid Orange 8, Acid Yellow 99, Amido Black 10B, Persian Orange, Reactive Orange 12, and a few other common colors are frequently detected in wastewater and effluent. To remove the colored chemicals from effluent water and wastewater, numerous physicochemical techniques have already been established. One of these techniques is photocatalytic degradation [53–56] separation by the membrane [57], coagulation process [58], electrolysis [59], liquid-liquid extraction [60], and adsorption [61, 62]. Since adsorption is a simpler and more effective technique of dye removal than other available technologies, it has found widespread use [63–67]. Nanoparticles with magnetic characteristics have been widely used in recent years for the adsorption of hazardous metal ions and organic contaminants. This is because of qualities including decreased toxicity, greater chemical stability, favorable magnetic characteristics, simplicity of synthesis, and superior recycling capacity. These tiny particles give the structure a great degree of mechanical robustness and surface area, which increases its ability to remove pollutants from the environment. With great accuracy and precision, poisonous dyes and heavy metals from aquatic environments are adsorbents based on magnetic nanoparticles [68–70]. However, in atmospheric conditions, the bare magnetic nanoparticles are easily oxidized. Researchers, therefore, investigated techniques to coat or functionalize magnetic nanoparticles to stabilize the nanoparticles in harsh environments and strengthen the functional groups to prevent easy oxidization and extend the life of nanoparticles. Due to its surface-attached, oxygenated functional groups of exfoliated graphite oxide and the high surface area-to-volume ratio, the association of graphene has resulted in a significant removal capacity of water contaminants. According to Chandra et al. rGO and GO-supporting materials have a greater binding capacity than bare nanoparticles [34]. In aqueous media, heavy metals and ionic dyes typically coexist, hence most researchers use techniques to remove each simultaneously. To explore the simultaneous removal of Cd(II) and ionic dyes like orange Gelb (OG) and methylene blue, Deng et al. developed MGO [12]. According to their findings, the maximal adsorption capacities of multi-walled carbon nanotubes/Fe_2O_3, magnetic rectorite/iron oxide, and graphene nanosheet/magnetite for the elimination of MB were 31.18, 43.82, and 42.30 mg/g, respectively. Additionally, they noted that the q_{max} of MGO towards OG (20.85 mg/g) and MB (64.23 mg/g) was significantly higher than that of exfoliated graphene oxide (5.98 for OG and 17.3 mg/g for MB) [71]. In the presence of GO substrate, Yang et al. synthesized reduced graphene oxide zero-valent nickel and NiAl-mixed metal oxide (rGO/Ni/MMO). The ability of this magnetic hybrid nanomaterial to remove MO from aqueous environments [72]. Abdi et al. (2018) developed a magnetic graphene-based composite (MMGO) and incorporated it in a polyether sulfone (PES) polymer, which serves as a membrane to study the copper removal and dye retention capacities [73]. They noticed that the dye (direct red) rejection for the MMGO hybrid inserted manufactured nanofiltration membrane was about 99%, and that this was true for the MMGO hybrid with variable concentrations. Eriochrome Black T (EBT) from dye wastewater was investigated by Khurana et al. using MGO in a batch procedure [74]. In this study, MGO was synthesized and impregnated with α-Fe_2O_3 to remove hazardous textile azodye EBT. The dye removal qmax was found to be 210.53 mg/g. Additionally, they investigated the effects of process variables like pH, contact time, and adsorbent dosage. They found that the maximum equilibrium adsorption capacity was achieved with MGO (5 mg) in a dye solution (50 mL), pH (2), and a contact time of 5 h. One of the most popular photocatalysts used is titanium dioxide (TiO2). GO has been found to improve the photocatalytic activity of immobilized TiO2 because GO sheets can help separate photo-induced electron-hole pairs, preventing charge recombination [75]. TiO_2 nanoparticle agglomeration prevention can be made by GO-TiO_2 nanocomposite formation at higher concentrations of TiO_2 by simple mixing or by sol-gel process [76]. A list of various graphene oxide/reduced graphene oxide-based nanocomposites that can significantly reduce various dyes are given in Table 2.

Table 2. List of various graphene-based material applied for the photodegradation of various dyes

Graphene based material	Properties	Ref
CdS-rGO composites	Photocatalytic reduction of Cr(VI)	[77]
$CoFe_2O_4$-functionalized graphene sheets	Adsorption of MO	[78]
$CoFe_2O_4$-graphene nanocomposite	Catalytic degradation of dyes (MB, RhB, MO, active red RGB, and active black BLG)	[79]
Graphene oxide	Adsorption of MB	[80]
Graphene-Fe_3O_4 composites	Removal of organic dyes	[81]
Graphene oxide-Fe_3O_4 hybrid	Adsorption of MB and cationic dyes	[82]
Graphene-TiO_2 nanocrystal hybrids	Photocatalytic degradation of RhB	[83]
Reduced graphene oxide	Adsorption of anionic dyes	[71]
$ZnFe_2O_4$-graphene nanocomposites	Photo-electrochemical degradation for organic molecules	[84]
ZnO/graphene nanocomposites	Photocatalytic degradation and filtered removal of RhB dye	[85]

Table 3. Compilation of different graphene oxide/reduced graphene oxide-based nanocomposites for removal of heavy metal

Adsorbent	Q_{max}(mg/g)	Ref
Graphene Oxide	Co(II) – 68.2 @ 30°C Cd(II) – 106.3 @ 30°C	[105]
Glycine functionalized GO	Cu(II)-72.6 Ni(II)-38.61	[106]
GO membrane	Cd(II)-83.8 Ni(II)-62.3	[107]
GO nanosheets coated with Fe_3O_4 nanoparticles	Cu(II)-11.339	[108]
Ethylene diamine tetraacetic acid functionalized MGO (EDTA-mGO)	Pb(II) - 508.4 @45°C Cu(II) - 301.2 @45°C Hg(II) - 268.4 @45°C	[109]
Ethylene diamine tri acetic acid graphene oxide (EDTA-GO)	Pb(II) – 479	[110]
Sulfonated magnetic graphene oxide composite (SMGO)	Cu(II) – 50.678 @ 10°C Cu(II) – 56.857 @ 30°C Cu(II) – 63.670 @ 50°C	[95]
Chitosan/graphene oxide ($CSGO_5$) composite CS:GO → 0.0263 g:0.5 g CS: GO → 0.0556 g:0.5 g CS: GO → 0.0882 g:0.5 g	Au(III) - 1076.6 Pd(II) - 216.9 Au(III) – 925.9 Pd(II) – 173.0 Au(III) – 869.6 Pd(II) – 160.0	[50]
Glutaraldehyde cross-linked chitosan (GCCS)	Au(III) – 990.1 Pd(II) – 180.2	[50]
sodium alginate/graphene oxide (SAGO) aerogel	Cu(II) – 98.0 Pb(II) – 267.4	[111]

Adsorbent	Q_{max}(mg/g)	Ref
Polypyrrole–reduced graphene oxide (PPy-RGO) composite	Hg(II) – 980.	[112]
Mn-dopedFe_3O_4 nanoparticle embedded graphene (GMIO)	Cd(II) – 87.2 @15°C Cd(II) – 101.1@30°C Cd(II) – 117.3@45°C Cd(II) – 127.1@60°C Cu(II) – 129.7@15°C Cu(II) – 133.2@30°C Cu(II) – 140.3@45°C Cu(II) – 144.3 @60°C	[113]
Fe_3O_4nanoparticles immobilized with silica	Cu(II) – 1.2649 Pb(II) – 2.0877 Cd(II) – 0.5282	[114]
Magnetic $CoFe_2O_4$ – rGO	Pb(II) – 299.4 Hg(II) – 157.9	[115]
Chitosan-magnetite	Pb(II) – 63.33 Ni(II) – 52.55	[116]
Triethylenetetramine-modified graphene oxide/chitosan composite (TGOCS)	Cr(VI) – 219.5	[117]
porous inverse spinel MGO composite	Pb(II) – 82.47 Cr(III) – 87.49	[118, 119]
porous inverse MGO nanocomposites (rMGO)	Pb(II) – 121.95 Cr(III) – 126.58 As(III) – 106.40 As(V) – 65.78	[120]

Heavy metals are known to be a significant contaminant in aqueous solutions. Humans may experience serious health risks if they consume heavy metals. Adsorption is a method for water filtration that is less expensive, more effective, and easier to use than traditional approaches [86, 87]. However, it has limitations because of adsorbent regeneration and filtration. The filtration and regeneration of adsorbents are two challenges that can be overcome by the use of magnetic materials for water purification, therefore many researchers have invented and used magnetic materials for the treatment of toxins [88, 89]. Magnetic inverse spinel ferrites and their derivatives are frequently used for water filtration due to their magnetic and chemical stability, as well as their very porous surface structure [90, 91]. Although, the nanometal ferrites exhibit low stability [92]. Hybrid materials have been developed using magnetic ferrites and GO using the hydrothermal process to get over these problems. Recent years have seen the usage of functionalized magnetic-carbon composites by researchers [93] and magnetic organic composites [94, 95] for the recovery and separation of precious metals. They can facilitate the adsorption of heavy metals due to the carboxylic, epoxy, and hydroxyl functionalities that are present at GO [96]. Additionally, the developed magnetic graphene oxide (MGO) composites have exhibited greater performance for the treatment of water [91, 97]. Chandra et al. (2010) developed a superparamagnetic rMGO composite with an average size of 10 nm through a chemical reaction. The adsorption capability of this composite exceeded 99.9% for the removal of both As(III) and As(V) [34]. Zhang et al. used a ferric hydroxide-GO composite for the remarkable absorption of arsenate from water [98]. The arsenate concentration of the polluted water was lowered from 20 to 0.5 ppm here due to significant arsenate adsorption over a pH range of (4-9) However, the adsorption effectiveness reduced when pH rose to >8. However, a leak of less than 1.0 ppm iron was found in the treated water. Using a copper-catalyzed azide-alkyne cycloaddition, water-soluble MGO nanocomposite was synthesized and used to adsorb Cd(II), Pb(II), and Cu(II) from an aqueous environment [99]. The results found that the nanocomposite has a higher surface area, superparamagnetic property, and excellent complex ability, with extraordinary removal capacity for heavy metals. RGO–MnO_2 composites exhibited excellent mercury uptake capacity [100]. Compared to its basic type, this composite exhibit increased adsorption capability. This procedure was also simple and environmentally beneficial because it involved the intrinsic reduction of RGO without the requirement of extra chemicals. Fe_3O_4-rGO-MnO composite was developed by Luo et al. for the adsorption of As(III) and As(V) [35]. The study showed that by decreasing MnO_2 and Fe_3O_4 aggregation, the synthesized composite enhanced its adsorption and produced more adsorption sites. Liu et al. effectively removed Co(II) using MGO in their study. The rate-limiting kinetics involved the adsorption of Co(II) on MGO, with inner-sphere surface complexation occurring at low pH. As the pH increased, simultaneous precipitation and inner-sphere surface complexation were linked to the removal of Co(II). Additionally, it was reported that the external ions had a considerable impact on the adsorption of Co(II) on MGO. An external magnetic field can be used to quickly remove, and regenerate Co(II)-loaded MGO from the aqueous environment [101]. The hydrothermal process has been used to produce hybrid materials using magnetic ferrites and GO. Recently, several researchers have employed magnetic organic composites [93] and functionalized magnetic-carbon composites [94, 102] to remove and separation of precious metal and metal ions. Heavy metal adsorption by graphene-based nanocomposite increases due to the carboxylic, hydroxy, and epoxy groups present at GO [96]. Moreover, the synthesized magnetic graphene oxide showed better water treatment efficiency due to its easy separation [91, 97]. To remove Cd(II),

Pb(II), and Cu(II) from the aqueous environment, copper catalyzed azide-alkyne cycloaddition was used to produce a water-soluble MGO nanocomposite by Zhang et al. [99]. Zhao et al. reported that the use of a layered GO based nanosheet for the adsorption of U(VI) ions resulted in a maximum adsorption capacity (q_m) of 97.5 mg/g. However, because of the aggregation of GO nanosheets in the water, the synthesized nanosheet adsorption ability was drastically reduced. The above restriction was overcome by the functionalization of GO nanosheets, which decreased aggregation. In this regard, it is possible to view the functionalization of GO by magnetic materials as a quick and efficient way to separate and defeat the aggregation of GO [103]. Fe_3O_4/GO was synthesized by Zong et al. from natural flake graphite to remove U(VI), and they reported a qmax of 69.49 mg/g [104]. The maximum adsorption capacities (Q_{max}) of the various types of GO/RGO based nanocomposites that have been reported for the remediation of heavy metals are listed in Table 3.

Conclusion and Future Prospects

Developing a reliable material for wastewater treatment is essential because freshwater resources are reducing and increasing worse every day. The chapter strongly endorses GO/RGO based adsorbent materials for the removal of various contaminants in water. Graphene and graphene oxide have been extensively used in desalination and the removal of organic contaminants applications to enhance membrane properties. Antimicrobial, antifouling, mechanical strength, selectivity, water flux, and thermal characteristics of GO are upgraded after fusion into membrane. To remove organic pollutants, photo-catalysts like zinc and titanium oxide can be used in association with graphene. Graphene oxide membranes can be used to filter out heavy metals and separate monovalent and divalent ions. Many techniques of preparing GO/RGO based nanocomposite have been developed. One of the most challenging projects right now is increasing the commercial production of ultrathin GO membranes and keeping a balance between expenses and the ease of synthesis processes. Future research projects involving the practical separation and regeneration of used GO-based adsorbents need to be more inventive. Additionally, pilot-scale and real field experiments must be conducted. GO membranes could be one of the most significant technologies for solving the impending global water issues.

Acknowledgement

Authors thank DST-FIST, New Delhi, India vide Ref. No. SR/FST/CS-I/2022/247 Dated December 19, 2022 for infrastructural support.

References

[1] S. Sharma, A. Bhattacharya, Drinking water contamination and treatment techniques, *Appl. Water Sci.* 7 (2017) 1043–1067. https://doi.org/10.1007/s13201-016-0455-7.

[2] M. S. Denny, J. C. Moreton, L. Benz, S. M. Cohen, Metal–organic frameworks for Membrane based seperation Nat. *Rev. Mater* 1 (2016) 16078. https://doi.org/10.1038/natrevmats.2016.78.

[3] N. Song, X. Gao, Z. Ma, X. Wang, Y. Wei, C. Gao, A review of graphene-based separation membrane: Materials, characteristics, preparation and applications, *Desalination*. 437 (2018). https://doi.org/10.1016/j.desal.2018.02.024.

[4] X. Feng, X. Ding, D. Jiang, Covalent organic frameworks, *Chem. Soc. Rev.* 41 (2012) 6010–6022. https://doi.org/10.1039/C2CS35157A.

[5] G. Yang, Z. Xie, M. Cran, D. Ng, C. D. Easton, M. Ding, H. Xu, S. Gray, Functionalizing graphene oxide framework membranes with sulfonic acid groups for superior aqueous mixture separation, *J. Mater. Chem. A*. 7 (2019) 19682–19690. https://doi.org/10.1039/C9TA04031E.

[6] Y. Kang, Y. Xia, H. Wang, X. Zhang, 2D Laminar Membranes for Selective Water and Transport, *Adv. Funct. Mater.* 29 (2019). https://doi.org/10.1002/adfm.201902014.

[7] J. Ran, P. Zhang, C. Chu, P. Cui, X. Ai, T. Pan, Y. Wu, T. Xu, Ultrathin lamellar MoS_2 membranes for organic solvent nanofiltration, *J. Memb. Sci.* 602 (2020) 117963. https://doi.org/https://doi.org/10.1016/j.memsci.2020.117963.

[8] G. Liu, W. Jin, N. Xu, Graphene-based membranes, *Chem. Soc. Rev.* 44 (2015) 5016–5030. https://doi.org/10.1039/C4CS00423J.

[9] M. Sun, J. Li, Graphene oxide membranes: Functional structures, preparation and environmental applications, *Nano Today*. 20 (2018) 121–137. http://dx.doi.org/10.1016/j.nantod.2018.04.007.

[10] Y. A. J. Al-Hamadani, B. M. Jun, M. Yoon, N. Taheri-Qazvini, S. A. Snyder, M. Jang, J. Heo, Y. Yoon, Applications of MXene-based membranes in water purification: A review., *Chemosphere*. 254 (2020) 126821. https://doi.org/10.1016/j.chemosphere.2020.126821.

[11] W. Wang, Y. Wei, J. Fan, J. Cai, Z. Lu, L. Ding, H. Wang, Recent progress of two-dimensional nanosheet membranes and composite membranes for separation applications, *Front. Chem. Sci. Eng*. 15 (2021) 793–819. https://doi.org/10.1007/s11705-020-2016-8.

[12] J. H. Deng, X. R. Zhang, G. M. Zeng, J. L. Gong, Q. Y. Niu, J. Liang, Simultaneous removal of Cd(II) and ionic dyes from aqueous solution using magnetic graphene

oxide nanocomposite as an adsorbent, *Chem. Eng. J.* 226 (2013) 189–200. https://doi.org/10.1016/j.cej.2013.04.045.

[13] M. Ding, H. Xu, W. Chen, G. Yang, Q. Kong, D. Ng, T. Lin, Z. Xie, 2D laminar maleic acid-crosslinked MXene membrane with tunable nanochannels for efficient and stable pervaporation desalination, *J. Memb. Sci.* 600 (2020) 117871. https://doi.org/10.1016/j.memsci.2020.117871.

[14] Z. Lu, Y. Wei, J. Deng, L. Ding, Z. K. Li, H. Wang, Self-Crosslinked MXene ($Ti_3C_2T_x$) Membranes with Good Anti swelling Property for Monovalent Metal Ion Exclusion, *ACS Nano.* 13 (2019) 10535–10544. https://doi.org/10.1021/acsnano.9b04612.

[15] J. Wang, Z. Zhang, J. Zhu, M. Tian, S. Zheng, F. Wang, X. Wang, L. Wang, Ion sieving by a two-dimensional $Ti_3C_2T_x$ alginate lamellar membrane with stable interlayer spacing, *Nat. Commun.* 11 (2020) 3540. https://doi.org/10.1038/s41467-020-17373-4.

[16] B. C. Brodie, II. On the atomic weight of graphite, *Proc. R. Soc. London.* (n.d.) 11–12. https://doi.org/10.1098/rstl.1859.0013.

[17] L. Staudenmaier, Ver Fahren zur Darstellung der Graphitsäure [Method for the representation of graphitic acid], *Eur. J. Inorg. Chem.* 31 (1898) 1481–1487. https://doi.org/10.1002/cber.18980310237.

[18] W. S. J. Hummers, R. E. Offeman, Preparation of Graphitic Oxide, *J. Am. Chem. Soc.* 80 (1958) 1339. https://doi.org/10.1021/ja01539a017.

[19] M. J. McAllister, J. L. Li, D. H. Adamson, H. C. Schniepp, A. A. Abdala, J. Liu, M. Herrera-Alonso, D. L. Milius, R. Car, R. K. Prud'homme, I. A. Aksay, Single Sheet Functionalized Graphene by Oxidation and Thermal Expansion of Graphite, *Chem. Mater.* 19 (2007) 4396–4404. https://doi.org/10.1021/cm0630800.

[20] W. Peng, H. Li, Y. Liu, S. Song, A review on heavy metal ions adsorption from water by graphene oxide and its composites, *J. Mol. Liq.* 230 (2017) 496–504. http://dx.doi.org/10.1016/j.molliq.2017.01.064.

[21] H. C. Schniepp, J. L. Li, M. J. McAllister, H. Sai, M. Herrera-Alonso, D. H. Adamson, R. K. Prud'homme, R. Car, D. A. Saville, I. A. Aksay, Functionalized Single Graphene Sheets Derived from Splitting Graphite Oxide, *J. Phys. Chem. B.* 110 (2006) 8535–8539. https://doi.org/10.1021/jp060936f.

[22] L. K. Wu, H. Wu, H. B. Zhang, H. Z. Cao, G. Y. Hou, Y. P. Tang, G. Q. Zheng, Graphene oxide/$CuFe_2O_4$ foam as an efficient absorbent for arsenic removal from water, *Chem. Eng. J.* 334 (2018) 1808–1819. https://doi.org/10.1016/j.cej.2017.11.096.

[23] S. Ye, W. Jin, Q. Huang, Y. Hu, Y. Li, J. Li, B. Li, Da-KGM based GO-reinforced FMBO-loaded aerogels for efficient arsenic removal in aqueous solution, *Int. J. Biol. Macromol.* 94 (2017) 527–534. https://doi.org/10.1016/j.ijbiomac.2016.10.059.

[24] C. Zhao, H. Wu, Density functional investigation of mercury and arsenic adsorption on nitrogen doped graphene decorated with palladium clusters: A promising heavy metal sensing material in farmland, *Appl. Surf. Sci.* 399 (2017) 55–66. https://doi.org/10.1016/j.apsusc.2016.12.084.

[25] J. Zhu, Z. Zhu, H. Zhang, H. Lu, W. Zhang, Y. Qiu, L. Zhu, S. Küppers, Calcined layered double hydroxides/reduced graphene oxide composites with improved photocatalytic degradation of paracetamol and efficient oxidation-adsorption of As(III), *Appl. Catal. B Environ.* 225 (2018) 550–562. https://doi.org/10.1016/j.apcatb.2017.12.003.

[26] D. Cai, M. Song, Preparation of fully exfoliated graphite oxide nanoplatelets in organic solvents, *J. Mater. Chem.* 17 (2007) 3678–3680. https://doi.org/10.1039/B705906J.

[27] C. K. Chua, M. Pumera, Chemical reduction of graphene oxide: a synthetic chemistry viewpoint, *Chem. Soc. Rev.* 43 (2014) 291–312. https://doi.org/10.1039/C3CS60303B.

[28] W. Peng, H. Li, Y. Hu, Y. Liu, S. Song, Does silicate mineral impurities in natural graphite affect the characteristics of synthesized graphene?, *Mater. Res. Bull.* 74 (2016) 333–339. https://doi.org/10.1016/j.materresbull.2015.10.036.

[29] S. You, S. M. Luzan, T. Szabó, A. V Talyzin, Effect of synthesis method on solvation and exfoliation of graphite oxide, *Carbon N. Y.* 52 (2013) 171–180. https://doi.org/10.1016/j.carbon.2012.09.018.

[30] J. Zhao, L. Liu, F. Li, Structural Characterization, in: *Graphene Oxide Phys. Appl.*, Springer Berlin Heidelberg, Berlin, Heidelberg, 2015: pp. 15–29. https://doi.org/10.1007/978-3-662-44829-8_2.

[31] P. Zhao, Y. Guo, W. Zhang, H. Chai, H. Xing, M. Xing, Neurotoxicity induced by arsenic in Gallus Gallus: Regulation of oxidative stress and heat shock protein response., *Chemosphere.* 166 (2017) 238–245. https://doi.org/10.1016/j.chemosphere.2016.09.060.

[32] Y. Zhu, S. Murali, W. Cai, X. Li, J. W. Suk, J. R. Potts, R. S. Ruoff, Graphene and graphene oxide: synthesis, properties, and applications., *Adv. Mater.* 22 (2010) 3906–3924. https://doi.org/10.1002/adma.201001068.

[33] S. Stankovich, D. A. Dikin, R. D. Piner, K. A. Kohlhaas, A. Kleinhammes, Y. Jia, Y. Wu, S. T. Nguyen, R. S. Ruoff, Synthesis of graphene-based nanosheets via chemical reduction of exfoliated graphite oxide, *Carbon N. Y.* 45 (2007) 1558–1565. https://doi.org/10.1016/j.carbon.2007.02.034.

[34] V. Chandra, J. Park, Y. Chun, J. W. Lee, I. C. Hwang, K. S. Kim, Water-Dispersible Magnetite-Reduced Graphene Oxide Composites for Arsenic Removal, *ACS Nano.* 4 (2010) 3979–3986. https://doi.org/10.1021/nn1008897.

[35] X. Luo, C. Wang, S. Luo, R. Dong, X. Tu, G. Zeng, Adsorption of As (III) and As (V) from water using magnetite Fe3O4-reduced graphite oxide–MnO2 nanocomposites, *Chem. Eng. J.* 187 (2012) 45–52. https://doi.org/10.1016/j.cej.2012.01.073.

[36] Y. Yoon, W. K. Park, T. M. Hwang, D. H. Yoon, W. S. Yang, J. W. Kang, Comparative evaluation of magnetite–graphene oxide and magnetite-reduced graphene oxide composite for As(III) and As(V) removal, *J. Hazard. Mater.* 304 (2016) 196–204. https://doi.org/10.1016/j.jhazmat.2015.10.053.

[37] F. Yu, S. Sun, J. Ma, S. Han, Enhanced removal performance of arsenate and arsenite by magnetic graphene oxide with high iron oxide loading, *Phys. Chem. Chem. Phys.* 17 (2015) 4388–4397. https://doi.org/10.1039/C4CP04835K.

[38] L. Yu, Y. Ma, C. N. Ong, J. Xie, Y. Liu, Rapid adsorption removal of arsenate by hydrous cerium oxide–graphene composite, *RSC Adv.* 5 (2015) 64983–64990. https://doi.org/10.1039/C5RA08922K.

[39] M. Khatamian, N. Khodakarampoor, M. Saket-Oskoui, Efficient removal of arsenic using graphene-zeolite based composites, *J. Colloid Interface Sci.* 498 (2017) 433–441. https://doi.org/10.1016/j.jcis.2017.03.052.

[40] V. Loryuenyong, K. Totepvimarn, P. Eimburanapravat, W. Boonchompoo, A. Buasri, Preparation and Characterization of Reduced Graphene Oxide Sheets via Water-Based Exfoliation and Reduction Methods, *Adv. Mater. Sci. Eng.* 2013 (2013) 923403. https://doi.org/10.1155/2013/923403.

[41] A. T. Habte, D. W. Ayele, Synthesis and Characterization of Reduced Graphene Oxide (rGO) Started from Graphene Oxide (GO) Using the Tour Method with Different Parameters, *Adv. Mater. Sci. Eng.* 2019 (2019) 5058163. https://doi.org/10.1155/2019/5058163.

[42] C. K. Chua, M. Pumera, Reduction of graphene oxide with substituted borohydrides, *J. Mater. Chem. A.* 1 (2013) 1892–1898. https://doi.org/10.1039/C2TA00665K.

[43] H. C. Vu, A. D. Dwivedi, T. T. Le, S. H. Seo, E. J. Kim, Y. S. Chang, Magnetite graphene oxide encapsulated in alginate beads for enhanced adsorption of Cr(VI) and As(V) from aqueous solutions: Role of crosslinking metal cations in pH control, *Chem. Eng. J.* 307 (2017) 220–229. https://doi.org/10.1016/j.cej.2016.08.058.

[44] S. Kumar, R. R. Nair, P. B. Pillai, S. N. Gupta, M. A. R. Iyengar, A. K. Sood, Graphene Oxide–$MnFe_2O_4$ Magnetic Nanohybrids for Efficient Removal of Lead and Arsenic from Water, *ACS Appl. Mater. Interfaces.* 6 (2014) 17426–17436. https://doi.org/10.1021/am504826q.

[45] H. Su, Z. Ye, N. Hmidi, High-performance iron oxide–graphene oxide nanocomposite adsorbents for arsenic removal, *Colloids Surfaces A Physicochem. Eng. Asp.* 522 (2017) 161–172. https://doi.org/10.1016/j.colsurfa.2017.02.065.

[46] D. Fu, Z. He, S. Su, B. Xu, Y. Liu, Y. Zhao, Fabrication of α-FeOOH decorated graphene oxide-carbon nanotubes aerogel and its application in adsorption of arsenic species, *J. Colloid Interface Sci.* 505 (2017) 105–114. https://doi.org/10.1016/j.jcis.2017.05.091.

[47] A. S. K. Kumar, S. J. Jiang, Chitosan-functionalized graphene oxide: A novel adsorbent an efficient adsorption of arsenic from aqueous solution, *J. Environ. Chem. Eng.* 4 (2016) 1698–1713. https://doi.org/10.1016/j.jece.2016.02.035.

[48] A. S. K. Kumar, S. J. Jiang, Synthesis of magnetically separable and recyclable magnetic nanoparticles decorated with β-cyclodextrin functionalized graphene oxide an excellent adsorption of As(V)/(III), *J. Mol. Liq.* 237 (2017) 387–401. https://doi.org/10.1016/j.molliq.2017.04.093.

[49] V. K. Sharma, T. J. McDonald, H. Kim, V. K. Garg, Magnetic graphene–carbon nanotube iron nanocomposites as adsorbents and antibacterial agents for water purification, *Adv. Colloid Interface Sci.* 225 (2015) 229–240. https://doi.org/10.1016/j.cis.2015.10.006.

[50] L. Liu, C. Li, C. Bao, Q. Jia, P. Xiao, X. Liu, Q. Zhang, Preparation and characterization of chitosan/graphene oxide composites for the adsorption of Au(III) and Pd(II), *Talanta*. 93 (2012) 350–357. https://doi.org/10.1016/j.talanta.2012.02.051.

[51] J. Fu, Z. Chen, M. Wang, S. Liu, J. Zhang, J. Zhang, R. Han, Q. Xu, Adsorption of methylene blue by a high-efficiency adsorbent (polydopamine microspheres): Kinetics, isotherm, thermodynamics and mechanism analysis, *Chem. Eng. J.* 259 (2015) 53–61. https://doi.org/10.1016/j.cej.2014.07.101.

[52] P. Mokhtari, M. Ghaedi, K. Dashtian, M. R. Rahimi, M. K. Purkait, Removal of methyl orange by copper sulfide nanoparticles loaded activated carbon: Kinetic and isotherm investigation, *J. Mol. Liq.* 219 (2016) 299–305. https://doi.org/10.1016/j.molliq.2016.03.022.

[53] D. He, Y. Chen, Y. Situ, L. Zhong, H. Huang, Synthesis of ternary g-C_3N_4/Ag/γ-FeOOH photocatalyst: An integrated heterogeneous Fenton-like system for effectively degradation of azo dye methyl orange under visible light, *Appl. Surf. Sci.* 425 (2017) 862–872. https://doi.org/10.1016/j.apsusc.2017.06.124.

[54] A. Mukherjee, P. Dhak, D. Dhak, *The solvothermal synthesis of a 3D rod-like Fe–Al bimetallic metal–organic-framework for efficient fluoride adsorption and photodegradation of water-soluble carcinogenic dyes*, (2022). https://doi.org/10.1039/D1VA00019E.

[55] B. Panda, D. Mondal, S. Mandal, J. Khatun, A. Mukherjee, D. Dhak, One-pot solution combustion synthesis of porous spherical-shaped magnesium zinc binary oxide for efficient fluoride removal and photocatalytic degradation of methylene blue and Congo red dye, *Environ. Sci. Pollut. Res.* 30 (2023) 81386–81402. https://doi.org/10.1007/s11356-022-22551-6.

[56] Mukherjee, S. Kundu, D. Chatterjee, D. Dhak, A Critical Review on Photoreduction Using Metal–Organic Frameworks: Kinetics, pH and Mechanistic Studies and Anthropogenic/Natural Sources of Cr(VI), *Chem. Africa.* 5 (2022) 1783–1796. https://doi.org/10.1007/s42250-021-00295-z.

[57] Bouazizi, M. Breida, B. Achiou, M. Ouammou, J. I. Calvo, A. Aaddane, S. A. Younssi, Removal of dyes by a new nano–TiO_2 ultrafiltration membrane deposited on low-cost support prepared from natural Moroccan bentonite, *Appl. Clay Sci.* 149 (2017) 127–135. https://doi.org/10.1016/j.clay.2017.08.019.

[58] H. Li, S. Liu, J. Zhao, N. Feng, Removal of reactive dyes from wastewater assisted with kaolin clay by magnesium hydroxide coagulation process, Colloids Surfaces A Physicochem. *Eng. Asp.* 494 (2016) 222–227. https://doi.org/10.1016/j.colsurfa.2016.01.048.

[59] H. Zou, Y. Wang, Azo dyes wastewater treatment and simultaneous electricity generation in a novel process of electrolysis cell combined with microbial fuel cell, *Bioresour. Technol.* 235 (2017) 167–175. https://doi.org/10.1016/j.biortech.2017.03.093.

[60] L. Bukman, N. R. C. Fernandes-Machado, W. Caetano, A. L. Tessaro, N. Hioka, Treatment of wastewater contaminated with ionic dyes: Liquid-liquid extraction induced by reversed micelle followed by photodegradation, *Sep. Purif. Technol.* 189 (2017) 162–169. https://doi.org/10.1016/j.seppur.2017.08.004.

[61] N. A. El Essawy, S. M. Ali, H. A. Farag, A. H. Konsowa, M. Elnouby, H. A. Hamad, Green synthesis of graphene from recycled PET bottle wastes for use in the adsorption of dyes in aqueous solution, *Ecotoxicol. Environ. Saf.* 145 (2017) 57–68. https://doi.org/10.1016/j.ecoenv.2017.07.014.

[62] J. Zhou, Q. F. Lü, J. J. Luo, Efficient removal of organic dyes from aqueous solution by rapid adsorption onto polypyrrole–based composites, *J. Clean. Prod.* 167 (2017) 739–748. https://doi.org/10.1016/j.jclepro.2017.08.196.

[63] M. A. Ahmed, A. A. brick, A. A. Mohamed, An efficient adsorption of indigo carmine dye from aqueous solution on mesoporous Mg/Fe layered double hydroxide nanoparticles prepared by controlled sol-gel route, *Chemosphere*. 174 (2017) 280–288. https://doi.org/10.1016/j.chemosphere.2017.01.147.

[64] G. B. Hong, Y. K. Wang, Synthesis of low-cost adsorbent from rice bran for the removal of reactive dye based on the response surface methodology, *Appl. Surf. Sci.* 423 (2017) 800–809. https://doi.org/10.1016/j.apsusc.2017.06.264.

[65] H. Jayasantha Kumari, P. Krishnamoorthy, T. K. Arumugam, S. Radhakrishnan, D. Vasudevan, An efficient removal of crystal violet dye from waste water by adsorption onto TLAC/Chitosan composite: A novel low cost adsorbent, *Int. J. Biol. Macromol.* 96 (2017) 324–333. https://doi.org/10.1016/j.ijbiomac.2016.11.077.

[66] S. Kundu, I. H. Chowdhury, M. K. Naskar, Synthesis of hexagonal shaped nanoporous carbon for efficient adsorption of methyl orange dye, *J. Mol. Liq.* 234 (2017) 417–423. https://doi.org/10.1016/j.molliq.2017.03.090.

[67] S. Wang, B. Yang, Y. Liu, Synthesis of a hierarchical SnS_2 nanostructure for efficient adsorption of Rhodamine B dye, *J. Colloid Interface Sci.* 507 (2017) 225–233. https://doi.org/10.1016/j.jcis.2017.07.053.

[68] K. P. Singh, S. Gupta, A. K. Singh, S. Sinha, Optimizing adsorption of crystal violet dye from water by magnetic nanocomposite using response surface modeling approach, *J. Hazard. Mater.* 186 (2011) 1462–1473. https://doi.org/10.1016/j.jhazmat.2010.12.032.

[69] L. Sun, C. Tian, L. Wang, J. Zou, G. Mu, H. Fu, Magnetically separable porous graphitic carbon with large surface area as excellent adsorbents for metal ions and dye, *J. Mater. Chem.* 21 (2011) 7232–7239. https://doi.org/10.1039/C1JM10470E.

[70] H. Wang, N. Yan, Y. Li, X. Zhou, J. Chen, B. Yu, M. Gong, Q. Chen, Fe nanoparticle-functionalized multi-walled carbon nanotubes: one-pot synthesis and their applications in magnetic removal of heavy metal ions, *J. Mater. Chem.* 22 (2012) 9230–9236. https://doi.org/10.1039/C2JM16584H.

[71] G. K. Ramesha, A. Vijaya Kumara, H. B. Muralidhara, S. Sampath, Graphene and graphene oxide as effective adsorbents toward anionic and cationic dyes, *J. Colloid Interface Sci.* 361 (2011) 270–277. https://doi.org/10.1016/j.jcis.2011.05.050.

[72] Z. Yang, S. Ji, W. Gao, C. Zhang, L. Ren, W. W. Tjiu, Z. Zhang, J. Pan, T. Liu, Magnetic nanomaterial derived from graphene oxide/layered double hydroxide hybrid for efficient removal of methyl orange from aqueous solution, *J. Colloid Interface Sci.* 408 (2013) 25–32. https://doi.org/10.1016/j.jcis.2013.07.011.

[73] G. Abdi, A. Alizadeh, S. Zinadini, G. Moradi, Removal of dye and heavy metal ion using a novel synthetic polyethersulfone nanofiltration membrane modified by magnetic graphene oxide/metformin hybrid, *J. Memb. Sci.* 552 (2018) 326–335. https://doi.org/10.1016/j.memsci.2018.02.018.

[74] Khurana, A. K. Shaw, Bharti, J. M. Khurana, P. K. Rai, Batch and dynamic adsorption of Eriochrome Black T from water on magnetic graphene oxide: Experimental and theoretical studies, *J. Environ. Chem. Eng.* 6 (2018) 468–477. https://doi.org/10.1016/j.jece.2017.12.029.

[75] L. M. Pastrana-Martínez, S. Morales-Torres, A. G. Kontos, N. G. Moustakas, J. L. Faria, J. M. Doña-Rodríguez, P. Falaras, A. M. T. Silva, TiO_2, surface modified TiO_2 and graphene oxide-TiO_2 photocatalysts for degradation of water pollutants under near-UV/Vis and visible light, *Chem. Eng. J.* 224 (2013) 17–23. https://doi.org/10.1016/j.cej.2012.11.040.

[76] M. Kumar, Z. Gholamvand, A. Morrissey, K. Nolan, M. Ulbricht, J. Lawler, Preparation and characterization of low fouling novel hybrid ultrafiltration membranes based on the blends of GO−TiO_2 nanocomposite and polysulfone for humic acid removal, *J. Memb. Sci.* 506 (2016) 38–49. https://doi.org/10.1016/j.memsci.2016.02.005.

[77] X. Liu, L. Pan, T. Lv, G. Zhu, Z. Sun, C. Sun, Microwave-assisted synthesis of CdS–reduced graphene oxide composites for photocatalytic reduction of Cr(vi), *Chem. Commun.* 47 (2011) 11984–11986. https://doi.org/10.1039/C1CC14875C.

[78] N. Li, M. Zheng, X. Chang, G. Ji, H. Lu, L. Xue, L. Pan, J. Cao, Preparation of magnetic $CoFe_2O_4$-functionalized graphene sheets via a facile hydrothermal method and their adsorption properties, *J. Solid State Chem.* 184 (2011) 953–958. https://doi.org/10.1016/j.jssc.2011.01.014.

[79] Y. Fu, H. Chen, X. Sun, X. Wang, Combination of cobalt ferrite and graphene: High-performance and recyclable visible-light photocatalysis, *Appl. Catal. B Environ.* 111–112 (2012) 280–287. https://doi.org/10.1016/j.apcatb.2011.10.009.

[80] S. T. Yang, S. Chen, Y. Chang, A. Cao, Y. Liu, H. Wang, Removal of methylene blue from aqueous solution by graphene oxide, *J. Colloid Interface* Sci. 359 (2011) 24–29. https://doi.org/10.1016/j.jcis.2011.02.064.

[81] Wang, C. Feng, Y. Gao, X. Ma, Q. Wu, Z. Wang, Preparation of a graphene-based magnetic nanocomposite for the removal of an organic dye from aqueous solution, *Chem. Eng. J.* 173 (2011) 92–97. https://doi.org/10.1016/j.cej.2011.07.041.

[82] F. He, J. Fan, D. Ma, L. Zhang, C. Leung, H. L. Chan, The attachment of Fe_3O_4 nanoparticles to graphene oxide by covalent bonding, *Carbon N. Y.* 48 (2010) 3139–3144. https://doi.org/10.1016/j.carbon.2010.04.052.

[83] Y. Liang, H. Wang, H. Sanchez Casalongue, Z. Chen, H. Dai, TiO_2 nanocrystals grown on graphene as advanced photocatalytic hybrid materials, *Nano Res.* 3 (2010) 701–705. https://doi.org/10.1007/s12274-010-0033-5.

[84] Y. Fu, X. Wang, Magnetically Separable $ZnFe_2O_4$–Graphene Catalyst and its High Photocatalytic Performance under Visible Light Irradiation, *Ind. Eng. Chem. Res.* 50 (2011) 7210–7218. https://doi.org/10.1021/ie200162a.

[85] T. Xu, L. Zhang, H. Cheng, Y. Zhu, Significantly enhanced photocatalytic performance of ZnO via graphene hybridization and the mechanism study, *Appl.*

Catal. B Environ. 101 (2011) 382–387. https://doi.org/10.1016/j.apcatb.2010.10.007.

[86] J. R. Koduru, Y. Y. Chang, I. S. Kim, Low-cost schizandra chinesis fruit peel for Co(II) removal from aqueous environment: Adsorption properties and mechanism, *Asian J. Chem.* 26 (2014) 289–297. https://doi.org/10.14233/ajchem.2014.16073.

[87] E. A. Oraby, J. J. Eksteen, The leaching of gold, silver and their alloys in alkaline glycine–peroxide solutions and their adsorption on carbon, *Hydrometallurgy.* 152 (2015) 199–203. https://doi.org/10.1016/j.hydromet.2014.12.015.

[88] D. H. K. Reddy, S. M. Lee, Application of magnetic chitosan composites for the removal of toxic metal and dyes from aqueous solutions, *Adv. Colloid Interface Sci.* 201–202 (2013) 68–93. https://doi.org/10.1016/j.cis.2013.10.002.

[89] D. H. K. Reddy, S. M. Lee, Three-Dimensional Porous Spinel Ferrite as an Adsorbent for Pb(II) Removal from Aqueous Solutions, *Ind. Eng. Chem. Res.* 52 (2013) 15789–15800. https://doi.org/10.1021/ie303359e.

[90] Li, H. Cao, J. Shao, M. Qu, J. H. Warner, Superparamagnetic Fe_3O_4 nanocrystals@graphene composites for energy storage devices, *J. Mater. Chem.* 21 (2011) 5069–5075. https://doi.org/10.1039/C0JM03717F.

[91] S. Xu, F. Zhang, Q. Kang, S. Liu, Q. Cai, The effect of magnetic field on the catalytic graphitization of phenolic resin in the presence of Fe–Ni, *Carbon N. Y.* 47 (2009) 3233–3237. https://doi.org/10.1016/j.carbon.2009.07.036.

[92] K. Maaz, S. Karim, A. Mumtaz, S. K. Hasanain, J. Liu, J. L. Duan, Synthesis and magnetic characterization of nickel ferrite nanoparticles prepared by co-precipitation route, *J. Magn. Magn. Mater.* 321 (2009) 1838–1842. https://doi.org/10.1016/j.jmmm.2008.11.098.

[93] M. I. Muravyov, N. V Fomchenko, A. V Usoltsev, E. A. Vasilyev, T. F. Kondrat'eva, Leaching of copper and zinc from copper converter slag flotation tailings using H2SO4 and biologically generated Fe2(SO4)3, *Hydrometallurgy.* 119–120 (2012) 40–46. https://doi.org/10.1016/j.hydromet.2012.03.001.

[94] S. Lv, X. Chen, Y. Ye, S. Yin, J. Cheng, M. Xia, Rice hull/MnFe2O4 composite: Preparation, characterization and its rapid microwave-assisted COD removal for organic wastewater, *J. Hazard. Mater.* 171 (2009) 634–639. https://doi.org/10.1016/j.jhazmat.2009.06.039.

[95] X. Hu, Y. Liu, H. Wang, A. Chen, G. Zeng, S. Liu, Y. Guo, X. Hu, T. Li, Y. Wang, L. Zhou, S. Liu, Removal of Cu(II) ions from aqueous solution using sulfonated magnetic graphene oxide composite, *Sep. Purif. Technol.* 108 (2013) 189–195. https://doi.org/10.1016/j.seppur.2013.02.011.

[96] J. Li, S. Guo, Y. Zhai, E. Wang, Nafion–graphene nanocomposite film as enhanced sensing platform for ultrasensitive determination of cadmium, *Electrochem. Commun.* 11 (2009) 1085–1088. https://doi.org/10.1016/j.elecom.2009.03.025.

[97] H. Koo, H. Lee, H. A. Go, Y. B. Lee, T. S. Bae, J. K. Kim, W. S. Choi, Graphene-based multifunctional iron oxide nanosheets with tunable properties., *Chemistry* (Easton). 17 4 (2011) 1214–1219. https://doi.org/10.1002/chem.201002252.

[98] K. Zhang, V. Dwivedi, C. Chi, J. Wu, Graphene oxide/ferric hydroxide composites for efficient arsenate removal from drinking water, *J. Hazard. Mater.* 182 (2010) 162–168. https://doi.org/10.1016/j.jhazmat.2010.06.010.

[99] W. Zhang, X. Shi, Y. Zhang, W. Gu, B. Li, Y. Xian, Synthesis of water-soluble magnetic graphene nanocomposites for recyclable removal of heavy metal ions, *J. Mater. Chem. A.* 1 (2013) 1745–1753. https://doi.org/10.1039/C2TA00294A.

[100] T. S. Sreeprasad, S. M. Maliyekkal, K. P. Lisha, T. Pradeep, Reduced graphene oxide–metal/metal oxide composites: Facile synthesis and application in water purification, *J. Hazard. Mater.* 186 (2011) 921–931. https://doi.org/10.1016/j.jhazmat.2010.11.100.

[101] M. Liu, C. Chen, J. Hu, X. Wu, X. Wang, Synthesis of Magnetite/Graphene Oxide Composite and Application for Cobalt(II) Removal, *J. Phys. Chem.* C. 115 (2011) 25234–25240. https://doi.org/10.1021/jp208575m.

[102] G. Zhao, X. Ren, X. Gao, X. Tan, J. Li, C. Chen, Y. Huang, X. Wang, Removal of Pb(ii) ions from aqueous solutions on few-layered graphene oxide nanosheets, *Dalt. Trans.* 40 (2011) 10945–10952. https://doi.org/10.1039/C1DT11005E.

[103] G. Zhao, T. Wen, X. Yang, S. Yang, J. Liao, J. Hu, D. Shao, X. Wang, Preconcentration of U(vi) ions on few-layered graphene oxide nanosheets from aqueous solutions, *Dalt. Trans.* 41 (2012) 6182–6188. https://doi.org/10.1039/C2DT00054G.

[104] P. Zong, S. Wang, Y. Zhao, H. Wang, H. Pan, C. He, Synthesis and application of magnetic graphene/iron oxides composite for the removal of U(VI) from aqueous solutions, *Chem. Eng. J.* 220 (2013) 45–52. https://doi.org/10.1016/j.cej.2013.01.038.

[105] G. Zhao, J. Li, X. Ren, C. Chen, X. Wang, Few-Layered Graphene Oxide Nanosheets As Superior Sorbents for Heavy Metal Ion Pollution Management, *Environ. Sci. Technol.* 45 (2011) 10454–10462. https://doi.org/10.1021/es203439v.

[106] F. Najafi, O. Moradi, M. Rajabi, M. Asif, I. Tyagi, S. Agarwal, V. K. Gupta, Thermodynamics of the adsorption of nickel ions from aqueous phase using graphene oxide and glycine functionalized graphene oxide, *J. Mol. Liq.* 208 (2015) 106–113. https://doi.org/10.1016/j.molliq.2015.04.033.

[107] L. Tan, J. Wang, Q. Liu, Y. Sun, X. Jing, L. Liu, J. Liu, D. Song, The synthesis of a manganese dioxide–iron oxide–graphene magnetic nanocomposite for enhanced uranium(vi) removal, *New J. Chem.* 39 (2015) 868–876. https://doi.org/10.1039/C4NJ01256A.

[108] J. Li, S. Zhang, C. Chen, G. Zhao, X. Yang, J. Li, X. Wang, Removal of Cu(II) and Fulvic Acid by Graphene Oxide Nanosheets Decorated with Fe3O4 Nanoparticles, *ACS Appl. Mater. Interfaces.* 4 (2012) 4991–5000. https://doi.org/10.1021/am301358b.

[109] L. Cui, Y. Wang, L. Gao, L. Hu, L. Yan, Q. Wei, B. Du, EDTA functionalized magnetic graphene oxide for removal of Pb(II), Hg(II) and Cu(II) in water treatment: Adsorption mechanism and separation property, *Chem. Eng. J.* 281 (2015) 1–10. https://doi.org/10.1016/j.cej.2015.06.043.

[110] C. J. Madadrang, H. Y. Kim, G. Gao, N. Wang, J. Zhu, H. Feng, M. Gorring, M.L. Kasner, S. Hou, Adsorption Behavior of EDTA-Graphene Oxide for Pb (II) Removal, *ACS Appl. Mater. Interfaces.* 4 (2012) 1186–1193. https://doi.org/10.1021/am201645g.

[111] Jiao, J. Xiong, J. Tao, S. Xu, D. Zhang, H. Lin, Y. Chen, Sodium alginate/graphene oxide aerogel with enhanced strength–toughness and its heavy metal adsorption study, *Int. J. Biol. Macromol.* 83 (2016) 133–141. https://doi.org/10.1016/j.ijbiomac.2015.11.061.

[112] V. Chandra, K. S. Kim, Highly selective adsorption of Hg2+ by a polypyrrole–reduced graphene oxide composite, *Chem. Commun.* 47 (2011) 3942–3944. https://doi.org/10.1039/C1CC00005E.

[113] Nandi, T. Basu, S. Debnath, A. K. Ghosh, A. De, U. C. Ghosh, Mechanistic Insight for the Sorption of Cd(II) and Cu(II) from Aqueous Solution on Magnetic Mn-Doped Fe(III) Oxide Nanoparticle Implanted Graphene, *J. Chem. Eng. Data.* 58 (2013) 2809–2818. https://doi.org/10.1021/je4005257.

[114] S. M. Lee, C. Laldawngliana, D. Tiwari, Iron oxide nano-particles-immobilized-sand material in the treatment of Cu(II), Cd(II) and Pb(II) contaminated waste waters, *Chem. Eng. J.* 195–196 (2012) 103–111. https://doi.org/10.1016/j.cej.2012.04.075.

[115] Y. Zhang, L. Yan, W. Xu, X. Guo, L. Cui, L. Gao, Q. Wei, B. Du, Adsorption of Pb(II) and Hg(II) from aqueous solution using magnetic CoFe2O4-reduced graphene oxide, *J. Mol. Liq.* 191 (2014) 177–182. https://doi.org/10.1016/j.molliq.2013.12.015.

[116] H. V. Tran, L. D. Tran, T. N. Nguyen, Preparation of chitosan/magnetite composite beads and their application for removal of Pb(II) and Ni(II) from aqueous solution, *Mater. Sci. Eng. C.* 30 (2010) 304–310. https://doi.org/10.1016/j.msec.2009.11.008.

[117] H. Ge, Z. Ma, Microwave preparation of triethylenetetramine modified graphene oxide/chitosan composite for adsorption of Cr(VI), *Carbohydr. Polym.* 131 (2015) 280–287. https://doi.org/10.1016/j.carbpol.2015.06.025.

[118] L. P. Lingamdinne, J. R. Koduru, Y. L. Choi, Y. Y. Chang, J. K. Yang, Studies on removal of Pb(II) and Cr(III) using graphene oxide based inverse spinel nickel ferrite nano-composite as sorbent, *Hydrometallurgy.* 165 (2016) 64–72. https://doi.org/10.1016/j.hydromet.2015.11.005.

[119] L. P. Lingamdinne, Y. L. Choi, I. S. Kim, J. K. Yang, J. R. Koduru, Y. Y. Chang, Preparation and characterization of porous reduced graphene oxide based inverse spinel nickel ferrite nanocomposite for adsorption removal of radionuclides, *J. Hazard. Mater.* 326 (2017) 145–156. https://doi.org/10.1016/j.jhazmat.2016.12.035.

[120] L. P. Lingamdinne, Y. L. Choi, I. S. Kim, Y. Y. Chang, J. R. Koduru, J. K. Yang, Porous graphene oxide based inverse spinel nickel ferrite nanocomposites for the enhanced adsorption removal of arsenic, *RSC Adv.* 6 (2016) 73776–73789. https://doi.org/10.1039/C6RA10134H.

About the Editors

Professor Nanda Gopal Sahoo earned his Ph.D. degree in material science from the Indian Institute of Technology (IIT) Kharagpur in 2004. After completing his Ph.D., he worked in several countries with eminent scientists for nine years, primarily in various fields of materials science, particularly focusing on carbon nanomaterials. In 2012-2013, he held the position of Scientist at the Institute of Materials Research and Engineering, A*Star, Singapore. Prior to this, he worked as a scientist at the Energy Research Institute at Nanyang Technological University (NTU) (ERI@N) from 2011 to 2012. In 2008, he was awarded the prestigious Lee Kuan Yew fellowship (2008-2011) in the School of Mechanical and Aerospace Engineering (MAE), NTU. Before the Lee Kuan Yew fellowship, he served as a Research Fellow at the School of Chemical and Biomedical Engineering, NTU, from 2006 to 2008 and at the Artificial Muscle Research Center, Konkuk University, South Korea, from 2005 to 2006.

After gaining extensive experience working with various eminent research groups, Professor Sahoo joined the Department of Chemistry at Kumaun University, Nainital, as an Associate Professor in 2013. He was later promoted to a full professor in 2016. At the university, he played a pivotal role in establishing the Nanoscience and Nanotechnology Centre. In 2014, he was selected as one of the BASE fellows sponsored by the Indo-US Science and Technological Forum, supporting his research in the field of solar cell development at Virginia Commonwealth University, USA. Professor Sahoo's contributions to research have been recognized with several awards, including National Awards for Technology Innovation from the Ministry of Chemicals and Fertilizers, Department of Chemicals and Petrochemicals, Government of

India, in 2018. He also received the prestigious Uttarakhand Governor's Best Research Award in 2017 and Uttarakhand Ratan in 2019. In 2021, he was recognized for Excellence in Research of the Year.

His international and extensive research experience has allowed him to make significant contributions to various areas of research and development, including solid waste management, fuel cells, nanopharmaceuticals, drug delivery, polymer nanocomposites, polymer chemistry, and particle technology. He has also worked as a visiting scientist at the Free University of Berlin, Berlin, Germany, Konkuk University, South Korea, Nanyang Technological University, Singapore and KIST, South Korea.

In terms of his academic output, to date, Professor Sahoo has supervised 12 Ph.D. students, published 142 research papers in high-quality international journals, holds 13 patents, edited 1 book and has presented 111 papers at international and national conferences and symposiums. According to Google Scholar, his work has received a total of 10,450 citations, with an h-index of 43.

Dr. Anirban Dandapat received his Ph. D. degree from Jadavpur University, Kolkata, India in 2013. After pursuing his Ph.D. degree, Dr. Dandapat perused postdoctoral research at Nanyang Technological University, Singapore (2013-2015) and The Hebrew University of Jerusalem, Israel (2015-2016). Then he moved to India and joined as Assistant Professor (INSPIRE Faculty) at Kumaun University, Nainital (2016-2022). Currently, he is working as Assistant Professor at University School of Automation and Robotics of Guru Gobind Singh Indraprastha University, New Delhi, India. His current research work is focused on the development of different nanomaterials for catalysis, photocatalysis, water purification and biomedical applications. He has published over forty journal papers in high-quality international journals, ten book chapters, 03 patents (granted 1 and filled 2) and presented over twenty papers in international and national conferences. His total citations according to Google Scholar are 1130 and h index is 19.

Index

L

M

N

O

P

R

S

T

U

V

W

Z